AN INTRODUCTION TO FUZZY LOGIC APPLICATIONS IN INTELLIGENT SYSTEMS

THE KLUWER INTERNATIONAL SERIES IN ENGINEERING AND COMPUTER SCIENCE

KNOWLEDGE REPRESENTATION, LEARNING AND EXPERT SYSTEMS

Consulting Editor

Tom Mitchell
Carnegie Mellon University

UNIVERSAL SUBGOALING AND CHUNKING OF GOAL HIERARCHIES, J. Laird, P. Rosenbloom, A. Newell, ISBN: 0-89838-213-0
MACHINE LEARNING: A Guide to Current Research, T. Mitchell, J. Carbonell, R. Michalski, ISBN: 0-89838-214-9
MACHINE LEARNING OF INDUCTIVE BIAS, P. Utgoff, ISBN: 0-89838-223-8
A CONNECTIONIST MACHINE FOR GENETIC HILLCLIMBING, D. H. Ackley, ISBN: 0-89838-236-X
LEARNING FROM GOOD AND BAD DATA, P. D. Laird, ISBN: 0-89838-263-7
MACHINE LEARNING OF ROBOT ASSEMBLY PLANS, A. M. Segre, ISBN: 0-89838-269-6
AUTOMATING KNOWLEDGE ACQUISITION FOR EXPERT SYSTEMS, S. Marcus, Editor, ISBN: 0-89838-294-7
MACHINE LEARNING, META-REASONING AND LOGICS, P. B. Brazdil, K. Konolige, ISBN: 0-7923-9047-4
CHANGE OF REPRESENTATION AND INDUCTIVE BIAS: D. P. Benjamin, ISBN: 0-7923-9055-5
KNOWLEDGE ACQUISITION: SELECTED RESEARCH AND COMMENTARY, S. Marcus, Editor, ISBN: 0-7923-9062-8
LEARNING WITH NESTED GENERALIZED EXEMPLARS, S. L. Salzberg, ISBN: 0-7923-9110-1
INCREMENTAL VERSION-SPACE MERGING: A General Framework for Concept Learning, H. Hirsh, ISBN: 0-7923-9119-5
COMPETITIVELY INHIBITED NEURAL NETWORKS FOR ADAPTIVE PARAMETER ESTIMATION, M. Lemmon, ISBN: 0-7923-9086-5
STRUCTURE LEVEL ADAPTATION FOR ARTIFICIAL NEURAL NETWORKS, T.C. Lee, ISBN: 0-7923-9151-9
CONNECTIONIST APPROACHES TO LANGUAGE LEARNING, D. Touretzky, ISBN: 0-7923-9216-7

AN INTRODUCTION TO FUZZY LOGIC APPLICATIONS IN INTELLIGENT SYSTEMS

Edited
by

Ronald R. Yager
Iona College

Lotfi A. Zadeh
University of California, Berkeley

KLUWER ACADEMIC PUBLISHERS
Boston/Dordrecht/London

Distributors for North America:
Kluwer Academic Publishers
101 Philip Drive
Assinippi Park
Norwell, Massachusetts 02061 USA

Distributors for all other countries:
Kluwer Academic Publishers Group
Distribution Centre
Post Office Box 322
3300 AH Dordrecht, THE NETHERLANDS

Library of Congress Cataloging-in-Publication Data

An Introduction to fuzzy logic applications in intelligent systems /
edited by Ronald R. Yager, Lotfi A. Zadeh.
p. cm. -- (The Kluwer International series in engineering and computer science ; SECS 165)
Includes bibliographical references and index.
ISBN: 0-7923-9191-8
1. Fuzzy sets. 2. Fuzzy systems. 3. Intelligent control systems.
I. Yager, Ronald R., 1941- . II. Zadeh, Lotfi Asker.
III. Series.
QA248.I568 1992
511.3'22--dc20 91-39272
CIP

Printed on acid-free paper.

Printed in the United States of America

CONTENTS

AN INTRODUCTION TO FUZZY LOGIC APPLICATIONS IN INTELLIGENT SYSTEMS

1

KNOWLEDGE REPRESENTATION IN FUZZY LOGIC

Lotfi A. Zadeh
Computer Science Division, Department of EECS
University of California, Berkeley, California 94720

ABSTRACT

The conventional approaches to knowledge representation, e.g., semantic networks, frames, predicate calculus and Prolog, are based on bivalent logic. A serious shortcoming of such approaches is their inability to come to grips with the issue of uncertainty and imprecision. As a consequence, the conventional approaches do not provide an adequate model for modes of reasoning which are approximate rather than exact. Most modes of human reasoning and all of commonsense reasoning fall into this category.

Fuzzy logic, which may be viewed as an extension of classical logical systems, provides an effective conceptual framework for dealing with the problem of knowledge representation in an environment of uncertainty and imprecision. Meaning representation in fuzzy logic is based on test-score semantics. In this semantics, a proposition is interpreted as a system of elastic constraints, and reasoning is viewed as elastic constraint propagation. Our paper presents a summary of the basic concepts and techniques underlying the application of fuzzy logic to knowledge representation and describes a number of examples relating to its use as a computational system for dealing with uncertainty and imprecision in the context of knowledge, meaning and inference.

INTRODUCTION

Knowledge representation is one of the most basic and actively researched areas of AI (Brachman, 1985,1988; Levesque, 1986, 1987; Moore, 1982, 1984; Negoita, 1985; Shapiro, 1987; Small, 1988). And yet, there are many important issues underlying knowledge representation which have not been adequately addressed. One such issue is that of the representation of knowledge which is lexically imprecise and/or uncertain.

As a case in point, the conventional knowledge representation techniques do not provide effective tools for representing the meaning of or inferring from the kind of everyday type facts exemplified by

(a) *Usually* it takes *about an hour* to drive from Berkeley to Stanford in *light* traffic.

(b) Unemployment is *not likely* to undergo a *sharp* decline during the next *few* months.

(c) *Most* experts believe that the likelihood of a *severe* earthquake in the *near* future is *very low*.

The italicized words in these assertions are the labels of fuzzy predicates, fuzzy quantifiers and fuzzy probabilities. The conventional approaches to knowledge representation lack the means for representing the meaning of fuzzy concepts. As a consequence, the approaches based on first order logic and classical probability theory do not provide an appropriate conceptual framework for dealing with the representation of commonsense knowledge, since such knowledge is by its nature both lexically imprecise and noncategorical (Moore, 1982, 1984; Zadeh, 1984).

The development of fuzzy logic was motivated in large measure by the need for a conceptual framework which can address the issues of uncertainty and lexical imprecision. The principal objective of this paper is to present a summary of some of the basic ideas underlying fuzzy logic and to describe their application to the problem of knowledge representation in an environment of uncertainty and imprecision. A more detailed discussion of these ideas may be found in Zadeh (1978a, 1978b, 1986, 1988a) and other entries in the bibliography.

ESSENTIALS OF FUZZY LOGIC

Fuzzy logic, as its name suggests, is the logic underlying modes of reasoning which are approximate rather than exact. The importance of fuzzy logic derives from the fact that most modes of human reasoning—and especially commonsense reasoning—are approximate in nature. It is of interest to note that, despite its pervasiveness, approximate reasoning falls outside the purview of classical logic largely because it is a deeply entrenched tradition in logic to be concerned with those and only those modes of reasoning which lend themselves to precise formulation and analysis.

Some of the essential characteristics of fuzzy logic relate to the following.

In fuzzy logic, exact reasoning is viewed as a limiting case of approximate reasoning.

In fuzzy logic, everything is a matter of degree.

Any logical system can be fuzzified.

In fuzzy logic, knowledge is interpreted a collection of elastic or, equivalently, fuzzy constraint on a collection of variables.

Inference is viewed as a process of propagation of elastic constraints.

Fuzzy logic differs from the traditional logical systems both in spirit and in detail. Some of the principal differences are summarized in the following (Zadeh, 1983b).

Truth: In bivalent logical systems, truth can have only two values: true or false. In multivalued systems, the truth value of a proposition may be an element of (a) a finite set; (b) an interval such as [0,1]; or (c) a boolean algebra. In fuzzy logic, the truth value of a proposition may be a fuzzy subset of any partially ordered set but usually it is assumed to be a fuzzy subset of the interval [0,1] or, more simply, a point in this interval. The so-called *linguistic* truth values expressed as *true, very true, not quite true*, etc. are interpreted as labels of fuzzy subsets of the unit interval.

Predicates: In bivalent systems, the predicates are crisp, e.g., *mortal, even, larger than*. In fuzzy logic, the predicates are fuzzy, e.g., *tall, ill, soon, swift, much larger than*. It should be noted that most of the predicates in a natural language are fuzzy rather than crisp.

Predicate Modifiers: In classical systems, the only widely used predicate modifier is the negation, *not*. In fuzzy logic, there is a variety of predicate modifiers which act as hedges, e.g., *very, more or less, quite, rather, extremely*. Such predicate modifiers play an essential role in the generation of the values of a linguistic variable, e.g., *very young, not very young, more or less young*, etc., (Zadeh, 1973).

Quantifiers: In classical logical systems there are just two quantifiers: *universal* and *existential*. Fuzzy logic admits, in addition, a wide variety of fuzzy quantifiers exemplified by *few, several, usually, most, almost always, frequently, about five*, etc. In fuzzy logic, a fuzzy quantifier is interpreted as a fuzzy number or a fuzzy proportion (Zadeh, 1983a).

Probabilities: In classical logical systems, probability is numerical or interval-valued. In fuzzy logic, one has the additional option of employing linguistic or, more generally, fuzzy probabilities exemplified by *likely, unlikely, very likely, around 0.8, high*, etc. (Zadeh 1986). Such probabilities may be interpreted as fuzzy numbers which may be manipulated through the use of fuzzy arithmetic (Kaufmann and Gupta, 1985).

In addition to fuzzy probabilities, fuzzy logic makes it possible to deal with fuzzy events. An example of a fuzzy event is: *tomorrow will be a warm day*, where *warm* is a fuzzy predicate. The probability of a fuzzy event may be a crisp or fuzzy number (Zadeh, 1968).

It is important to note that from the frequentist point of view there is an interchangeability between fuzzy probabilities and fuzzy quantifiers or, more generally, fuzzy measures. In this perspective, any proposition which contains labels of fuzzy probabilities may be expressed in an equivalent from which contains fuzzy quantifiers rather than fuzzy probabilities.

Possibilities: In contrast to classical modal logic, the concept of possibility in fuzzy logic is graded rather than bivalent. Furthermore, as in the case of probabilities, possibilities may be treated as linguistic variables with values such as *possible, quite possible, almost impossible*, etc. Such values may be interpreted as labels of fuzzy subsets of the real line.

A concept which plays a central role in fuzzy logic is that of a possibility distribution (Zadeh, 1978a; Dubois and Prade, 1988; Klir, 1988). Briefly, if X is a variable taking values in a universe of discourse U, then the *possibility distribution* of X, Π_X, is the fuzzy set of all possible values of X. More specifically, let $\pi_X(u)$ denote the possibility that X can take the value u, $u \in U$. Then the membership function of X is numerically equal to the *possibility distribution function* $\pi_X(u)$: U->[0, 1], which associates with each element $u \in U$ the possibility that X may take u as its value. More about possibilities and possibility distributions will be said at a later point in this paper.

It is important to observe that in every instance fuzzy logic adds to the options which are available in classical logical systems. In this sense, fuzzy logic may be viewed as an extension of such systems rather than as system of reasoning which is in conflict with the classical systems.

Before taking up the issue of knowledge representation in fuzzy logic, it will be helpful to take a brief look at some of the principal modes of reasoning in fuzzy logic. These are the following, with the understanding that the modes in question are not necessarily disjoint.

1. *Categorical Reasoning*

In this mode of reasoning, the premises contain no fuzzy quantifiers and no fuzzy probabilities. A simple example of categorical reasoning is:

Carol is slim
Carol is very intelligent

Carol is slim and very intelligent

In the premises, *slim* and *very intelligent* are assumed to be fuzzy predicates. The fuzzy predicate in the conclusion, *slim and very intelligent*, is the conjunction of *slim* and *intelligent.*

Another example of categorical reasoning is:

Mary is young
John is much older than Mary

John is (much older young).

where (*much_older young*) represents the composition of the binary fuzzy predicate *much_older* with the unary fuzzy predicate *young*. More specifically, let π_{much_older} and π_{young} denote the possibility distribution functions associated with the fuzzy predicates *much_older* and *young*, respectively. Then, the possibility distribution function of John's age may be expressed as (Zadeh, 1978a)

$$\pi_{Age(John)}(u) = \vee_v (\pi_{much\_older}(u, v) \wedge \pi_{young}(nu))$$

where $\vee$ and $\wedge$ stand for max and min, respectively.

2. *Syllogistic Reasoning*

In contrast to categorical reasoning, syllogistic reasoning relates to inference from premises containing fuzzy quantifiers (Zadeh, 1985; Dubois and Prade, 1978a). A simple example of syllogistic reasoning is the following

most Swedes are blond
most blond Swedes are tall

$most^2$ Swedes are blond and tall

where the fuzzy quantifier *most* is interpreted as a fuzzy proportion and *$most^2$* is the square of *most* in fuzzy arithmetic (Kaufmann and Gupta, 1985).

3. *Dispositional Reasoning*

In dispositional reasoning the premises are dispositions, that is, propositions which are preponderantly but necessarily always true (Zadeh, 1987). An example of dispositional reasoning is:

heavy smoking is a leading cause of cancer

---

to avoid lung cancer avoid heavy smoking

Note that in this example the conclusion is a maxim which may be interpreted as a dispositional command. Another example of dispositional reasoning is:

usually the probability of failure is not very low
usually the probability of failure is not very high

(2 *usually* ⊖ 1) *the probability of failure is not very low and not very high*

In this example, *usually* is a fuzzy quantifier which is interpreted as a fuzzy proportion and 2 *usually* ⊖ 1 is a fuzzy arithmetic expression whose value may be computed through the use of fuzzy arithmetic. (⊖ denotes the operation of subtraction in fuzzy arithmetic.) It should be noted that the concept of *usuality* plays a key role in dispositional reasoning (Zadeh, 1985, 1987), and is the concept that links together the dispositional and syllogistic modes of reasoning. Furthermore, it underlies the theories of nonmonotonic and default reasoning (McCarthy, 1980; McDermott, 1980, 1982; Reiter, 1983).

4. *Qualitative Reasoning*

In fuzzy logic, the term *qualitative reasoning* refers to a mode of reasoning in which the input-output relation of a system is expressed as a collection of fuzzy if-then rules in which the antecedents and consequents involve linguistic variables (Zadeh, 1975, 1989). In this sense, qualitative reasoning in fuzzy logic bears some similarity to—but is not coextensive with—qualitative reasoning in AI (de Kleer, 1984; Forbus, 1989; Kuipers, 1986).

A very simple example of qualitative reasoning is:

volume is small if pressure is high
volume is large if pressure is low

---

volume is $(w1 \wedge high + w2 \wedge large)$ *if pressure is medium*

where + should be interpreted as infix max; and

$$w1 = sup\ (high \wedge medium)$$

and

$$w2 = sup\ (low \wedge medium)$$

are weighting coefficients which represent, respectively, the degrees to which the antecedents *high* and *low* match the input *medium*. In w1, the conjunction *high* $\wedge$ *medium* represents the intersection of the possibility distributions of *high* and *low*, and the suprenum is taken over the domain of *high* and *medium*. The same applies to w2.

Qualitative reasoning underlies many of the applications of fuzzy logic in the realms of control and systems analysis (Sugeno, 1985; Pospelov, 1987; Togai, 1986). In this connection, it should be noted that fuzzy Prolog provides an effective knowledge representation language for qualitative reasoning (Baldwin, 1984, 1987; Mukaidono, 1987; Zadeh, 1989).

MEANING AND KNOWLEDGE REPRESENTATION

In a general setting, knowledge may be viewed as a collection of propositions, e.g.,

Mary is young
Pat is much taller than Mary
overeating causes obesity
most Swedes are blond
tomatoes are red unless they are unripe
usually high quality goes with high price
if pressure is high then volume is low

To constitute knowledge a proposition must be understood. In this sense, meaning and knowledge are closely interrelated. In fuzzy logic, meaning representation—and thus knowledge representation—is based on test-score semantics (Zadeh, 1978a, 1986).

A basic idea underlying test-score semantics is that a proposition in a natural language may be viewed as a collection of elastic, or, equivalently, fuzzy constraints. For example, the proposition *Mary is tall* represents an elastic constraint on the height of Mary. Similarly, the proposition *Jean is blonde* represents an elastic constraint on the color of Jean's hair. And, the proposition *most tall men are not very agile* represents an elastic constraint on the proportion of men who are not very agile among tall men.

In more concrete terms, representing the meaning of a proposition, p, through the use of test-score semantics involves the following steps.

1. Identification of the variables $X_1, \ldots, X_n$ whose values are constrained by the proposition. Usually, these variables are implicit rather than explicit in p.
2. Identification of the constraints $C_1, \ldots, C_m$ which are induced by p.
3. Characterization of each constraint, C_i, by describing a testing procedure which associates with C_i a test score τ_i representing the degree to which C_i is satisfied. Usually τ_i is expressed as a number in the interval [0,1]. More generally, however, a test score may be a probability/possibility distribution over the unit interval.
4. Aggregation of the partial test scores $\tau_1, \ldots, \tau_m$ into a smaller number of test scores $\bar{\tau}_1, \ldots, \bar{\tau}_k$, which are represented as an *overall vector test score* $\tau = (\bar{\tau}_1, \ldots, \bar{\tau}_k)$. In most cases $k = 1$, so that the overall test scores is a scalar. We shall assume that this is the case unless an explicit statement to the contrary is made.

It is important to note that, in test-score semantics, the meaning of p is represented not by the overall test score τ but by the procedure which leads to it. Viewed in this perspective, test-score semantics may be regarded as a generalization of truth-conditional, possible-world and model-theoretic semantics. However, by providing a computational framework for dealing with uncertainty and dispositionality—which the conventional semantical systems disregard—test-score semantics achieves a much higher level of expressive power and thus provides a basis for representing the meaning of a much wider variety of propositions in a natural language.

In test-score semantics, the testing of the constraints induced by p is performed on a collection of fuzzy relations which constitute an *explanatory database*, or ED for short. A basic assumption which is made about the explanatory database is that it is comprised of relations whose meaning is known to the addressee of the meaning-representation process. In an indirect way, then, the testing and aggregation procedures in test-score semantics may be viewed as a description of a process by which the meaning of p is composed from the meanings of the constituent relations in the explanatory database. It is this explanatory role of the relations in ED that motivates its description as an *explanatory database*.

As will be seen in the sequel, in describing the testing procedures we need not concern ourselves with the actual entries in the constituent relations. Thus, in general, the description of a test involves only the frames of the constituent relations, that is, their names, their variables (or attributes) and the domain of each variable.

As a simple illustration of the concept of a test procedure, consider the proposition $p \triangleq$ *Maria is young and attractive*. The ED in this case will be assumed to consist of the following relations:

$$ED \triangleq POPULATION\ [Name; Age; \mu Attractive] + YOUNG\ [Age; \mu]\ , \quad (3.1)$$

in which + should be read as "and," and $\triangleq$ stands for "denotes."

The relation labeled POPULATION consists of a collection of triples whose first element is the name of an individual; whose second element is the age of that individual; and whose third element is the degree to which the individual in question is attractive. The relation YOUNG is a collection of pairs whose first element is a value of the variable *Age* and whose second element is the degree to which that value of

Age satisfies the elastic constraint characterized by the fuzzy predicate *young*. In effect, this relation serves to calibrate the meaning of the fuzzy predicate *young* in a particular context by representing its denotation as a fuzzy subset, YOUNG, of the interval [0,100].

With this ED, the test procedure which computes the overall test score may be described as follows:

1. Determine the age of Maria by reading the value of *Age* in POPULATION, with the variable *Name* bound to Maria. In symbols, this may be expressed as

$$Age\ (Maria) = {}_{Age}\ POPULATION\ [Name = Maria]\ .$$

   In this expression, we use the notation ${}_Y R\,[X = a]$ to signify that X is bound to a in R and the resulting relation is projected on Y, yielding the values of Y in the tuples in which $X = a$.

2. Test the elastic constraint induced by the fuzzy predicate *young*:

$$\tau_1 = {}_{\mu}\, YOUNG\, [Age = Age\,(Maria)]\ .$$

3. Determine the degree to which Maria is attractive:

$$\tau_2 = {}_{\mu Attractive}\ POPULATION\,[Name = Maria]\ .$$

4. Compute the overall test score by aggregating the partial test scores τ_1 and τ_2. For this purpose, we shall use the min operator $\wedge$ as the aggregation operator, yielding

$$\tau = \tau_1 \wedge \tau_1\ , \tag{3.2}$$

   which signifies that the overall test score is taken to be the smaller of the operands of $\wedge$. The overall test score, as expressed by (3.2), represents the compatibility of $p \triangleq$ *Maria is young and attractive* with the data resident in the explanatory database.

In testing the constituent relations in ED, it is helpful to have a collection of standardized translation rules for computing the test score of a combination of elastic constraints $C_1, \ldots, C_k$ from the knowledge of the test scores of each constraint considered in isolation. For the most part, such rules are *default* rules in the sense that they are intended to be used in the absence of alternative rules supplied by the user.

For purposes of knowledge representation, the principal rules of this type are the following.

1. Rules pertaining to modification

If the test score for an elastic constraint C in a specified context is τ, then in the same context the test score for

(a) *not* C is $1-\tau$ (*negation*)

(b) *very* C is τ^2 (*concentration*)

(c) *more or less* C is $\tau^{\frac{1}{2}}$ (*diffusion*) .

2. Rules pertaining to composition

If the test scores for elastic constraints C_1 and C_2 in a specified context are τ_1 and τ_2, respectively, then in the same context the test score for

(a) C_1 *and* C_2 is $\tau_1 \wedge \tau_2$ (*conjunction*), where $\wedge \triangleq$ *min*.

(b) C_1 *or* C_2 is $\tau_1 \vee \tau_2$ (*disjunction*), **where** $\vee \triangleq max$.

(c) *If* C_1 *then* C_2 is $1 \wedge (1-\tau_1+\tau_2)$ (*implication*) .

3. Rules pertaining to quantification

The rules in question apply to propositions of the general form *Q A's are B's* , where *Q* is a fuzzy quantifier, e.g., *most, many, several, few,* etc, and *A* and *B* are fuzzy sets, e.g., *tall men, intelligent men,* etc. As was stated earlier, when the fuzzy quantifiers in a proposition are implied rather than explicit, their suppression may be placed in evidence by referring to the proposition as a *disposition.* In this sense, the proposition *overeating causes obesity* is a disposition which results from the suppression of the fuzzy quantifier *most* in the proposition *most of those who overeat are obese.*

To make the concept of a fuzzy quantifier meaningful, it is necessary to define a way of counting the number of elements in a fuzzy set or, equivalently, to determine its cardinality.

There are several ways in which this can be done (Zadeh, 1978a; Dubois and Prade, 1985; Yager, 1980). For our purposes, it will suffice to employ the concept of a *sigma-count*, which is defined as follows:

Let F *be a fuzzy subset of* $U = \{u_1, \ldots, u_n\}$

expressed symbolically as

$$F = \mu_1/u_1 + \cdots + \mu_n/u_n = \Sigma_i \mu_i/u_i$$

or, more simply, as

$$F = \mu_1 u_1 + \cdots + \mu_n u_n \ ,$$

in which the term μ_i/u_i, $i = 1, \ldots, n$, signifies that μ_i is the grade of membership of u_i in F, and the plus sign represents the union.

The sigma-count of F is defined as the arithmetic sum of the μ_i , i.e.,

$$\Sigma\, Count(F) \triangleq \Sigma_i \mu_i \ , i = 1, \ldots, n \ ,$$

with the understanding that the sum may be rounded, if need be, to the nearest integer. Furthermore, one may stipulate that the terms whose grade of membership falls below a specified threshold be excluded from the summation. The purpose of such an exclusion is to avoid a situation in which a large number of terms with low grades of membership become count-equivalent to a small number of terms with high membership.

The *relative sigma-count*, denoted by $\Sigma\, Count(F/G)$, may be interpreted as the proportion of elements of F which are in G. More explicitly,

$$\Sigma\, Count(F/G) = \frac{\Sigma\, Count(F \cap G)}{\Sigma\, Count(G)} \ ,$$

where $F \cap G$, the intersection of F and G, is defined by

$$\mu_{F \cap G}(u) = \mu_F(u) \wedge \mu_G(u) \ , \quad u \in U \ .$$

Thus, in terms of the membership functions of F and G, the the relative sigma-count of F in G is given by

$$\Sigma\, Count(F/G) = \frac{\Sigma_i \mu_F(u_i) \wedge \mu_G(u_i)}{\Sigma_i \mu_G(u_i)} \ .$$

The concept of a relative sigma-count provides a basis for interpreting the meaning of propositions of the form *Q A's are B's*, e.g., *most young men are healthy*. More specifically, if the focal variable (i.e., the constrained variable) in the proposition in question is taken to be the proportion of *B's* in *A's*, then the corresponding translation rule may be expressed as

$$Q\ A\text{'s are } B\text{'s} \rightarrow \Sigma Count(B/A) \text{ is } Q \ .$$

As an illustration, consider the proposition $p \triangleq$ *over the past few years Naomi earned far more than most of her close friends*. In this case, we shall assume that the constituent relations in the explanatory database are:

ED $\triangleq$ INCOME [Name; Amount; Year] +
FRIEND [Name; μ] +
FEW [Number; μ] +
FAR.MORE [Income1; Income2; μ] +
MOST [Proportion; μ] .

Note that some of these relations are explicit in *p*; some are not; and that most of the constituent words in *p* do not appear in ED.

In what follows, we shall describe the process by which the meaning of *p* may be composed from the meaning of the constituent relations in ED. Basically, this process is a test procedure which tests, scores and aggregates the elastic constraints which are induced by *p*.

1. Find Naomi's income, IN_i, in $Year_i$, i = 1, 2, 3,..., counting backward from present. In symbols,

$$IN_i \triangleq {}_{Amount}\, INCOME\,[Name = Naomi; Year = Year_i],$$

   which signifies that *Name* is bound to Naomi, *Year* to $Year_i$, and the resulting relation is projected on the domain of the attribute *Amount*, yielding the value of *Amount* corresponding to the values assigned to the attributes *Name* and *Year*.

2. Test the constraint induced by FEW:

$$\mu_i \triangleq {}_{\mu}FEW\ [Year = Year_i] \ ,$$

   which signifies that the variable *Year* is bound to $Year_i$ and the corresponding value of μ is read by projecting on the domain of μ.

3. Compute Naomi's total income during the past few years:

$$TIN \triangleq \Sigma_i \mu_i IN_i ,$$

   in which the μ_i play the role of weighting coefficients. Thus, we are tacitly assuming that the total income earned by Naomi during a fuzzily specified interval of time is obtained by weighting Naomi's income in year $Year_i$ by the degree to which $Year_i$ satisfies the constraint induced by FEW and summing the weighted incomes.

4. Compute the total income of each $Name_j$ (other than Naomi) during the past few years:

$$TIName_j = \Sigma_i \mu_i IName_{ji} ,$$

   where $IName_{ji}$ is the income of $Name_j$ in $Year_i$.

5. Find the fuzzy set of individuals in relation to whom Naomi earned far more. The grade of membership of $Name_j$ in this set is given by

$$\mu_{FM}(Name_j) = {}_{\mu}FAR.MORE\,[Income\,1{=}TIN;\,Income\,2{=}TIName_j].$$

6. Find the fuzzy set of close friends of Naomi by intensifying (Zadeh, 1978a) the relation FRIEND:

$$CF \triangleq CLOSE.FRIEND \triangleq {}^{2}FRIEND,$$

which implies that

$$\mu_{CF}(Name_j) = ({}_{\mu}FRIEND\,[Name{=}Name_j])^2,$$

where the expression

$${}_{\mu}FRIEND\,[Name{=}Name_j]$$

represents $\mu_F(Name_j)$, that is, the grade of membership of $Name_j$ in the set of Naomi's friends.

7. Count the number of close friends of Naomi. On denoting the count in question by Σ Count (CF), we have:

$$\Sigma Count(CF) = \Sigma_j \mu^2{}_{FRIEND}(Name_j).$$

8. Find the intersection of FM with CF. The grade of membership of $Name_j$ in the intersection is given by

$$\mu_{FM \cap CF}(Name_j) = \mu_{FM}(Name_j) \wedge \mu_{CF}(Name_j),$$

where the min operator $\wedge$ signifies that the intersection is defined as the conjunction of its operands.

9. Compute the sigma-count of $FM \cap CF$:

$$\Sigma Count(FM \cap CF) = \Sigma_j \mu_{FM}(Name_j) \wedge \mu_{CF}(Name_j).$$

10. Compute the relative sigma-count of FM in CF, i.e., the proportion of individuals in $FM \cap CF$ who are in CF:

$$\rho \triangleq \frac{\Sigma Count(FM \cap CF)}{\Sigma Count(CF)}$$

11. Test the constraint induced by MOST:

$$\tau \triangleq {}_{\mu}MOST\,[Proportion{=}\rho],$$

which expresses the overall test score and thus represents the compatibility of p with the explanatory database.

In application to the representation of dispositional knowledge, the first step in the representation of the meaning of a disposition involves the process of *explicitation*, that is, making explicit the implicit quantifiers. As a simple example, consider the disposition

$$d \triangleq \textit{young men like young women}$$

which may be interpreted as the proposition

$$p \triangleq \textit{most young men like mostly young women.}$$

The candidate ED for p is assumed to consist of the following relations:

$$\text{ED} \triangleq \begin{array}{l} \text{POPULATION [Name; Sex; Age]} + \\ \text{LIKE [Name1; Name2; } \mu \text{]} + \\ \text{MOST [Proportion; } \mu \text{],} \end{array}$$

in which μ in LIKE is the degree to which *Name1* likes *Name2*.

To represent the meaning of p, it is expedient to replace p with the semantically equivalent proposition

$$q \triangleq \textit{most young men are } P \ ,$$

where P is the fuzzy *dispositional* predicate

$$P \triangleq \textit{likes mostly young women} \ .$$

In this way, the representation of the meaning of p is decomposed into two simpler problems, namely, the representation of the meaning of P, and the representation of the meaning of q knowing the meaning of P.

The meaning of P is represented by the following test procedure.

1. Divide POPULATION into the population of males, M.POPULATION, and population of females, F.POPULATION:

   $$\text{M.POPULATION} \triangleq {}_{Name,Age}\text{POPULATION [Sex = Male]}$$
   $$\text{F.POPULATION} \triangleq {}_{Name,Age}\text{POPULATION [Sex = Female]} \ ,$$

   where ${}_{Name,Age}$POPULATION denotes the projection of POPULATION on the attributes *Name* and *Age*.

2. For each $Name_j$, $j = 1, \ldots, l$, in F.POPULATION, find the age of $Name_j$:

   $$A_j \triangleq {}_{Age}\, F.POPULATION\,[Name = Name_j] \ .$$

3. For each $Name_j$, find the degree to which $Name_j$ is young:

   $$\alpha_j \triangleq {}_{\mu}\, YOUNG\,[Age = A_j] \ ,$$

   where α_j may be interpreted as the grade of membership of $Name_j$ in the fuzzy set, *YW*, of young women.

4. For each $Name_i$, $i = 1, \ldots, k$, in M.POPULATION, find the age of $Name_i$:

   $$B_i \triangleq {}_{Age}\, M.POPULATION\,[Name = Name_i] \ .$$

5. For each $Name_i$, find the degree to which $Name_i$ is young:

   $$\delta_i \triangleq {}_{\mu}\, YOUNG\,[Age = B_i] \ ,$$

   where δ_i may be interpreted as the grade of membership of $Name_i$ in the fuzzy set, *YM*, of young men.

6. For each $Name_j$, find the degree to which $Name_i$ likes $Name_j$:

   $$\beta_{ij} \triangleq {}_{\mu}\, LIKE\,[Name1 = Name_i\, ; Name2 = Name_j] \ ,$$

   with the understanding that β_{ij} may be interpreted as the grade of membership of $Name_j$ in the fuzzy set, WL_i, of women whom $Name_i$ likes.

7. For each $Name_j$ find the degree to which $Name_i$ likes $Name_j$ and $Name_j$ is young:

$$\gamma_{ij} \triangleq \alpha_j \wedge \beta_{ij} \ .$$

Note: As in previous examples, we employ the aggregation operator min ($\wedge$) to represent the effect of conjunction. In effect, γ_{ij} is the grade of membership of $Name_j$ in the intersection of the fuzzy sets WL_i and YW.

8. Compute the relative sigma-count of young women among the women whom $Name_i$ likes:

$$\rho_i \triangleq \Sigma Count(YW / WL_i)$$
$$= \frac{\Sigma Count(YW \cap WL_i)}{\Sigma Count(WL_i)}$$
$$= \frac{\Sigma_j \gamma_{ij}}{\Sigma_j \beta_{ij}}$$
$$= \frac{\Sigma_j \alpha_j \wedge \beta_{ij}}{\Sigma_j \beta_{ij}} \ .$$

9. Test the constraint induced by MOST:

$$\tau_i \triangleq {}_{\mu}MOST[Proportion = \rho_i] \ .$$

This test score, then, represents the degree to which $Name_i$ has the property expressed by the predicate

$$P \triangleq \textit{likes mostly young women} \ .$$

Continuing the test procedure, we have:

10. Compute the relative sigma-count of men who have property P among young men:

$$\rho \triangleq \Sigma\, Count(P / YM)$$
$$= \frac{\Sigma\, Count(P \cap YM)}{\Sigma\, Count(YM)}$$
$$= \frac{\Sigma_i\, \tau_i \wedge \delta_i}{\Sigma_i\, \delta_i} \ .$$

11. Test the constraint induced by MOST:

$$\tau = {}_{\mu}MOST[Proportion = \rho] \ .$$

This test score represents the overall test score for the disposition *young men like young women.*

THE CONCEPT OF A CANONICAL FORM AND ITS APPLICATION TO THE REPRESENTATION OF MEANING

When the meaning of a proposition, p, is represented as a test procedure, it may be hard to discern in the description of the procedure the underlying structure of the process through which the meaning of p is constructed from the meanings of the constituent relations in the explanatory database.

A concept which makes it easier to perceive the logical structure of p and thus to develop a better understanding of the meaning representation process, is that of a canonical form of p, abbreviated as $cf(p)$ (Zadeh, 1978b, 1986).

The concept of a canonical form relates to the basic idea which underlies test-score semantics, namely, that a proposition may be viewed as a system of elastic constraints whose domain is a collection of relations in the explanatory database. Equivalently, let $X_1, \ldots, X_n$ be a collection of variables which are constrained by p. Then, the canonical form of p may be expressed as

$$cf(p) \triangleq X \textit{ is } F \quad , \tag{4.1}$$

where $X = (X_1, \ldots, X_n)$ is the constrained variable which is usually implicit in p, and F is a fuzzy relation, likewise implicit in p, which plays the role of an elastic (or fuzzy) constraint on X. The relation between p and its canonical form will be expressed as

$$p \rightarrow X \textit{ is } F \quad , \tag{4.2}$$

signifying that the canonical form may be viewed as a representation of the meaning of p.

In general, the constrained variable X in $cf(p)$ is not uniquely determined by p, and is dependent on the focus of attention in the meaning-representation process. To place this in evidence, we shall refer to X as the *focal* variable.

As a simple illustration, consider the proposition

$$p \triangleq \textit{Anne has blue eyes} \quad . \tag{4.3}$$

In this case, the focal variable may be expressed as

$$X \triangleq \textit{Color}\,(\textit{Eyes}\,(\textit{Anne})) \quad ,$$

and the elastic constraint is represented by the fuzzy relation BLUE. Thus, we can write

$$p \rightarrow \textit{Color}\,(\textit{Eyes}\,(\textit{Anne})) \textit{ is } \textit{BLUE} \quad . \tag{4.4}$$

As an additional illustration, consider the proposition

$$p \triangleq \textit{Brian is much taller than Mildred.} \tag{4.5}$$

Here, the focal variable has two components, $X = (X_1, X_2)$, where

$$X_1 = \textit{Height}\,(\textit{Brian})$$

$$X_2 = \textit{Height}\,(\textit{Mildred})\,;$$

and the elastic constraint is characterized by the fuzzy relation MUCH.TALLER [$Height1$; $Height2$; μ], in which μ is the degree to which $Height1$ is *much taller* than $Height2$. In this case, we have

$$p \rightarrow (\textit{Height}\,(\textit{Brian}), \textit{Height}\,(\textit{Mildred})) \textit{ is } \textit{MUCH.TALLER} \quad . \tag{4.6}$$

In terms of the possibility distribution of X, the canonical form of p may be interpreted as the assignment of F to Π_X. Thus, we may write

$$p \rightarrow X \textit{ is } F \rightarrow \Pi_X = F \quad , \tag{4.7}$$

in which the equation

$$\Pi_X = F \tag{4.8}$$

is termed the *possibility assignment equation* (Zadeh 1978b). In effect, this equation signifies that the canonical form $cf(p) \triangleq X \textit{ is } F$ implies that

$$Poss\{X = u\} = \mu_F(u)\ , \quad u \in U\ , \tag{4.9}$$

where μ_F is the membership function of F. It is in this sense that F, acting as an elastic constraint on X, restricts the possible values which X can take in U. An important implication of this observation is that a proposition, p, may be interpreted as an implicit assignment statement which characterizes the possibility distribution of the focal variable in p.

As an illustration, consider the disposition

$$d \triangleq overeating\ causes\ obesity\ , \tag{4.10}$$

which upon explicitation becomes

$$p \triangleq most\ of\ those\ who\ overeat\ are\ obese\ . \tag{4.11}$$

If the focal variable in this case is chosen to be the relative sigma-count of those who are obese among those who overeat, the canonical form of p becomes

$$\Sigma\, Count\ (OBESE/OVEREAT)\ is\ MOST\ , \tag{4.12}$$

which in virtue of (4.9) implies that

$$Poss\{\Sigma\, Count\ (OBESE/OVEREAT) = u\} = \mu_{MOST}(u)\ , \tag{4.13}$$

where μ_{MOST} is the membership function of $MOST$. What is important to note is that (4.13) is equivalent to the assertion that the overall test score for p is expressed by

$$\tau = \mu_{MOST}(\Sigma\, Count(OBESE/OVEREAT))\ , \tag{4.14}$$

in which OBESE, OVEREAT and MOST play the roles of the constituent relations in ED.

It is of interest to observe that the notion of a semantic network may be viewed as a special case of the concept of a canonical form. As a simple illustration, consider the proposition

$$p \triangleq Richard\ gave\ Cindy\ a\ red\ pin\ . \tag{4.15}$$

As a semantic network, this proposition may be represented in the standard form:

$$\begin{aligned} &Agent\ (GIVE) = Richard \\ &Recipient\ (GIVE) = Cindy \\ &Time\ (GIVE) = Past \\ &Object\ (GIVE) = Pin \\ &Color\ (Pin) = Red\ . \end{aligned} \tag{4.16}$$

Now, if we identify X_1 with *Agent* (*GIVE*), X_2 with *Recipient* (*GIVE*), etc., the semantic network representation (4.16) may be regarded as a canonical form in which $X = (X_1, \ldots, X_5)$, and

$$\begin{aligned} &X_1 = Richard \\ &X_2 = Cindy \\ &X_3\ \text{is}\ Past \\ &X_4\ \text{is}\ Pin \\ &X_5\ \text{is}\ Red\ . \end{aligned} \tag{4.17}$$

More generally, since any semantic network may be expressed as a collection of triples of the form (Object, Attribute, Attribute Value), it can be transformed at once into a canonical form. However, since a canonical form has a much greater expressive power than a semantic network, it may be difficult to transform a canonical form into a semantic network.

INFERENCE

The concept of a canonical form provides a convenient framework for representing the rules of inference in fuzzy logic. Since the main concern of the paper is with knowledge representation rather than with inference, our discussion of the rules of inference in fuzzy logic in this section has the format of a summary.

In the so-called categorical rules of inference, the premises are assumed to be in the canonical form *X is A* or the conditional canonical form *X is A if Y is B*, where *A* and *B* are fuzzy predicates (or relations). In the syllogistic rules, the premises are expressed as *Q A's are B's*, where *Q* is a fuzzy quantifier and *A* and *B* are fuzzy predicates (or relations).

The rules in question are the following

CATEGORICAL RULES

$X, Y, Z, \cdots \triangleq$ variables taking values in $U, V, W, \cdots$

Examples

$X \triangleq Age(Mary),\ Y = Distance(P1, P2)$

$A, B, C, \cdots =$ fuzzy predicates (relations)

Examples

$A = small,\ B = much\ larger$

ENTAILMENT RULE

X *is* A

$A \subset B \rightarrow \mu_A(u) \leq \mu_B(u),\ u \in U$

X *is* B

Example

Mary is very young
very young $\subset$ *young*

Mary is young

CONJUNCTION RULE

X is A
X is B

X is A $\cap$ *B* $\rightarrow$ $\mu_{A \cap B}(u) = \mu_A(u) \wedge \mu_B(u)$

$\cap$ = intersection (conjunction)

Example

pressure is not very high
pressure is not very low

---

pressure is not very high and not very low

DISJUNCTION RULE

X is A
or *X is B*

X is A $\cup$ *B* $\rightarrow$ $\mu_{A \cup B}(u) = \mu_A(u) \vee \mu_B(u)$

$\cup$ = union (disjunction)

PROJECTION RULE

(X,Y) is R

X is ${}_X R$ $\rightarrow$ $\mu_{{}_X R}(u) = sup_v \, \mu_R(u, v)$

${}_X R \triangleq$ projection of R on U

Example

(X,Y) is close to (3,2)

X is close to 3

COMPOSITIONAL RULE

(X,Y) is R $\rightarrow$ *binary predicate*
Y is B

X is A$\circ$*R* $\rightarrow$ $\mu_{A \circ R}(u) = sup_v (\mu_R(u, v) \wedge \mu_B(v))$

Example

X is much larger than Y
Y is large

X is much larger $\circ$ *large*

NEGATION RULE

$$\frac{not\ (X\ is\ A)}{X\ is\ \neg A} \quad \rightarrow \quad \mu_{\neg A}(u) = 1 - \mu_A(u)$$

$\neg \triangleq$ negation

Example

$$\frac{not\ (Mary\ is\ young)}{Mary\ is\ not\ young}$$

EXTENSION PRINCIPLE

$$\frac{X\ is\ A}{f(X)\ is\ f(A)}$$

$$A = \mu_1/u_1 + \mu_2/u_2 + \cdots + \mu_n/u_n$$

$$f(A) = \mu_1/f(u_1) + \mu_2/f(u_2) + \cdots + \mu_n/f(u_n)$$

Example

$$\frac{X\ is\ small}{X^2\ is\ {}^2small}$$

$${}^2small \triangleq \text{very small}, \mu_{very\ small} = (\mu_{small})^2$$

It should be noted that the use of the canonical form of in these rules stands in sharp contrast to the way in which the rules of inference are expressed in classical logic. The advantage of the canonical form is that it places in evidence that inference in fuzzy logic may be interpreted as a propagation of elastic constraints. This point of view is particularly useful in the applications of fuzzy logic to control and decision analysis (*Proc. of the 2nd IFSA Congress*, 1987, *Proc. of the International Workshop*, Iizuka, 1988).

As was pointed out already, it is the qualitative mode of reasoning that plays a key role in the applications of fuzzy logic to control. In such applications, the input-output relations are expressed as collections of fuzzy if-then rules (Mamdani and Gaines, 1981).

For example, if X and Y are input variables and Z is the output variable, the relation between X, Y, and Z may be expressed as

$$Z\ is\ C_1 \text{ if } X\ is\ A_1 \text{ and } Y\ is\ B_1$$
$$Z\ is\ C_2 \text{ if } X\ is\ A_2 \text{ and } Y\ is\ B_2$$
$$\cdot \quad \cdot \quad \cdot \quad \cdot \quad \cdot \quad \cdot \quad \cdot$$
$$Z\ is\ C_n \text{ if } X\ is\ A_n \text{ and } Y\ is\ B_n$$

where C_i, A_i, and B_i, $i=1,\ldots,n$ are fuzzy subsets of their respective universes of discourse. For example,

Z is small if X is large and Y is medium
Z is not large if X is very small and Y is not large

Given a characterization of the dependence of Z on X and Y in this form, one can employ the compositional rule of inference to compute the value of Z given the values of X and Y. This is what underlies the Togai-Watanbe fuzzy logic chip (Togai, 1986) and the operation of fuzzy logic controllers in industrial process control (Sugeno, 1985).

In general, the applications of fuzzly logic in systems and process control fall into two categories. First, there are those applications in which, in comparison with traditional methods, fuzzy logic control offers the advantage of greater simplicity, greater robustness, and lower cost. The cement kiln control pioneered by the F.L. Smidth Company falls into this category.

Second, are the applications in which the traditional methods provide no solution. The self-parking fuzzy car conceived by Sugeno (Sugeno, 1985) is a prime example of what humans can do so easily and is so difficult to emulate by the traditional approaches to systems control.

SYLLOGISTIC RULES

In its generic form, a fuzzy syllogism may be expressed as the inference schema

$$Q_1 A\text{'s are } B\text{'s}$$
$$\underline{Q_2 C\text{'s are } D\text{'s}}$$
$$Q_3 E\text{'s are } F\text{'s}$$

in which A, B, C, D, E and F are interrelated fuzzy predicates and Q_1, Q_2 and Q_3 are fuzzy quantifiers.

The interrelations between A,B,C,D,E and F provide a basis for a classification of fuzzy syllogisms. The more important of these syllogisms are the following

(a) *Intersection/product syllogism:*

$$C=A \wedge B\,,\, E=A\,,\, F=C \wedge D$$

(b) *Chaining syllogism:*

$$C=B\,,\, E=A\,,\, F=D$$

(c) *Consequent conjunction syllogism:*

$$A=C=E\,,\, F=B \wedge D$$

(d) *Consequent disjunction syllogism:*

$$A=C=E\,,\, F=B \vee D$$

(e) *Antecedent conjunction syllogism:*

$$B=D=F\,,\, E=A \wedge C$$

(f) *Antecedent disjunction syllogism:*

$$B=D=F\,,\, E=A \vee C$$

In the context of expert systems, these and related syllogisms provide a set of inference rules for combining evidence through conjunction, disjunction and chaining (Zadeh, 1983b).

One of the basic problems in fuzzy syllogistic reasoning is the following: Given A, B, C, D, E and F, find the maximally specific (i.e., most restrictive) fuzzy quantifier Q_3 such that the proposition Q_3 *E's are F's* is entailed by the premises. In the case of (a), (b) and (c), this leads to the following syllogisms:

INTERSECTION/PRODUCT SYLLOGISM.

$$Q_1 A\text{'s are } B\text{'s} \tag{5.1}$$

$$\underline{Q_2 (A \text{ and } B)\text{'s are } C\text{'s}}$$

$$(Q_1 \otimes Q_2) A\text{'s are } (B \text{ and } C)\text{'s}$$

where $\otimes$ denotes the product in fuzzy arithmetic (Kaufmann and Gupta, 1985). It should be noted that (5.1) may be viewed as an analog of the basic probabilistic identity

$$p(B,C/A) = p(B/A)p(C/A,B)$$

A concrete example of the intersection/product syllogism is the following:

most students are young (5.2)

<u>*most young students are single*</u>

most^2 *students are young* and *single*

where *most*^{2} denotes the product of the fuzzy quantifier *most* with itself.

CHAINING SYLLOGISM.

$$Q_1 A\text{'s are } B\text{'s}$$

$$\underline{Q_2 B\text{'s are } C\text{'s}}$$

$$(Q_1 \otimes Q_2)\ A\text{'s are } C\text{'s}$$

This syllogism may be viewed as a special case of the intersection product syllogism. It results when $B \subset A$ and Q_1 and Q_2 are monotone increasing, that is, $\geq Q_1 = Q_1$, and $\geq Q_2 = Q_2$,where $\geq Q_1$ should be read as *at least* Q_1, Q_2. A simple example of the chaining syllogism is the following:

most students are undergraduates

<u>*most undergraduates are single*</u>

most^2 *students are single*

Note that *undergraduates* $\subset$ *students* and that in the conclusion F = *single*, rather than *young and single*, as in (5.2).

CONSEQUENT CONJUNCTION SYLLOGISM.

The consequent conjunction syllogism is a example of a basic syllogism which is not a derivative of the intersection/product syllogism. Its statement may be expressed as follows:

$$Q_1 A\text{'s are } B\text{'s} \tag{5.3}$$

$$\underline{Q_2 A\text{'s are } C\text{'s}}$$

$$Q\ A\text{'s are } (B \text{ and } C)\text{'s},$$

where Q is a fuzzy quantifier which is defined by the inequalities

$$0 \circledvee (Q_1 \otimes Q_2 \ominus 1) \leq Q \leq Q_1 \leq \circledwedge Q_2 \tag{5.4}$$

in which $\circledvee, \circledwedge, \otimes$ and $\ominus$ are the operations of $\vee$ (max), $\wedge$ (min), + and - in fuzzy arithmetic.

An illustration of (5.3) is provided by the example

most students are young

<u>most students are single</u>

Q students are single and *young*

where

$$2\,most \circledvee 1 \leq Q \leq most.$$

This expression for Q follows from (5.4) by noting that

$$most \circledvee most = most$$

and

$$0 \circledvee (2most \ominus 1) = 2most \ominus 1.$$

The three basic syllogisms stated above are merely examples of a collection of fuzzy syllogisms which may be developed and employed for purposes of inference from commonsense knowledge. In addition to its application to commonsense reasoning, fuzzy syllogistic reasoning may serve to provide a basis for combining uncertain evidence in expert systems (Zadeh, 1983b).

CONCLUDING REMARKS

One of the basic aims of fuzzy logic is to provide a computational framework for knowledge representation and inference in an environment of uncertainty and imprecision. In such environments, fuzzy logic is effective when the solutions need not be precise and/or it is acceptable for a conclusion to have a dispositional rather than categorical validity. The importance of fuzzy logic derives from the fact that there are many real world applications which fit these conditions, especially in the realm of knowledge-based systems for decision-making and control.

REFERENCES AND RELATED PUBLICATIONS

Baldwin, J.F. "FRIL—A fuzzy relational inference language," *Fuzzy Sets and Systems 14*, 155-174, 1984.

Baldwin, J.F., Martin, T.P., and Pilsworth, B.W., "Implementation of FPROG—a fuzzy Prolog interpreter," *Fuzzy Sets and Systems 23*, 119-129, 1987.

Bezdek, J.C. (ed.), *Analysis of Fuzzy Information—Vol 1,2, and 3: Applications in Engineering and Science*, CRC Press, Boca Raton, FL., 1987

Brachman, R.J. and Levesque, H.J. *Readings in Knowledge Representation*, Morgan Kaufmann Publishers, Inc., Los Altos, Calif., 1985.

Brachman, R.J. "The basics of knowledge representation and reasoning," *AT&T Technical Journal 67*, 25-40, 1988.

de Kleer, J., and J. Brown, "A qualitative physics based on confluences," *Artificial Intelligence, 24*, 7-84, 1984.

Doyle, J. "A truth-maintenance system," *Artificial Intelligence 12*, 231-272, 1979.

Dubois, D. and Prade, H., *Fuzzy Sets and Systems: Theory and Applications*. Academic Press, New York, 1980.

Dubois, D. and Prade, H. "Fuzzy cardinality and the modeling of imprecise quantification," *Fuzzy Sets and Systems 16*, 199-230, 1985.

Dubois, D. and Prade, H., *Possibility Theory—An Approach to Computerized Processing of Uncertainty*. Plenum Press, New York, 1988.

Dubois, D. and Prade, H. "On fuzzy syllogisms," *Computational Intelligence 14*, 171-179, 1988a.

Dubois, D. and Prade, H. "The treatment of uncertainty in knowledge-based systems using fuzzy sets and possibility theory," *Int. J. Intelligent Systems 3*, 141-165, 1988b.

Farreny, H. and Prade, H. "Dealing with the vagueness of natural languages in man-machine communication," *Applications of Fuzzy Set Theory in Human Factors*, W. Karvowski and A. Mital (eds.), Elsevier Science Publ., Amsterdam, 71-85, 1986.

Forbus, K., "Qualitative physics: past, present, and future," *Exploring Artificial Intelligence*, H. Shrobe, ed., Morgan Kaufman, Los Altos, CA, 1989.

Fujitec, "Artificial intelligence type elevator group control system," *JETRO*, 26, 1988.

Goguen, J.A., "The logic of inexact concepts," *Synthese 19*, 325-373, 1969.

Goodman, I.R. and Nguyen, H.T., *Uncertainty Models for Knowledge-Based Systems*. North-Holland, Amsterdam, 1985.

Gupta, M.M. and Yamakawa, T. (eds.). *Fuzzy Logic in Knowledge-Based Systems*. North-Holland, Amsterdam, 1988.

Isik, C., "Inference engines for fuzzy rule-based control," *International Jour. of Approximate Reasoning 2*, 122-187, 1988.

Johnson-Laird, P.N. "Procedural semantics," *Cognition 5*, 189-214, 1987.

Kacprzyk, J. and Yager, R.R. (eds.), *Management Decision Support Systems Using Fuzzy Sets and Possibility Theory. Interdisciplinary Systems Research Series*, vol. 83, Verlag TUV Rheiland, Koln, 1985.

Kacprzyk, J. and Orlovski, S.A. (eds.), *Optimization Models Using Fuzzy Sets and Possibility Theory*. D. Reidel, Dordrecht, 1987.

Kasai Y. and Y. Morimoto, "Electronically controlled continuously variable transmission," *Proc. Int. Congress on Transportation Electronics*, Dearborn, Michigan, 1988.

Kaufmann, A. and Gupta, M.M., *Introduction to Fuzzy Arithmetic*. Van Nostrand, New York, 1985.

Kaufmann, A. and Gupta, M.M., *Fuzzy Mathematical Models with Applications to Engineering and Management Science*. North-Holland, Amsterdam, 1988.

Kinoshita, M., and T. Fukuzaki, T. Satoh, and M. Miyake, "An automatic operation method for control rods in BWR plants," *Proc. Specialists' Meeting on In-core Instrumentation and Reactor Core Assessment*, Cadarache, France, 1988.

Kiszka, J.B., M.M. Gupta, and P.N. Nikiforuk, "Energetistic stability of fuzzy dynamic systems," *IEEE Transactions on Systems, Man and Cybernetics SMC-15*, 1985.

Klir, G.J. and Folger, T.A., *Fuzzy Sets, Uncertainty and Information*. Prentice Hall, Englewood Cliffs, N.J., 1988.

Kuipers, P., "Qualitative simulation," *Artificial Intelligence 29*, 289- 338, 1986.

Levesque, H.J. "Knowledge representation and reasoning," *Annual Reviews of Computer Science 1*, Annual Review, Inc., Palo Alto, Calif., 255-287, 1986.

Levesque, H.J. and Brachman, R. "Expressiveness and tractability in knowledge representation and reasoning," *Computational Intelligence 3*, 78-93, 1987.

Mamdani, E.H. and Gaines, B.R. (eds.), *Fuzzy Reasoning and its Applications*. Academic Press, London, 1981.

McCarthy, J., "Circumscription: non-monotonic inference rule," *Artificial Intelligence 13*, 27-40, 1980.

McDermott, D.V., "Non-monotonic logic, I," *Artificial Intelligence 13*, 41-72, 1980.

McDermott, D.V., "Non-monotonic logic, II: non-monotonic modal theories," *Journal of the Association for Computer Machinery 29*, 33-57, 1982.

Moore, R.C., "The role of logic in knowledge representation and commonsense reasoning," *Proceedings of the National Conference on Artificial Intelligence*, 428-433, 1982.

Moore, R.C. and Hobbs, J.C. (eds.), *Formal Theories of the Commonsense World*. Ablex Publishing, Harwood, N.J., 1984.

Mukaidono, M., Z. Shen, and L. Ding, "Fuzzy Prolog," *Proc. 2nd IFSA Congress*, Tokyo, Japan, 452-455, 1987.

Negoita, C.V., *Expert Systems and Fuzzy Systems*. Benjamin/Cummings, Menlo Park, CA 1985.

Nilsson, N., "Probabilistic logic," *Artificial Intelligence 20*, 71-87, 1986.

Peterson, P., "On the logic of *few, many,* and *most*," *Notre Dame Journal of Formal Logic 20*, 155-179, 1979.

Pospelov, G.S., "Fuzzy set theory in the USSR," *Fuzzy Sets and Systems 22*, 1-24, 1987.

Proceedings of the Second Congress of the International Fuzzy Systems Association, Tokyo, Japan, 1987.

Proceedings of the International Workshop on Fuzzy Systems Applications, Kyushu Institute of Technology, Iizuka, Japan, 1988.

Reiter, R. and Criscuolo, G., "Some representational issues in default reasoning," *Computers and Mathematics 9*, 15-28, 1983.

Shapiro, J.C. (ed.), *Enclyclopedia of Artificial Intelligence*. John Wiley & Sons, New York, 1987.

Small, S.L., Cottrell, G.W., and Tanenhaus, M.K. (eds.), *Lexical Ambiguity Resolution*. Morgan Kaufman Publishers, Los Altos, CA, 1988.

Sugeno, M., ed., *Industrial Applications of Fuzzy Control*, North Holland, Amsterdam, 1985.

Talbot, C.J., "Scheduling TV advertising: an expert systems approach to utilising fuzzy knowledge," *Proc. of the Fourth Australian Conference on Applications of Expert Systems*, Sydney, Australia, 1988.

Togai, M., and H. Watanabe, "Expert systems on a chip: an engine for real-time approximate reasoning," *IEEE Expert 1*, 55-62, 1986.

Wilensky, R. "Some problems and proposals for knowledge representation," Technical Report 87/351, Computer Science Division, University of California, Berkeley, 1987.

Yager, R.R. "Quantified propositions in a linguistic logic," *Proceedings of the 2nd International Seminar of Fuzzy Set Theory*, Klement, E.P. (ed.), Johannes Kepler University, Linz, Austria, 1980.

Yager, R.R. "Reasoning with fuzzy quantified statements—I," *Kybernetes 14*, 233-240, 1985.

Yasunobu, S. and G. Hasegawa, "Evaluation of an automatic container crane operation system based on predictive fuzzy control," *Control Theory and Advanced Technology*, vol. 2, no. 3, 1986.

Yasunobu, S. and S. Myamoto, "Automatic train operation by predictive fuzzy control," *Industrial Applications of Fuzzy Control*, M. Sugeno, ed., North Holland, Amsterdam, 1985.

Zadeh, L.A. "Probability measures of fuzzy events," *Jour. Math. Anal. and Applications 23*, 421-427, 1968.

Zadeh, L.A. "Outline of a new approach to the analysis of complex systems and decision processes," *IEEE Trans. on Systems, Man and Cybernetics SMC-3*, 28-44, 1973.

Zadeh, L.A. "The Concept of a Linguistic Variable and its Application to Approximate Reasoning," Part I; *Inf. Science 8*, 199-249; Part II *Inf. Science 8*, 301-357; Part III *Inf. Science 9*, 43-80, 1975.

Zadeh, L.A. "Fuzzy sets as a basis for a theory of possibility," *Fuzzy Sets and Systems 1*, 3-28, 1978a.

Zadeh, L.A. "PRUF—A meaning representation language for natural languages," *Int. J. Man—Machine Studies 10*, 395-460, 1978b..

Zadeh, L.A. "A fuzzy-set-theoretic approach to fuzzy quantifiers in natural languages," *Computers and Mathematics 9*, 149-184, 1983a.

Zadeh, L.A. "The role of fuzzy logic in the management of uncertainty in expert systems," *Fuzzy Sets and Systems 11*, 199-227, 1983b.

Zadeh, L.A. "A theory of commonsense knowledge," in *Aspects of Vagueness*, H.J. Skala, S. Termini and E. Trillas, (eds.), Reidel, Dordrecht, 1984.

Zadeh, L.A. "Syllogistic reasoning in fuzzy logic and its application to reasoning with dispositions, " *IEEE Trans. on Systems, Man and Cybernetics SMC-15*, 754-763, 1985.

Zadeh, L.A. "Test-Score Semantics as a Basis for a Computational Approach to the Representation of Meaning," *Literary and Linguistic Computing 1*, 24-35, 1986.

Zadeh, L.A., "A computational theory of dispositions," *Int. J. of Intelligent Systems 2*, 39-63, 1987.

Zadeh, L.A., "Fuzzy logic," *Computer 1*, 83-93, 1988a.

Zadeh, L.A., "Dispositional logic," *Appl. Math. Lett.*, 95-99, 1988b.

Zadeh, L.A. "QSA/FL—Qualitative systems analysis based on fuzzy logic," *Proc. AAAI Symposium*, Stanford University, 1989.

Zemankova-Leech, M. and Kandel, A., *Fuzzy Relational Data Bases - A Key to Expert Systems*. Verlag TUV Rheinland, Cologne, 1984.

Zimmerman, H.J., *Fuzzy Set Theory and its Applications*. Kluwer, Nijhoff, Dordrecht, 1987.

2

EXPERT SYSTEMS USING FUZZY LOGIC

Ronald R. Yager
Machine Intelligence Institute
Iona College
New Rochelle, NY 10801

ABSTRACT

We show how the theory of approximate reasoning developed by L.A. Zadeh provides a natural format for representing the knowledge and performing the inferences in rule based expert systems. We extend the representational ability of these systems by providing a new structure for including rules which only require the satisfaction to some subset of the antecedent conditions. This is accomplished by the use of fuzzy quantifiers. We also provide a methodology for the inclusion of a form of uncertainty in the expert systems associated with the belief attributed to the data and production rules.

INTRODUCTION

In [1] Buchanan and Duda provide an excellent introduction to the principles of rule-based expert systems. In [2] Buchanan provides a bibliography on expert systems. A particularly well cited example of a rule based expert system is MYCIN [3,4]. In [5] Van Melle has abstracted the basic structure of the MYCIN system and provided a language for the development of prototypical rule based expert systems called EMYCIN.

A rule based expert system is essentially an example of a production system consisting of the following components [1]:

1. A rule or knowledge base - This consists of the experts knowledge in the form of conditional type statements. Each conditional statement consisting of an antecedent portion and consequent portion. Typical rules are of the form

if antecedent then consequent.

2. A problem or global database - This consists of a set of facts or assertions about the current problem.

3. A rule interpreter- This consists of the portion of the system that carries out the problem solving. The rule interpreter can be considered to have two components. The first component consists of the inference mechanism. This helps to determine when a particular rule is valid and what is the effect of applying this rule. The second component consists of some meta-rules which helps determine in which order the rule base is to be searched for applicability of rules.

The information in the problem database as well as the antecedent and consequents of the rules are of the form *the (attribute) of (object) is (value)*, an optional certainty measure can be assigned to these propositions.

The expert system is generally activated by the introduction of a problem to be solved in the form of a goal to be satisfied as well as the insertion in the problem data base of data about the current problem. In many cases the expert system is a pattern directed or forward chaining production system. In this type of situation the problem is initiated with the insertion of the goal state and the data rule base is then searched to find which rules can be applied based upon the information in the problem database. A rule is applicable if the information in the global database satisfies the antecedent portion of the rule. If a rule is fired the appropriate information is added to the global database forming a new augmented database. One sees this in essence as being an application of modens pollens. The rule base is then again searched for fireable rules using this new augmented global database again adding new information to the database. The meta-knowledge in the rule interpreter is used to help direct the search for fireable rules. The determination of good heuristics for searching the rule base plays a significant role in the intelligent aspects of the system. The process of firing rules continues until no new information can be added or a goal state is reached.

In this paper we are concerned with the question of representation of the propositions forming the information in the global database and the antecedents and consequents in rules, as well as the inference mechanism used to infer the consequence of fired rules. We shall also provide a format for the representation of complex rules in which only some of the antecedents need to be satisfied. Furthermore, we shall provide a mechanism for the inclusion of some forms of uncertainty. In particular we shall suggest that the theory of approximate reasoning based upon fuzzy subsets developed by L.A. Zadeh provides a very robust methodology for representing the propositions and implementing the appropriate inference [6-10]. We note that Yager [11,12] has provided an approach for querying large knowledge bases of the type found in expert systems where the information is described in terms of fuzzy propositions.

REPRESENTATION OF DATA AND RULES

As noted by Buchanan and Duda [1] the fundamental building block for the information in both the database and the rule base of an expert system are propositional statements of the form:

the (attribute) of (object) is (value)

For example

The height of John is 6 feet.

The temperature of the patient is 102.

One can combine the ideas of attribute and object into a concept called a variable. Thus in the above examples John's height and the patient's temperature can be considered variables. In this notation the fundamental building blocks of the rule-based expert systems would be

$$V \text{ is } A,$$

where V is a variable, attribute (object) and A is its current value.

It is at this point we diverge from the current representational approach to expert systems knowledge. In the current systems, such as MYCIN, the values of the variables are left as symbols, words or values with no meaning. That is, the data *Temperature is high* is left in this form, no attempt is made to give any meaning to the value high. That is, the values are considered as atomic items with no further attempts at understanding their meaning. The matching used to determine the fireability of rules is carried out at this level of semantics. Using the values at this level of detail provokes some important questions. When two people use the same word, such as the designer of a system and the user, do they mean the same thing ? Secondly, if a rule has a certain value for a variable in its antecedent can we still learn something about the consequent variable if we only know that the value of antecedent is close to the value in the rule ? The ability to handle these types of problems requires us to provide a deeper semantics for the values associated with variables. Just as the predicate logic refines and improves upon the propositional logic by further decomposing the atomic statements the theory of approximate reasoning [6-10] further refines the meaning of the values associated with variables.

The approach we suggest is based upon idea of fuzzy subsets introduced by Zadeh [13]. Assume X is a set of objects. A fuzzy subset A of X is a subset in which the membership grade for each $x \in X$ is a element in the unit interval [0,1]. We denote this membership function A(x). In our approach a proposition such as

Age is old

has the effect of associating with the variable age a possibility distribution [9].

Assume we have the proposition

V is A

where A is some value. We can express A as a fuzzy subset of a base set, the set values the variable can assume. For example, if A is old we can express A as a fuzzy subset if interval of ages [0,150]. In particular, X is the set of all values that V can assume. The statement in turn induces a possibility distribution, π_V over the set X such that

$$\Pi_V(x) = A(x),$$

where A(x) is the membership grade of x in A. In particular $\Pi_V(x)$ is seen to be the possibility that V = x given the data V is A.

In a rule based expert system the fundamental component of the rules are conditional statements of the form

$$\textit{if } V_1 \textit{ is } A \textit{ then } V_2 \textit{ is } B.$$

As suggested by Zadeh [8] propositions of this type can also be seen to generate possibility distributions. In particular if V_1 and V_2 have as their base sets the sets X and Y respectively then

if V_1 is A then V_2 is B

induces a conditional possibility distribution $\pi_{v_1|v_2}$ over X × Y such that

$$\Pi_{v_1|v_2}(x, y) = \text{Min}(1, 1 - A(x) + B(y)).$$

an alternative definition is

$$\Pi_{v_1|v_2}(x, y) = \text{Max}(1 - A(x), B(y)).$$

Thus in this approach the effect of both data statements and rules are to introduce possibility distributions.

More complex forms of rules can easily be represented in this approach. If $V_1, V_2, \ldots V_n$ are variables taking values in the base sets $X_1, X_2, \ldots X_n$ respectively then the statement

V_1 is A_1 and V_2 is A_2 and ... and V_n is A_n

is seen to induce the joint possibility distribution $\Pi_{V_1, V_2, V_3, \quad V_n}$ over $X_1 \times X_2 \ldots \times X_n$ such that

$$\Pi_{V_1, V_2, V_3, \quad V_n}(x_1, x_2, \ldots x_n) = \text{Min}_i [A_i(x_i)]$$

The statement $V = A_1$ or $V_2 = A_2$ or $V_n = A_n$ induces the joint possibility distribution $\Pi_{V_1, V_2, V_3, \quad V_n}$ over $X_1 \times X_2 \ldots \times X_n$ such that

$$\Pi_{V_1, V_2, V_3, \quad V_n}(x_1, x_2, \ldots x_n) = \text{Max}_i [A_i(x_i)].$$

With these ideas we can easily represent more complicated rules. Let $V_1, V_2 \ldots V_n$ be variables with base sets $X_1 \ldots X_n$ respectively and let $U_1, U_2, \ldots U_p$ be variables with base sets $Y_1, Y_2, \ldots Y_n$ respectively. Consider the rule

if V_1 is A_1 and V_2 is A_2 ... and V_n is A_n then U_1 is B_1.

This induces a condition possibility distribution $\Pi_{u_1|v_1, v_2, \ldots v_n}$ over $X_1 \times X_2 \ldots X_n \times Y_1$ such that

$$\Pi_{u_1|v_1,v_2,\ldots v_n}(x_1, x_2, \ldots x_n, y_1) = \text{Min}(1, 1\text{-}H(x_1, x_2, \ldots x_n) + B_1(y_1)),$$

where $H(x_1, x_2, \ldots x_n) = \text{Min}_i [A_i(x_i)]$.

Consider the rule

if V_1 is A_1 and V_2 is A_2 ... and V_n is A_n

then

U_1 is B_1 or U_2 is B_2 or U_p is B_p.

This generates the conditional possibility distribution $\Pi_{u_1,u_2,\ldots u_p|v_1,v_2,\ldots v_n}$ over the set $X_1 \times X_2, \ldots \times X_n \times Y_1 \times Y_2 \times Y_3 \ldots \times X_p$ such that

$$\Pi_{U_1, U_2, \ldots\ldots U_p/V1, V2, \ldots Vp}(x_1, x_2, x_n, y_1, y_2, \ldots yp) = \text{Min}(1, 1\text{-}H(x_1, x_2, \ldots x_n) + G(y_1, y_2, \ldots y_p))$$

where

$$H(x_1, x_2, \ldots x_n) = \mathrm{Min}_i\,(A_i\,(x_i)\,)$$

and

$$G(y_1, y_2, \ldots y_p) = \mathrm{Max}_i\,(B_i\,(y_i)\,).$$

Other complex rules can be expressed in a similar manner.

INFERENCES FROM THE SYSTEM

The ability to use the database to search the rule base to infer further data in this approach is based upon the inference laws of the theory of approximate reasoning. The essential laws for this purpose are the conjunction principle, and the entailment principle. These laws are related respectively to the laws of adjunction, law of simplification, law of modens pollens and the law of addition in the classic binary propositional logic.

The conjunction principle states that if we have two pieces of data about some variable, for example

V is $A \Rightarrow \Pi_V(x) = A(x)$

V is $B \Rightarrow \Pi_V(x) = B(x)$

then we can conjunct these distributions getting the proposition V is C where

$$C(x) = A(x) \wedge B(x)$$
$$C(x) = \mathrm{Min}\,(A(x), B(x)\,).$$

The projection principle allows us to project out marginal possibility distributions from joint distributions. Assume $\Pi_{v_1,\, v_2}$ is a joint possibility distribution over the base set $X_1 \times X_2$, then this the projection principle allows us to infer that

$$\Pi_{v_1}(x) = \underset{y \in X_2}{\overset{\mathrm{Max}}{\mathrm{all}}}\ [\Pi_{v_1, v_2}\,(x, y)].$$

The law of fuzzy compositional inference which combines conjunction and projection plays a role similar to modens pollens in binary logic. Consider the data

$$V \text{ is } A_1$$

and the rule

$$\text{if } V \text{ is } A_2 \text{ then } U \text{ is } B$$

The proposition *V is A_1* induces the possibility distribution

$$\Pi_V = A_1(x)$$

over X. The rule *if V is A_2 then U is B* induces the conditional possibility distribution

$$\Pi_{u|v}(x,y) = \mathrm{Min}\,[1, 1 - A_2(x) + B(y)]$$

over $X \times Y$.

The law of fuzzy compositional inference says from these two pieces of information we can infer that

$$\Pi_u(y) = \mathrm{Max}_x\,[\Pi_v(x) \wedge \Pi_{u|v}\,(x, y)].$$

Consider the situation where there is more than one element in the antecedent

$$\text{if } V_1 \text{ is } A_1 \text{ and } V_2 \text{ is } A_2 \text{ then } U \text{ is } B.$$

Let our data be

$$V_1 \text{ is } C_1$$
$$V_2 \text{ is } C_2$$

First the rule "if V_1 is A_1 and V_2 is A_2 then U is B" induces the conditional possibility distribution

$$\Pi_{u|v_1, v_2}(y, x_1, x_2) = \text{Min}\,[1, 1 - (A_1(x_1) \wedge A_2(x_2)) + B(y)]$$

In order to obtain U from this via fuzzy composition inference, we need Π_{v_1, v_2}. This can be obtained from our data as

$$\Pi_{v_1, v_2}(x_1, x_2) = \text{Min}\,[C_1(x_1), C_2(x_2)]$$

then

$$\Pi_u(y) = \text{Max}_{(x_1,x_2)}\,[\Pi_{v_1, v_2}(x_1, x_2) \wedge \Pi_{u|v_1, v_2}(y, x_1, x_2)]$$

Consider next the situation in which we have two elements in the consequent of our rule:

if V is A then U_1 is B_1 or U_2 is B_2.

This induces the conditional possibility distribution

$$\Pi_{u_1,u_2|v}(y_1, y_2, x) = \text{Min}\,[1, 1 - A(x) + (B_1(y_1) \vee B_2(y_2))]$$

using the data

$$V \text{ is } C \; (\Pi_v(x) = C(x))$$

we can apply fuzzy composition inference to obtain

$$\Pi_{u_1, u_2}(y_1, y_2) = \text{Max}_x\,[\Pi_v(x) \wedge \Pi_{u_1, u_2|v}(x, y_1, y_2)]$$

The projection principle can now be applied to get either Π_{u_1} or Π_{u_2}. For example

$$\Pi_{u_1}(y_1) = \text{Max}_{y_2}[\Pi_{u_1, u_2}(y_1, y_2)]$$

The entailment principle implies that from the datum V is A, we can infer V is B, where B is any fuzzy subset such as the A C B.

We now can see the applicability of this theory to rule-based expert systems. Our global data base consists of information of the form V_i is A_i, our rule base consists of rules of the type "if V_i is B_i then U_j is C_j" by application of the laws of inference especially compositional fuzzy inference we can obtain new information to add to our global data base.

QUANTIFIERS IN THE ANTECEDENT

In this section we shall provide an extension of the ideas presented in the previous part to allow for the representation of more sophisticated rules in our rule base, such as:

"if **most** of the conditions V_1 is A_1, V_2 is A_2,. . .V_n is A_n are satisfied then U is B"

"if **at least half** of the conditions V_1 is A_1, V_2 is A_2, . . .V_n is A_n are satisfied then U is B"

The ability to represent such rules will greatly enhance the ability of any expert system to capture the types of rules used by experts.

We shall provide a methodology for representing such rules in a manner consistent with the rest of our formulation and one which allows inferences to be made about the value of the consequence using the rule and observed values about the variables in the antecedent. This methodology is based upon Zadeh's [14] representation of quantifiers and Yager's procedure for evaluating quantified statements [15].

The class of rules we are concerned with can be described to consist of the following components, an antecedent and a consequence. The antecedent component consists of a collection of requirements specified in the form of proposition of the type V_i *is* A_i is a variable and A_i is a fuzzy subset of the base set X. In addition the antecedent consists of a quantifier, Q, such as most, all, almost all, at least one, at least half, etc. The consequent consists of a proposition of the type *U is B*.

The rule than reflects the fact that if Q of the antecedent conditions, the V_i is A_i's, are satisfied than U is B can be added to our knowledge base. The fundamental difference between this type of rule and the types studied in the previous section is that rather than requiring all the antecedent conditions to be satisfied, only Q of them need be satisfied.

Like the other types of conditional rules, these rules also induce a conditional possibility distribution $\Pi_{U|V_1, V_2, \ldots V_n}$ over the set $X_1 \times X_1 \times \,.\, X_n \times Y$. In particular for any point $(x_1, x_2, \ldots x_n, y)$ where $x_i \in X_i$ and $y \in Y$

$$\Pi_{U|V_1, V_2, \ldots V_n}(x_1, x_2, \ldots x_n, y) = \text{Min}\,[1, 1 - H(x_1, x_2, \ldots x_n) + B(y)].$$

The essential difference lies in the determination of the joint possibility $H(x_1, \ldots x_n)$, the component due to the antecedent. The method for determining this H is based upon ideas developed by Yager [15].

As suggested by Zadeh [14], a linguistic quantifier can be expressed as a fuzzy subset. In particular there exists three kinds of quantifiers, the first two of which are of interest to us. A kind one quantifier or absolute quantifier such as "about 5", "at least seven" and a kind two or relative quantifier is exemplified by values such as "almost all" and "at least half." As suggested by Zadeh, a kind one quantifier can be expressed as a fuzzy subset of the non-negative reals whereas a kind two quantifier can be expressed as a fuzzy subset of the unit interval. For example, if Q_1 is the kind one quantifier "at least 5", then for each $x \in R^+$, Q(x) indicates the degree to which x satisfied the concept "at least 5". Similarly, if Q_2 is a kind two quantifier, "most", then for any $x \in I$, Q(x) indicates the degree to which the proposition x satisfies the concept "most".

Let Q be a quantifier either kind I or kind II, with base set W, for kind I, $W = R^+$ and for kind II, $W = [0, 1]$. Then Q is said to be monotonically non-decreasing if for any $w_1, w_2 \in W$ such that $u_2 > u_1$ then $Q(u_2) \geq Q(u_1)$. We shall restrict ourselves to these monotonically non-decreasing quantifiers as they appear to be the types that naturally appear in the rules used in expert systems.

We can now describe the procedure for obtaining H, from the antecedent of Q and the conditions V_i is A_i. We shall initially assume Q is a kind I quantifier.

For any point $(x_1, x_2, \ldots x_n) \in X_1 \times X_2 \times \ldots X_n$, where X_i is the base set of A_i we obtain $H(x_1, x_2, \ldots x_n)$ in the following manner.

Let $D(x_1, x_2, \ldots x_n) = \{A_1(x_1), A_2(x_2), \ldots A_n(x_n)\}$ and let $D_i(x_1, x_2, \ldots x_n)$ be the ith largest element in the set $D(x_1, x_2, \ldots x_n)$.

For any absolute quantifier Q_1

$$H(x_1, \ldots x_n) = \mathrm{Max}_i\, [Q_1(i) \wedge D_i(x_1, \ldots x_n)].$$

If Q_1 is a relative quantifier then we replace $Q_1(i)$ by $Q_1(i|n)$.

Having obtained $H(x_1, \ldots x_n)$ for every $(x_1, \ldots x_n) \in X_1 \times X_2 \times \ldots X_n$ we obtain

$$\Pi_{u|V_1,V_2,\ldots V_n}(x_1, \ldots x_n) = \mathrm{Min}\, [\, 1 - H(x_1, \ldots x_n) + B(y)]$$

This then becomes the induced possibility distribution from the rule,

if Q of V_1 is A_1, V_2 is A_2, . .V_n is A_n are satisfied then U is B.

If in addition we have in our database the values V_1 is C_1, V_2 is C_2, . . . V_n is C_n, where C_i is a fuzzy subset of Xi then we can obtain a value for U as

U is M

where

$$M(y) = \mathrm{Max}_{(x_1,x_2,..x_n)}\, [\Pi_{U|v_1,v_2,.\,v_n}(x_1,x_2,..\,x_n) \wedge \Pi_{v_1,v_2,.\,v_n}(x_1,x_2,..x_n)$$

where $\Pi_{v_1,v_2,.\,v_n}(x_1,x_2,..x_n) = \mathrm{Min}_i C_i(x_i)$.

A simple example will illustrate this procedure. Assume V_1, V_2, V_3, U are variables with base sets

$$X_1 = \{a, b\}$$
$$X_2 = \{c, d\}$$
$$X_3 = \{e, f\}$$
$$Y = \{g, h\}$$

Let Q be the kind II quantifier, *most* defined by

$$Q(0) = 0, Q(1/3) = 0, Q(2/3) = 1/2, Q(1) = 1.$$

Assume our rule is

If Q of [V_1 is A_1, V_2 is A_2, V_3 is A_3] are satisfied then U is B

where

$$A_1 = a = \{1/a, 0/b\}$$
$$A_2 = c = \{1/c, 0/d\}$$
$$A_3 = f = \{0/e, 1/f\}$$
$$B = g = \{1/g, 0/h\}$$

We shall first obtain H. Consider the point (a, c, e)

$$D(a, c, e) = \{1, 1, 0\}$$

hence

$$D_1\ (a, c, e) = 1$$
$$D_2 = (a,c,e) = 1$$
$$D_3\ (a,c,e) = 0.$$

Using this we get

$$H(a,c,e) = \text{Max}_i\ [Q(i/3) \wedge Di\ (a,c,e)] = [0 \wedge 1,\ 1/2 \wedge 1,\ 1 \wedge 0] = 1/2$$

The following table provides the formulation of H, $H = \text{Max}[0 \wedge d_1, 1/2 \wedge d_2, 1 \wedge d_3]$

x_1	x_2	x_3	$A(x_1)$	$A(x_2)$	$A(x_3)$	d_1	d_2	d_3	H
a	c	e	1	1	0	1	1	0	.5
a	c	f	1	1	1	1	1	1	1
a	d	e	1	0	0	1	0	0	1
a	d	f	1	0	1	1	1	0	.5
b	c	e	0	1	0	1	0	0	0
b	c	f	0	1	1	1	1	0	.5
b	d	e	0	0	0	0	0	0	0
b	d	f	0	0	1	1	0	0	0

Since

$$\Pi_{U|V_1,V_2,V_3}(x_1, x_2, x_3, y) = \text{Min}\ (1, 1 - H\ (x_1, x_2, x_3) + B(y))$$

the following table expresses $\pi_u|V_1,V_2,V_3$

x_1	x_2	x_3	y	$\pi_u|V_1,V_2,V_3\ (x_1, x_2, x_3, y)$
a	c	e	g	$1 \wedge (1 - ½ + 1) = 1$
a	c	e	h	$1 \wedge (1 - ½ + 0) = ½$
a	c	f	g	$1 \wedge (1 - 1 + 1) = 1$
a	c	f	h	$1 \wedge (1 - 1 + 0) = 0$
a	d	e	g	$1 \wedge (1 - 0 + 1) = 1$
a	d	e	h	$1 \wedge (1 - 0 + 0) = 1$
a	d	f	g	$1 \wedge (1 - ½ + 1) = 1$
a	d	f	h	$1 \wedge (1 - ½ + 0) = ½$
b	c	e	g	$1 \wedge (1 - 0 + 1) = 1$
b	c	e	h	$1 \wedge (1 - 0 + 0) = 1$
b	c	f	g	$1 \wedge (1 - ½ + 1) = 1$
b	c	f	h	$1 \wedge (1 - ½ + 0) = ½$
b	d	e	g	$1 \wedge (1 - 0 + 1) = 1$
b	d	e	h	$1 \wedge (1 - 0 + 0) = 1$
b	d	f	g	$1 \wedge (1 - 0 + 1) = 1$
b	d	f	h	$1 \wedge (1 - 0 + 0) = 1$

Assume we have the data

$V_1 = \{1/a, 0/b\} = a = C_1$

$V_2 = \{1/c, 0/d\} = c = C_2$

$V_3 = \{0/e, 1/f\} = f = C_3$

then to obtain $\Pi_u\ (y)$ we see that

$\Pi_u(y) = \text{Max}_{(x_1, x_2, x_3)} \Pi_{u|V_1,V_2,V_3}(x_1, x_2, x_3, y) \wedge C_1(x_1) \wedge C_2(x_2) \wedge C_3(x_3)]$

hence

$\Pi_u(g) = \text{Max}\ [1\wedge1\wedge1\wedge0,\ 1\wedge1\wedge1\wedge1,\ 1\wedge1\wedge0\wedge0,\ 1\wedge1\wedge0\wedge1,\ 1\wedge0\wedge1\wedge0,\ 1\wedge0\wedge1\wedge1,$
$1\wedge0\wedge0\wedge0,\ 1\wedge0\wedge0\wedge1] = \text{Max}\ [0,1,0,0,0,0,0,0] = 1$

$\Pi_u(h) = \text{Max}\ [1/2\wedge1\wedge1\wedge0,\ 0\wedge1\wedge1\wedge1,\ 1\wedge1\wedge0\wedge0,\ 1/2\wedge1\wedge0\wedge1,\ 1\wedge0\wedge1\wedge0,$
$1/2\wedge0\wedge1\wedge1\wedge1,\ 1\wedge0\wedge0\wedge0,\ 1\wedge0\wedge0\wedge1] = \text{Max}\ [0,0,0,....0] = 0$

hence as we would have anticipated

$$U = \{1/g, 0/h\} = g$$

Consider the next situation where

$$V_1 = \{0/a, 1/b\} = b$$
$$V_2 = \{1/c, 0/d\} = c$$
$$V_3 = \{0/e, 1/f\} = f$$

$\Pi_u(g) = \text{Max}\ [1\wedge0\wedge1\wedge0,\ 1\wedge0\wedge1\wedge1,\ 1\wedge0\wedge0\wedge0,\ 1\wedge0\wedge0\wedge1,\ 1\wedge1\wedge1\wedge0,\ 1\wedge1\wedge1\wedge1,$
$1\wedge1\wedge0\wedge0,\ 1\wedge1\wedge0\wedge1] = \text{Max}\ [0,0,0,0,0,1,0,0] = 1$

$\Pi_u(h) = \text{Max}\ [1/2\wedge0\wedge1\wedge0,\ 0\wedge0\wedge1\wedge1,\ 1\wedge0\wedge0\wedge0,\ 1/2\wedge0\wedge0\wedge1,\ 1\wedge1\wedge1\wedge0,\ 1/2\wedge1\wedge1\wedge1,$
$1\wedge1\wedge0\wedge0,\ 1\wedge1\wedge0\wedge1] = \text{Max}\ [0,0,0,0,0,1/2,0,0] = 1/2$

hence

$$U = \{1/g, .5/h\}$$

In the situation where

$V_1 = \{1/a, 1/b\}$ = "I don't know"
$V_2 = \{1/c, 0/d\}$ = c
$V_3 = \{0/e, 1/f\}$ = f

we can show that again

$$U = \{1/g, .5/h\}$$

In the case where

$$V_1 = \{0/a, 1/b\} = b$$
$$V_2 = \{0/c, 1/d\} = d$$
$$V_3 = \{0/e, 1/f\} = f$$

we can show that

$$U = \{1/g, 1/h\} \text{ (Unkown).}$$

The following theorem shows that the conjunction of antecedent conditions is one of the quantifiers.

Theorem: When Q is the quantifier *all* then the rule

If Q [Vi is Ai] are satisfied then U is B I

is equivalent to the proposition

if V_1 *is* X_1 *and* V_2 *is* X_2 *and . . .* V_n *is* X_n *then* $U = B$ II .

Proof: For rule II we have

if V is H the U is B

where $V = (V_1, V_2, V_3, \ldots, V_n)$ and

$$H(x_1, \ldots x_n) = A_1(x_1) \wedge A_2(x_2) \ldots \wedge A_n(x_n).$$

For rule I we have

if V is G the U is B

where

$$G(x_1, \ldots X_n) = \mathrm{Max}_i\,[Q(i) \wedge D(i)]$$

where D(i) = i^{th} largest element in the set $\mathcal{Q} = \{A_1(x_1), A_2(x_2), \ldots A_n(x_n)$.

When Q is the quantifier *all*, then

$$Q(i) = 1 \qquad i = n$$
$$Q(i) = 0 \qquad i \neq n.$$

In this case

$$G(x_1, \ldots x_n) = 1 \wedge D(n) = n^{th} \text{ largest element in } \mathcal{Q}$$

hence $G(x_1, \ldots x_n) = A_1(x_1) \wedge A_2(x_2) \wedge \ldots \wedge A_n(x_n) = H(x_1, \ldots x_n)$.

Theorem: When Q is the quantifier *at least one* then the rule

if Q [V_i is A_i] then U is B **I**

is equivalent to the proposition

if V_1 is A_1 or V_2 is A_2 or V_n is A_n then U is B. **III**

Proof: For rule III we have

if V is H the U is B

where V = $(V_1, V_2, V_3, \ldots, V_n)$ and

$$H(x_1, \ldots x_n) = A_1(x_1) \vee A_2(x_2) \vee \ldots . A_n(x_n)$$

For rule I we have

if V is G the U is B

where

$$G(x_1, \ldots X_n) = \mathrm{Max}_i\,[Q(i) \wedge D(i)]$$

where D(i) = i^{th} largest element in the set $\mathcal{Q} = \{A_1(x_1), A_2(x_2), \ldots A_n(x_n)$.

When Q is the quantifier *at least one*, then

$$Q(i) = 1 \text{ for all } i \geq 1.$$

Thus

$$G(x_1, \ldots . x_n) = A_1(x_1) \vee A_2(x_2) \vee A_n(x_n).$$

CERTAINTY QUALIFICATION

In providing information to the database and rule base of an expert system, as discussed by Buchanan and Duda [1], a person may not be completely confident as to the value he is providing for a variable. Thus a user of a system may provide the information that

V is A *with confidence (or certainty)* **a**.

In the above the quantity **a**, which is a number in the unit interval, expresses the degree to which the informant believes that this information is valid.

We would like to provide a mechanism to include these types of qualified statements into our system. In the spirit of keeping the very powerful structure which we have developed the approach will be to assume that a statement

V is A with **a** *confidence*

is equivalent to an unqualified statement of the form

V is B.

This new statement implicitly implies a confidence of one. Thus we see that the statement

V is B with 1 confidence

is equivalent to V is B. Thus all our previous work can be seen to have been done with the implicit certainty one.

We impose the condition that the statement

V is A with *zero* confidence

should be equivalent to the proposition *V is X*, where X is the base set of V. Thus zero confidence is equivalent to saying *I don't know anything about the value of V*

We should note that this is different than probability, for in probability we should have

V is A *zero* probability $\Rightarrow$ V is $\overline{A}$ with 1 probability

In general we see that an informant usually makes a tradeoff in providing information between the specificity of the information and the confidence. That is, the more specific he is required to provide the information the less confident he can be about it.

In [17] Yager has suggested a mechanism for transforming statements of the form

V is A with confidence a

into statements of the form

V is B

with implied confidence one. In particular if A and B are fuzzy subsets of X then

V is A with confidence a

can be transformed into the equivalent proposition V is B where for any $x \in X$

$$B(x) = (a \wedge A(x)) + (1-a)$$

NOTE: For the statement *V is A with confidence 1*, then we get

V is A.

Proof: $B(x) = (a \wedge A(x)) + (1-a) = 1 \wedge A(x) + 0 = A(x)$.

NOTE: For the statement *V is A with confidence 0*, then we get the unqualified proposition V is X.

Proof: $B(x) = (a \wedge A(x)) + (1-a) = 0 \wedge A(x) + (1-0) = 1$.

In [18] Yager has introduced a measure of specificity associated with a fuzzy subset. Assume F is a fuzzy subset of the finite set X, then the specificity of F, S(F) is defined as

$$S(F) = \int_0^{\alpha_{max}} \frac{1}{\text{card } F_\alpha} d\alpha$$

where $F_\alpha = \{x \mid F(x) \geq \alpha\}$, Card F_α is the number of elements in F_α and α_{max} is the largest membership grade in F. For the case where F is normal, then

$$S(F) = \int_0^1 \frac{1}{\text{card } F_\alpha} d\alpha$$

Yager [18] has shown for the case of normal fuzzy subsets if $F \subset G$, that is for $F(x) \leq G(x)$ for all $x \in X$, then

$$S(F) \geq S(G).$$

The following theorems reinforce our observations about the tradeoff between specificity and certainty.

Lemma : If A is normal, then the transformation of the proposition *V is A with a certainty* into the proposition *V is B* will yield B as a normal set.

Proof: Let x be such that $A(x) = 1$, then

$$B(x) = a \wedge A(x) + (1\text{-}a) = a \wedge 1 + (1\text{-}a) = a + (1\text{-}a) = 1.$$

In the following theorems A is assumed normal.

Theorem: Assume the proposition *V is A with* **a** *certainty* transforms into the proposition *V is B* then

$$S(A) \geq S(B).$$

Proof: We shall first show that for each $x \in X$, $B(x) \geq A(x)$ from the definition

$$B(x) = (a \wedge A(x)) + (1\text{-}a).$$

Assume $a \geq A(x)$ then

$$B(x) = a \wedge A(x) + (1\text{-}a) = A(x) + (1\text{-}a) \geq A(x).$$

Assume $a < A(x)$ then

$$B(x) = a + (1\text{-}a) = 1 \geq A(x).$$

Since $B(x) \geq A(x)$ for each x, then it follows that $S(A) \geq S(B)$.

Thus we see that the act of qualifying a proposition by a certainty has the effect of reducing the specificity of its unqualified equivalent.

Theorem: Assume *V is A with* $\mathbf{a_1}$ *certainty* transforms into *V is* B_1 and *V is A with* $\mathbf{a_2}$ *certainty* transforms into *V is* B_2 if $a_1 > a_2$, then

$$S(B_1) \geq S(B_2).$$

Proof: $B_1(x) = (a_1 \wedge A(x)) + (1\text{-}a_1)$ and $B_2(x) = (a_2 \wedge A(x)) + (1\text{-}a_2)$. There are three possibly situations: **1.** $A(x) \leq a_2 \leq a_1$. In this case

$$B_1(x) = A(x) + (1\text{-}a_1)$$
$$B_2(x) = A(x) + (1\text{-}a_2),$$

since $a_1 > a_2$, then $(1\text{-}a_1) \leq (1\text{-}a_2)$ and hence $B_2(x) \geq B_1(x)$.

2. $a_2 \leq A(x) \leq a_1$. In this case

$$B_1(x) = A(x) + (1\text{-}a_1)$$

$$B_2(x) = a_2 + (1\text{-}a_2) = 1 \geq B_1(x).$$

3. $a_2 \leq a_1 \leq A(x)$. In this case

$$B_1(x) = a_1 + (1-a_1) = 1$$

$$B_2(x) = a_2 + (1-a_2) = 1 \geq B_1(x)$$

Since $B_1(x) \leq B_2(x)$ for all x, $S(B_1) \geq S(B_2)$.

It should be noted that this approach to certainty qualification can easily be applied to rules in the expert system. Consider the rule

if V is A then U is B with α *certainty*

where A and B are fuzzy subsets of X and Y respectively. This transforms into the possibility distribution,

$$\Pi_{u|v}(x,y) = (H(x,y) \wedge \alpha) + (1-\alpha)$$

where

$$H(x,y) = \mathrm{Min}(1, 1-A(x) + B(y)).$$

REPRESENTATION OF DEFAULT KNOWLEDGE

The construction of useful knowledge based systems requires the representation and manipulation of so called commonsense knowledge . Commonsense knowledge is very often characterized by pieces of knowledge that are usually true but not necessarily always true. The essential feature of this is the assumption of a piece of knowledge without conclusive evidence of its truth. Within this approach one assumes some piece of commonsense knowledge as valid if it is consistent or possible within the framework of what we already know.

It is the use of the absence of contradictory evidence which strongly characterizes the process of commonsense reasoning. That it, classic reasoning systems require the certainty of propositions before asserting its truth. In commonsense reasoning systems some facts are asserted as true it there exists a possibility of it being true, nothing contradicts it. Knowledge of this type is often called defeasible because we want the option of withdrawing it if contradictory evidence subsequently appears.

Systems which allow for the inference of information based upon the lack of some contradictory fact are faced with the problem that their reasoning process is nonmonotonic. In particular some proposition that was inferred may cease to be inferable with the acquisition of further knowledge.

In [16] Yager introduced a reasoning system which we shall call fuzzy default reasoning. This system is rooted in the theory of approximate reasoning [17]. He order to discuss this system we need introduce some additional concepts.

Intuitively speaking the statement

V is A

says that the value of V lies in the subset A. Knowledge that V is A can be used to help determine viability of other statements. If

V is B

is a second statement we define

$$\text{Poss}[V \text{ is } B/V \text{ is } A] = \text{Max}_x[A(x) \wedge B(x)].$$

Formally this definition captures a measure of the degree of intersection between the two sets A and B. Pragmatically, this measure provides an upper bound on the truth of the statement V is B given V is A. That is if A and B intersect then it is possible that V lies in B, V is B is true, given that V is A. We shall see this is a measure of consistency of the two statements. A second closely related definition is

$$\text{Cert}[V \text{ is } B/V \text{ is } A] = 1 - \text{Poss}[\bar{B}/A]$$

We note that an equivalent formulation is

$$\text{Cert}[V \text{ is } B/V \text{ is } A] = \text{Min}_x[\bar{A}(x) \vee B(x)].$$

Formally this definition captures the degree to which A is contained in B. Pragmatically this measure provides a lower bound on the truth of V is B given V is A. In general

$$\text{Cert}[V \text{ is } B/V \text{ is } A] \le \text{Poss}[V \text{ is } B/V \text{ is } A]$$

In binary reasoning systems we require that

$$\text{Cert}[V \text{ is } B/KB] = 1,$$

where KB is our knowledge base, to infer that V is B is true. We shall see that in the commonsense environment we essentially allow an inference of a commonsense piece of knowledge to occur if

$$\text{Poss}[V \text{ is } B/KB] = 1$$

Recalling that a rule is of the form

if V is A then U is B,

where A and B are fuzzy subsets of the base sets X and Y. The above statement gets translated into a joint canonical statement

(V, U) is D

where D is a fuzzy subset of $X \times Y$ such that

$$D(x, y) = (1 - A(x)) \vee B(y).$$

If we have two pieces of knowledge

If V is A then U is B

V is E

then we can conjunct these to get

(U, V) is H.

Here H is a subset of the cartesian space $X \times Y$ such that

$$H = E \cap D = (\bar{A} \cup B) \cap E = (\bar{A} \cap E) \cup (B \cap E).$$

The inferred value of V, denoted G, can represent this as a

V is G

where

$$G(y) = \text{Max}_x[\bar{A}(x) \wedge E(x)] \vee B(y) = \text{Poss}(\bar{A}/E) \vee B(y) = (1 - \text{Cert}(A/E)) \vee B(y).$$

We see that if we are certain that A occurs given E, effectively $E \subset A$, then we get $G = B$. If $\text{Poss}(\bar{A}/E) = 1$ then we get $G = Y =$ "unknown", hence no inference is made.

In [16,18,19] Yager has suggested that we can use possibility qualification as a basis for the implementation of many different kinds of commonsense knowledge. A possibility qualified statement is of the form

V is A is possible.

This statement characterizes a piece of information that says our knowledge of the value of V is such that it is possible (or consistent) with it to assume that V lies in the set A. Note that it doesn't specifically say V lies in A. Formally this statement gets translated into

$$V \text{ is } A^+$$

where A^+ is a subset of the power set of the base set X. In particular for any subset G of X

$$A^+(G) = \text{Poss}[A/G] = \text{Max}_x[A(x) \wedge G(x)]$$

Essentially A^+ is made up of the subsets of X which intersect, are consistent, with A.

Closely related to possibility qualification is certainty qualification. A statement

V is A is certain

translates into

$$V \text{ is } A^{\nabla}$$

where A^{∇} is a subset of the power set of the base set of A, X, such that for any subset F of X

$$A^{\nabla}(F) = \text{Cert}(A/F) = 1 - \text{Poss}\,(A^{@}/F)$$

We shall now describe the representation of some primary types of commonsense knowledge by the possibilistic reasoning approach.

We shall initially consider the statement

typically V is A.

The interpretation of "typically V is A" afforded by Reiter's default reasoning system[20] is to say "if we have not established V is $\neg$A then assume V is A. Thus we can translate the above into

if V is A is possible then V is A.

Using our translation rules we get

$$\text{if } V \text{ is } A^+ \text{ then } V \text{ is } A.$$

This translate into

$$V \text{ is } \neg(A^+) \cup A.$$

We shall denote $\neg(A^+)$ as A^*, hence we get

$$V \text{ is } (A^* \cup A).$$

Furthermore assume that our knowledge base consists simply of the fact that

V is B.

Combining this with our typical knowledge we get V is D where

$$D = (A^* \cap B) \cup (A \cap B).$$

Furthermore as discussed in [16,18,19] this becomes

$$D(x) = (B(x) \wedge (1 - \text{Poss}[A/B]) \vee (A(x) \wedge B(x)).$$

Two extremal cases should be noted. If our typical value A is completely inconsistent with our known value, $A \cap B = \phi$, then $\text{Poss}[A/B] = 0$

and thus $D(x) = B(x)$ and hence

$$V \text{ is } B$$

We have discounted our typical information when it conflicted with our knowledge-base.

On the other hand if A has some consistency with B, $A \cap B \neq \phi$, thus $Poss[A/B] = 1$ then we get

$$D(x) = B(x) \wedge A(x)$$

and hence

$$V \text{ is } A \cap B.$$

Thus when our typical knowledge doesn't contradict our firm knowledge we conjunct these sources of knowledge. In the special case when B is unknown, $B = X$, then we get

$$V \text{ is } A.$$

CONCLUSION

We have discussed the applicability of the theory of approximate reasoning to rule based expert systems. The novel aspects of this work concerns our introduction of an approach in this framework for the inclusion of complex rules and the ability to introduce certainty qualification into our system.

REFERENCES

(1) Buchanan, B.G. and Duda, R.O., "Principles of rule-based expert systems," Fairchild Technical Report No. 626, Lab. for Artificial Intelligence Research, Fairchild Camera, Palo Alto, Ca., 1982.

(2) Buchanan, B.G., "Partial bibliography of work on expert systems," Sigart Newsletter, No. 84, 45-50, 1983.

(3) Shortliffe, E.H., Computer Based Medical Consultations: MYCIN, American Elsevier, New York 1976.

(4) Davis, R., Buchanan, B.G. & Shortliffe, E.H., "Production rules as a representation of a knowledge-based consultation program," Artificial Intelligence 8, 15-45, 1977.

(5) Van Melle, W., "A domain independent system that aids in constructing knowledge-based consultation program," Ph.D. dissertation, Stanford University Computer Science Dept., Stanford CS-80-820, 1980.

(6) Zadeh, L.A., "Fuzzy logic and approximate reasoning," Synthese 30, 407-428, 1975.

(7) Zadeh, L.A., "The concept of a linguistic variable and its application to approximate reasoning," Information Science 8 and 9, 199-249, 301-357, 43-80, 1975.

(8) Zadeh, L.A., "A theory of approximate reasoning," in Hayes, J.E., Michie, D. and Kulich, L.I., (eds) Machine Intelligence 9, 149-194, John Wiley & Sons, New York, 1979.

(9) Zadeh, L.A., "Fuzzy sets as a basis for a theory of possibility," Fuzzy Sets and Systems 1, 3–28, 1978.

(10) Zadeh, L.A., "PRUF-a meaning representation language for natural languages," Int. J. of Man-Machine Studies 10, 395-460, 1978.

(11) Yager, R.R., "Querying knowledge base systems with linguistic information via knowledge trees," Int. J. Man- Machine Studies 19, 1983.

(12) Yager, R. R., "Knowledge trees in complex knowledge bases," Fuzzy Sets and Systems 15, 45-64, 1985.

(13) Zadeh, L.A., "Fuzzy sets," Information and Control 8, 338-353, 1965.

(14) Zadeh, L.A., "A computational approach to fuzzy quantifiers in natural languages," Com. 8 Maths. with Appl. 9, 149-184, 1983.

[15]. Yager, R. R., "Quantifiers in the formulation of multiple objective decision functions," Information Sciences 31, 107-139, 1983.

[16]. Yager, R. R., "Default and approximate reasoning," Proc. 2nd IFSA Conference, Tokyo, 690-692, 1987.

[17]. Yager, R. R., Ovchinnikov, S., Tong, R. and Nguyen, H., Fuzzy Sets and Applications: Selected Papers by L. A. Zadeh, John Wiley & Sons: New York, 1987.

[18]. Yager, R. R., "Nonmonotonic inheritance systems," IEEE Transactions on Systems, Man and Cybernetics 18, 1028-1034, 1988.

[19]. Yager, R. R., "On the representation of commonsense knowledge by possibilistic reasoning," Int. J. of Man-Machine Systems 31, 587-610, 1989.

[20]. Reiter, R., "A logic for default reasoning," Artificial Intelligence 13, 81-132, 1980.

3

Fuzzy rules in knowledge-based systems – Modelling gradedness, uncertainty and preference –

Didier DUBOIS Henri PRADE

Institut de Recherche en Informatique de Toulouse
Université Paul Sabatier, 118 route de Narbonne
31062 Toulouse Cedex (France)

The paper starts with ideas of possibility qualification and certainty qualification for specifying the possible range of a variable whose value is ill-known. The notion of possibility which is used for that purpose is not the standard one in possibility theory, although the two notions of possibility can be related. Based on these considerations four distinct types of rules with different semantics involving gradedness and uncertainty are then introduced. The combination operations which appear for taking advantage of the available knowledge are all derived from the intended semantics of the rules. The processing of these four types of rules is studied in detail. Fuzzy rules modelling preference in decision processes are also discussed.

1. INTRODUCTION

The applications of fuzzy set and possibility theories to rule-based expert systems have been mainly developed along two lines in the eighties : i) the generalization of the certainty factor approach introduced in MYCIN (Buchanan and Shortliffe, 1984) by enlarging the possible operations to be used for combining the uncertainty coefficients ; ii) the handling of vague predicates in the expression of the expert rules or of the available information. The first line of research is exemplified by the inference system RUM (Bonissone et al., 1987) where a control layer chooses the triangular norm operation governing the propagation of uncertainty, or by the inference system MILORD (Godo et al., 1988) where the combination and propagation operations associated with each rule reflect the expert knowledge. The second trend has motivated a huge amount of literature especially for discussing the multiple-valued logic implication connective → to be used in the modelling of a rule of the form "if X is A then Y is B" by means of a fuzzy relation R (defined by $\mu_R(x,y) = \mu_A(x) \rightarrow \mu_B(y)$). The choice of the implication function has been investigated from an algebraic point of view by classifying the implications according to axiomatic properties, and from a deduction-oriented perspective by requiring some prescribed kind of results for the generalized modus ponens applied to fuzzy "if... then..." rules (e.g. Mizumoto and Zimmermann (1982), Dubois and Prade (1984), Trillas and Valverde (1985), Bouchon (1987), Smets and Magrez (1987)). Although the available results indeed enable us to jointly choose an implication function and the conjunction to be used for combining the two premisses "X is A' " and "if X is A then Y is B" in order to obtain an expected behavior for the generalized modus ponens, these approaches do not really consider the intended semantics of the rules. See (Dubois and Prade, 1990d) and (Dubois, Lang and Prade, 1990) for an extensive overview and a discussion of the generalized modus ponens and of the certainty factor approaches respectively.

In this paper, extending recently obtained results (Dubois and Prade, 1989a, 1990b, d), we show how the choice of the implication operation is induced by the type of rule we have to model in the framework of possibility theory. The approach which is proposed formalizes ideas which have been more empirically studied by Bouchon (1988), Després (1989) about the role of different kinds of modifiers in the expression and the intended meaning of fuzzy rules and can be also somewhat related to recent works about possibility and necessity qualifications (Magrez and Smets, 1989 ; Dubois and Prade, 1990a ; Fonck, 1990 ; Yager, 1990).

We first discuss two distinct ways of specifying a possibility distribution, either by possibility or by certainty qualification. This can be regarded as a new approach in possibility theory. The consequences of the mode of qualification on the manipulation of the pieces of knowledge which are thus specified, are emphasized. The notion of possibility which is used in possibility qualification do not correspond to the standard notion of possibility measure in possibility theory ; the links between the two concepts are clarified in Section 3. Using the ideas of Section 2, Section 4 introduces four different types of rules which are closely related to particular types of fuzzy truth-values (or, if we prefer, of modifiers). Section 5 discusses the behavior of these rules in the generalized modus ponens and when used in parallel. Section 6 is devoted to another kind of fuzzy rules expressing preference.

2. TWO WAYS OF SPECIFYING A POSSIBILITY DISTRIBUTION

2.1. The Concept of a Possibility Distribution

A possibility distribution is a function π_x, attached to a variable x, from a so-called universe of discourse U to the real interval [0,1] which aims at representing our current view of the feasible, or epistemically possible, or admissible values of a single-valued variable x whose domain is U. Depending on the interpretations, $\pi_x(u)$ estimates the degree of ease, the degree of unsurprizingness or of expectedness, the degree of acceptability or of preference attached to the proposition "the value of x is u", i.e. x = u. The possibility distribution π_x is just a way of specifying an ordering among the elements of U, which expresses that the closer to 1 (resp. to 0), the more (resp. the less) feasible, epistemically possible, or admissible, according to the interpretation, the value u is for x. In the following we shall use the neutral term "possible", saying that $\pi_x(u)$ estimates the extent to which u is possible for x, when it is not interesting to put forward any specific interpretation. Thus the interval [0,1] is just considered here as an ordinal scale where 1 stands for complete possibility and 0 for complete impossibility. As soon as U entirely covers the domain of the variable x, it is natural to require that there exists at least one element u of U which can be considered as completely possible for x, i.e. such that $\pi_x(u) = 1$; then π_x is said to be normalized.

2.2. Specifications by Means of Ordinary Subsets

A possibility distribution is not usually specified as such, but by the qualification of subsets of U. Let A be an ordinary subset of U. It can be qualified either in terms of possibility or in terms of certainty in order to specify a possibility distribution π over A. Namely

i) if A is a (completely) possible range for x, it means that $\forall\, u \in A,\ \pi_x(u) = 1$ and π_x remains unspecified outside of A, or equivalently, that

"A is possible" is translated by $\forall\, u \in U,\ \mu_A(u) \le \pi_x(u)$ (1)

where μ_A is the $\{0,1\}$-valued characteristic function of A.

ii) if it is (completely) certain that the value of x lies in A, it means that any value outside A is (completely) impossible, i.e. $\forall\, u \notin A,\ \pi_x(u) = 0$ and π_x is unspecified over A, or if we prefer

"A is certain" is translated by $\forall\, u \in U,\ \pi_x(u) \le \mu_A(u)$. (2)

Thus, let A_c and A_s be two ordinary subsets of U satisfying (1) and (2) respectively for a possibility distribution π_x, then we have

$$\forall\, u \in U,\ \mu_{A_c}(u) \le \pi_x(u) \le \mu_{A_s}(u) \qquad (3)$$

which expresses that A_c is included in the core of π_x, i.e. $\{u \in U,\ \pi_x(u) = 1\}$ while A_s contains the support of π_x, i.e. $\{u \in U,\ \pi_x(u) > 0\}$.

Let A_1 and A_2 be two subsets of U which both satisfy (1) for the same possibility distribution π_x, then we see that

"A_1 is possible" and "A_2 is possible" $\Rightarrow \forall u \in U,\ \max(\mu_{A_1}(u),\mu_{A_2}(u)) \le \pi_x(u)$ (4)

while if A_1 and A_2 both satisfy (2), we have

"A_1 is certain" and "A_2 is certain" $\Rightarrow \forall\, u \in U,\ \pi_x(u) \le \min(\mu_{A_1}(u),\mu_{A_2}(u))$. (5)

We observe that pieces of knowledge which are simultaneously qualified in terms of possibility are combined by means of max operation in a union-like manner, while pieces of knowledge which are simultaneously qualified in terms of certainty are combined by means of min operation in an intersection-like manner.

Let us consider cases of qualification where the possibility or the certainty is not complete but corresponds to an intermediary level α in the scale [0,1]. It leads to the two following generalizations of (1) and (2)

i) the statement "A is a possible range for x at least at the degree α" will be understood as $\forall u \in A,\ \pi_x(u) \ge \alpha$, which leads to

"A is α-possible" is translated by $\forall$ u $\in$ U, $\min(\mu_A(u), \alpha) \leq \pi_x(u)$. (6)

Note that for $\alpha = 1$, (1) is recovered.

ii) the statement "it is certain at least at the degree α that the value of x is in A", will be interpreted as any value outside A is at most possible at the complementary degree, namely $1 - \alpha$, i.e. $\forall u \notin A$, $\pi_x(u) \leq 1 - \alpha$, which leads to

"A is α-certain" is translated by $\forall$ u $\in$ U, $\pi_x(u) \leq \max(\mu_A(u), 1 - \alpha)$ (7)

Note that for $\alpha = 1$, (2) is recovered. Certainty qualification was first discussed by Yager (1984) and Prade (1985) ; in this latter reference, (7) already appears with equality (i.e. the less restrictive possibility distribution compatible with (7) is chosen). Possibility-qualification goes back to Zadeh (1978b) and Sanchez (1978).

Clearly (4) and (5) can be straightforwardly extended to "A_i is α_i-possible" and to "A_i is α_i-certain", for i = 1,2 using (6) and (7) respectively.

2.3. Specifications by Means of Fuzzy Subsets

We now consider the more general case where the subset A which is qualified is fuzzy. It is well-known that a fuzzy (sub)set A can be represented in terms of a collection of ordinary subsets, namely the α-cuts $A_\alpha = \{u \in U, \mu_A(u) \geq \alpha\}$ of A. We have (Zadeh, 1971)

$$\forall u, \mu_A(u) = \sup_{\alpha \in (0,1]} \min(\mu_{A_\alpha}(u), \alpha) \quad (8)$$

Then we immediately notice that, if we interpret "A is possible" (where A is fuzzy) as the conjunctive collection of possibility-qualified non-fuzzy statements of the form "A_α is α-possible", $\forall \alpha \in (0,1]$, we obtain, using (6) for the possibility-qualification and (4) for the max-combination (here extended to a sup-combination since the collection may be not finite)

$$\forall u \in U, \sup_{\alpha \in (0,1]} \min(\mu_{A_\alpha}(u), \alpha) \leq \pi_x(u)$$

i.e.

$$\forall u \in U, \mu_A(u) \leq \pi_x(u)$$

which clearly generalizes (1) to a fuzzy set A.

From (8), by taking the complement of A, i.e. the complement to 1 of its membership function, we can obtain another representation formula, namely

$$\forall u \in U, \mu_{\overline{A}}(u) = 1 - \mu_A(u) = \inf_{\alpha \in (0,1]} \max(\mu_{\overline{A_{1-\alpha}}}(u), 1 - \alpha) \quad (9)$$

where the overbar on a subset denotes the complementation and we use the identity $\overline{(A_\alpha)} = (\overline{A})_{\overline{1-\alpha}}$, with $B_{\overline{\beta}}$ denoting the *strong* β-cut of a fuzzy set B, namely $\{u \in U, \mu_B(u) > \beta\}$ (i.e. '$\geq$' is changed into '>' in the definition of the level cut). Clearly (9) applies to any A and thus (9) still holds changing $\overline{A}$ into A, which gives

$$\forall u \in U, \mu_A(u) = \inf_{\alpha \in (0,1]} \max(\mu_{A_{\overline{1-\alpha}}}(u), 1 - \alpha) \quad (10)$$

Then, if we interpret "A is certain" (where A is fuzzy) as the conjunctive combination of certainty-qualified non-fuzzy statements of the form $A_{\overline{1-\alpha}}$ is α-certain, $\forall \alpha \in (0,1]$, we obtain using (7) and the min-combination (5) (extended to an inf-combination)

$$\forall u \in U, \pi_x(u) \le \inf_{\alpha \in (0,1]} \max(\mu_{\overline{A_{\overline{1-\alpha}}}}(u), 1-\alpha)$$

i.e.

$$\forall u \in U, \pi_x(u) \le \mu_A(u) \qquad (11)$$

which clearly generalizes (2) to a fuzzy set A. The interpretation of "A is certain" (i.e. we are completely certain that the possible values of x are restricted by μ_A), as the conjunction of statements of the form "$A_{\overline{1-\alpha}}$ is (at least) α-certain" is quite natural. Indeed we are completely certain ($\alpha = 1$) that the support $A_{\overline{0}} = \{u \in U, \mu_A(u) > 0\}$ of A contains the value of x, while the certainty that the strong β-cut of A includes the value of x decreases when β increases, because the β-cut becomes smaller due to the nestedness property $\beta \le \beta' \Rightarrow A_{\overline{\beta}} \supseteq A_{\overline{\beta'}}$ (here $\beta = 1 - \alpha$). Note also that "$A_{\overline{1-\alpha}}$ is at least α-certain", according to (7), means that any value in $\overline{A_{\overline{1-\alpha}}} = (\overline{A})_\alpha$ is at most possible at the degree $1 - \alpha$ (or if we prefer at least impossible at the degree α).

An immediate consequence of (9) and (11) is that if the fuzzy set A is both a completely possible and completely certain fuzzy range for the value of x in the above sense, then we should have

$$\forall u \in U, \pi_x(u) = \mu_A(u) \qquad (12)$$

i.e. the equality with which Zadeh (1978a) starts the introduction of possibility theory.

We now generalize (6) and (7) to a fuzzy set A thus introducing gradedness in possibility and certainty qualification of fuzzy subsets. "A is possible" has been interpreted as "A_β is at least β-possible" $\forall \beta \in (0,1]$. Then it is natural to interpret "A is α-possible" as $\forall \beta \ge \alpha$, "A_β is at least α-possible" and $\forall \beta < \alpha$, "A_β is at least β-possible". In other words, the smallest β-cuts with β close to 1 are only assigned a minimal degree of possibility equal to α (instead of β). Then using the max-combination (4), we get

$$\forall u \in U, \max(\sup_{\beta \ge \alpha} \min(\mu_{A_\beta}(u), \alpha), \sup_{\beta < \alpha} \min(\mu_{A_\beta}(u), \beta)) \le \pi_x(u)$$

i.e.

$$\max(\min(\sup_{\beta \ge \alpha} \min(\mu_{A_\beta}(u), \beta), \alpha), \sup_{\beta < \alpha} \min(\mu_{A_\beta}(u), \beta)) \le \pi_x(u)$$

i.e.

$$\forall u \in U, \min(\sup_{\beta \in (0,1]} \min(\mu_{A_\beta}(u), \beta), \alpha) \le \pi_x(u) \text{ (since } \sup_{\beta < \alpha} \min(\mu_{A_\beta}(u), \beta) \le \alpha)$$

i.e.

$$\forall u \in U, \min(\mu_A(u), \alpha) \le \pi_x(u) \qquad (13)$$

which extends (6) to the case where A is fuzzy. Interestingly enough, (13) was already discussed by Zadeh (1978b) and Sanchez (1978) for possibility-qualification purposes.

Similarly, "A is certain" has been interpreted as "$A_{\overline{1-\beta}}$ is at least β-certain", $\forall \beta \in (0,1]$. Then "A is α-certain" will be interpreted as "$A_{\overline{1-\beta}}$ is at least $\min(\alpha,\beta)$

certain". Then using the min-combination (5), we get (Dubois and Prade, 1990a, d) :

$$\forall u \in U, \pi_X(u) \leq \inf_{\beta \in (0,1]} \max(\mu_{A_{\overline{1-\beta}}}(u), 1 - \min(\alpha,\beta))$$

i.e.

$$\forall u \in U, \pi_X(u) \leq \max(\inf_{\beta \in (0,1]} \max(\mu_{A_{\overline{1-\beta}}}(u), 1 - \beta), 1 - \alpha)$$

i.e.

$$\forall u \in U, \pi_x(u) \leq \max(\mu_A(u), 1 - \alpha) \qquad (14)$$

which extends (7) to the case where A is fuzzy.

Before applying the above model, especially (13) and (14), to the representation of different types of fuzzy rules in Section 4, it is important to clarify the relationship between the notion of possibility qualification introduced in this section, and possibility theory as developed until now (Zadeh, 1978a ; Dubois and Prade, 1988).

3. TWO COMPLEMENTARY NOTIONS OF POSSIBILITY

The idea of possibility which has been used for qualification purposes in the preceding section is not the same as the one underlying the definition of a possibility measure. Indeed when "A is possible" is modelled by the inequality

$$\forall u \in U, \mu_A(u) \leq \pi_x(u)$$

it means, when A is an ordinary subset, that

$$\forall u \in A, \pi_x(u) = 1 \qquad (15)$$

while the measure of possibility Π induced by π_x, of a non-fuzzy event A is defined by (Zadeh, 1978a)

$$\Pi(A) = \sup_{u \in A} \pi_x(u) \qquad (16)$$

and clearly, when A has a bounded support,

$$\Pi(A) = 1 \Leftrightarrow \exists u \in A, \pi_x(u) = 1 \qquad (17)$$

$\Pi(A) = 1$ only says that the statement "x is in A" is consistent with the available information described by π_x. The discrepancy between (15) and (17) is obvious and is expressed by the difference between the logical quantifiers '$\forall$' and '$\exists$'. This discrepancy between possibility measures and the notion of "possible" appearing in possibility-qualification was noticed by Zadeh (1978b), but no attempt had been made to define the set-function underlying possibility-qualification.

The notion of possibility introduced in Section 2 rather relates to the following quantity, called "everywhere-possibility" or for short E-possibility of A

$$\Delta(A) = \inf_{u \in A} \pi_x(u) \qquad (18)$$

which is such that

$$\text{"A is possible" in the sense of (1)} \Leftrightarrow \Delta(A) = 1 \qquad (19)$$

Clearly the set functions Π and Δ correspond respectively to a weak and a strong requirement of possibility, and indeed $\forall A, \Pi(A) \geq \Delta(A)$, or if we prefer $\forall \alpha, \Delta(A) \geq \alpha \Rightarrow \Pi(A) \geq \alpha$. In practice it corresponds to the distinction between saying that "it is possible that A is true" which means that there exists at least one value u in A which is completely possible for x (i.e. $\Pi(A) = 1$), and saying that "A is possible" is short for "the range A is (completely) possible for x" whose intended meaning is really that all the values in A are possible for the variable x. This latter notion of possibility is particularly important, as advocated in this paper, for the specification of possibility distributions in general and more particularly of fuzzy rules.

The notion of E-possibility seems to have been largely ignored in the fuzzy set literature. However its counterpart in Shafer (1976)'s evidence theory is well-known ; it is the commonality function Q, which, by the way, is mainly used for technical reasons and does not seem to have received any practical interpretation until now. Indeed starting with a basic probability assignment m such that $\Sigma_A\, m(A) = 1$, the commonality of A is defined by $Q(A) = \Sigma_{B \supseteq A}\, m(A)$; it can be easily checked that the following analogue of (19) holds

$$Q(A) = 1 \Leftrightarrow \forall\, u \in A, Pl(\{u\}) = 1$$

where the plausibility function Pl is defined by $Pl(C) = \Sigma_{C \cap B \neq \emptyset}\, m(B)$. This results from $Q(A) = 1 \Leftrightarrow \forall B$ such that $m(B) > 0$, $B \supseteq A \Leftrightarrow \forall\, u \in A$, $\forall B$ such that $m(B) > 0$, $u \in B \Leftrightarrow \forall\, u \in A, Pl(\{u\}) = 1$. Moreover $\Delta(A)$ can be put under a form which looks analogous to Q(A). Indeed introducing the fuzzy set F such that $\mu_F = \pi_x$, we have $\Delta(A) = \inf_{u \in A} \mu_F(u)$; hence $\forall\, u \in A$, $\mu_F(u) \geq \Delta(A)$ or equivalently $A \subseteq F_{\Delta(A)}$, and more generally $\forall\, \alpha \leq \Delta(A)$, $A \subseteq F_\alpha$. Besides $\nexists \varepsilon > 0$, $A \subseteq F_{\Delta(A)+\varepsilon}$ (from the definition of $\Delta(A)$). Hence

$$\Delta(A) = \sup\{\alpha \in (0,1], F_\alpha \supseteq A\} \quad \text{(with the convention } \sup \mathcal{E} = 0 \text{ if } \mathcal{E} = \emptyset)$$

A possibility distribution π_x such that $\{\pi_x(u) \in (0,1], u \in U\}$ is an ordered finite set $M = \{\alpha_1, \ldots, \alpha_n\}$ with $\alpha_1 = 1 > \ldots > \alpha_n > \alpha_{n+1} = 0$, is equivalent to the basic probability assignment (Dubois and Prade, 1982) defined by

$$\forall i,\ m(F_{\alpha_i}) = \alpha_i - \alpha_{i+1}$$
$$\forall\, B \neq F_{\alpha_i},\ m(B) = 0$$

Then, using the nestedness property $\alpha < \beta \Rightarrow F_\alpha \supset F_\beta$, it can be easily seen that $\Delta(A) = Q(A)$, where Q is defined from the function m above, i.e. the two definitions coincide. More generally, for A remaining non-fuzzy, it is easy to see that

$$\text{A is (at least) } \alpha\text{-possible} \Leftrightarrow \forall\, u \in A,\ \pi_x(u) \geq \alpha \Leftrightarrow \Delta(A) \geq \alpha \qquad (20)$$

which generalizes (19). The definition of the E-possibility can be extended to fuzzy sets still preserving the equivalence

$$\text{A is (at least) } \alpha\text{-possible} \Leftrightarrow \forall\, u \in U,\ \pi_x(u) \geq \min(\mu_A(u), \alpha) \Leftrightarrow \Delta(A) \geq \alpha \qquad (21)$$

This is satisfied by taking

$$\Delta(A) = \inf_{u \in U} \mu_A(u) \rightarrow \pi_x(u) \qquad (22)$$

where $a \rightarrow b = \begin{cases} 1 \text{ if } a \leq b \\ b \text{ if } a > b \end{cases}$ is a multiple-valued logic implication connective, known as Gödel's implication. It is easy to see that (22) reduces to (18) when A is an ordinary subset. Moreover (21) is ensured by the equivalence $a \to b \geq c \Leftrightarrow b \geq \min(a,c)$. By contrast, a lower bound α on the extension of the possibility measure Π (defined by (16)) to a fuzzy event A, i.e.

$$\Pi(A) = \sup_{u \in U} \min(\mu_A(u), \pi_x(u)) \geq \alpha \qquad (23)$$

is equivalent to $\forall \beta < \alpha, \Pi(A_\beta) \geq \alpha$, i.e. $\forall \beta < \alpha, \exists u \in A_\beta, \pi_x(u) \geq \alpha$ (see, e.g., Dubois and Prade, 1990a), which clearly departs from $\Delta(A) \geq \alpha$, the latter means that $\forall \beta \geq \alpha, \forall u \in A_\beta, \pi_x(u) \geq \alpha$ and $\forall \beta < \alpha, \forall u \in A_\beta, \pi_x(u) \geq \beta$.

We now identify to what evaluation of A is associated the certainty qualification presented above. "A is α-certain" is represented, in the general case where A is fuzzy, by (14), i.e.

$$\forall u \in U, \pi_x(u) \leq \max(\mu_A(u), 1 - \alpha)$$

Since $c \leq \max(a, 1 - b) \Leftrightarrow (1 - a) \to (1 - c) \geq b$, where $\to$ denotes Gödel's implication, (14) is equivalent to (Dubois and Prade, 1989a)

$$\mathcal{N}(A) = \inf_{u \in U} (1 - \mu_A(u)) \to (1 - \pi_x(u)) \geq \alpha \qquad (24)$$

i.e.

$$\text{"A is } \alpha\text{-certain"} \Leftrightarrow \mathcal{N}(A) \geq \alpha.$$

When A is an ordinary subset, (24) reduces to

$$\mathcal{N}(A) = 1 - \Pi(\bar{A}) = \inf_{u \notin A} (1 - \pi_x(u)) \qquad (25)$$

where the overbar denotes the complementation. It means that certainty qualification, when A is non-fuzzy is in complete agreement with the necessity measure based on π_x and defined by duality with respect to the possibility measure. However when A is fuzzy, the duality relation $\mathcal{N}(A) = 1 - \Pi(\bar{A})$, where Π is extended to fuzzy events by (23), is no longer satisfied. When A is fuzzy, since (14) is equivalent to "$A_{\overline{1-\beta}}$ is at least $\min(\alpha,\beta)$-certain", we get

$$\mathcal{N}(A) \geq \alpha \Leftrightarrow \forall \beta < 1, \mathcal{N}(A_{\bar{\beta}}) \geq \min(\alpha, 1 - \beta) \qquad (26)$$

which expresses the relation between certainty-qualification and the measure of necessity of the β-cuts of A. See (Dubois and Prade, 1990a) for a discussion about certainty-qualification in possibilistic logic with fuzzy predicates (which is in full agreement with $\mathcal{N}$ defined by (24)) and (Dubois and Prade, 1989a, 1990d) for the distinct uses of $\mathcal{N}(A)$ and $1 - \Pi(\bar{A})$, the former in certainty-qualification, the latter in fuzzy pattern matching. More particularly, as already said in (Dubois and Prade, 1989a), the statement "A is certain" may either mean that we are certain that the possible values of x are inside A, i.e. $\pi_x \leq \mu_A$ (which is captured by $\mathcal{N}(A) = 1$), or that we are certain that the value of x is among the elements of U which *completely* belong to A, i.e. $\text{support}\pi_x = \{u \in U, \pi_x(u) > 0\} \subseteq \text{core}(A) = \{u \in U, \mu_A(u) = 1\}$ which is captured by $1 - \Pi(\bar{A}) = 1$). This latter interpretation which is more demanding is clearly related to the fuzzy filtering of fuzzily-known objects ; see (Dubois, Prade and Testemale, 1988).

4. REPRESENTATION OF DIFFERENT KINDS OF FUZZY RULES

We now apply the results of Section 2 on possibility and certainty qualifications to the specification of fuzzy rules relating a variable x ranging on U to a variable y ranging on V.

Possibility rules : A first kind of fuzzy rule corresponds to statements of the form "the more x is A, the more possible B is a range for y". If we interpret this rule as "$\forall$u, if x = u, B is a range for x is at least μ_A(u)-possible", a straightforward application of (13), yields the following constraint on the conditional possibility distribution $\pi_{y|x}(\cdot, u)$ representing the rule when x = u

$$\forall u \in U, \forall v \in V, \min(\mu_A(u), \mu_B(v)) \leq \pi_{y|x}(v,u). \qquad (27)$$

Certainty rules : A second kind of fuzzy rule corresponds to statements of the form "the more x is A, the more certain y lies in B". Interpreting the rule as "$\forall$u, if x = u, y lies in B is at least μ_A(u)-certain", by application of (14) we get the following constraint for the conditional possibility distribution modelling the rule

$$\forall u \in U, \forall v \in V, \pi_{y|x}(v,u) \leq \max(\mu_B(v), 1 - \mu_A(u)). \qquad (28)$$

In the particular case where A is an ordinary subset and where we know that, if x is in A, B is both a possible and a certain range for y, (27) and (28) yield

$$\begin{cases} \forall u \in A, & \pi_{y|x}(v,u) = \mu_B(v) \\ \forall u \notin A, & \pi_{y|x}(v,u) \text{ is completely unspecified.} \end{cases} \qquad (29)$$

This corresponds to the usual modelling of a fuzzy rule with a non-fuzzy condition part. Note that B may be any kind of fuzzy set in (27), (28) and then in (29). Thus B may itself includes some uncertainty ; for instance the membership function of B may be of the form $\mu_B = \max(\mu_{B*}, 1 - \beta)$ in order to express that when x is A, B* is the (fuzzy) range of y with a certainty β (any value outside the support of B* remains a possible value for y with a degree equal to $1 - \beta$) ; we may even have an unnormalized possibility distribution which can be put under the form $\mu_B = \min(\mu_{B*},\alpha)$ if the possibility that y takes its value in V is bounded from above by α (i.e. there is a possibility $1 - \alpha$ that y has no value in V, when x takes its value in A).

Gradual rules : This third kind of fuzzy rule has been discussed in (Dubois and Prade, 1989a, 1990b, d). Gradual rules correspond to statements of the form "the more x is A, the more y is B". Statements involving "the less" in place of "the more" are easily obtained by changing A or B into their complements $\bar{A}$ and $\bar{B}$ due to the equivalence between "the more x is A" and "the less x is $\bar{A}$" (with $\mu_{\bar{A}} = 1 - \mu_A$). More precisely, the intended meaning of a gradual rule can be understood in the following way : "the greater the degree of membership of the value of x to the fuzzy set A and the more the value of y is considered to be in relation (in the sense of the rule) with the value of x, the greater the degree of membership to B should be for this value of y", i.e.

$$\forall u \in U, \min(\mu_A(u), \pi_{y|x}(v,u)) \leq \mu_B(v) \qquad (30)$$

or, using the equivalence $\min(a,t) \leq b \Leftrightarrow t \leq a \rightarrow b$ where $\rightarrow$ denotes Gödel's implication,

$$\forall u \in U, \pi_{y|x}(v,u) \leq \mu_A(u) \rightarrow \mu_B(v) = \begin{cases} 1 & \text{if } \mu_A(u) \leq \mu_B(v) \\ \mu_B(v) & \text{if } \mu_A(u) > \mu_B(v) \end{cases} \qquad (31)$$

(31) can be equivalently written

$$\forall u \in U, \pi_{y|x}(v,u) \leq \max(\mu_{[\mu_A(u),1]}(\mu_B(v)), \mu_B(v)) = \mu_{[\mu_A(u),1] \cup T}(\mu_B(v)) \qquad (32)$$

where $\mu_{[\mu_A(u),1]}$ is the characteristic function of the interval $[\mu_A(u),1]$ and where T is the fuzzy set of [0,1], defined by $\forall t \in [0,1], \mu_T(t) = t$, which models the fuzzy truth-value '**true**' in fuzzy logic (Zadeh, 1978b). If we remember that "x is A is τ-true", where τ is a fuzzy truth-value modelled by the fuzzy set τ of [0,1], is represented by the possibility distribution (Zadeh, 1978b)

$$\forall u \in U, \pi_x(u) = \mu_\tau(\mu_A(u)) \qquad (33)$$

(note that τ = T yields the basic assignment (12)), we can interpret the meaning of gradual rules in the following way using (32) : $\forall u \in U$, if x = u then y is B is at least $\mu_A(u)$-true. The membership function of the fuzzy truth-value 'at least α-true' is pictured in Figure 1.a. As it can be seen, it is not a crisp "at least α-true" (which would correspond to the ordinary subset $[\alpha,1]$), but a fuzzy one in agreement with truth-qualification in the sense of (33), indeed "at least 1-true" corresponds to the fuzzy truth-value T.

If we are only looking for the *crisp* possibility distributions $\pi_{y|x}$ (i.e. the {0,1}-valued ones) which satisfy (30), because we assume that it is a crisp relation between y and x which underlies the rule "the more x is A, the more y is B", then we obtain the constraint

$$\forall u \in U, \pi_{y|x}(v,u) \leq \begin{cases} 1 \text{ if } \mu_A(u) \leq \mu_B(v) \\ 0 \text{ if } \mu_A(u) > \mu_B(v) \end{cases} = \mu_{[\mu_A(u),1]}(\mu_B(v)) \qquad (34)$$

which expresses that the rule is now viewed as meaning : $\forall u \in U$, if x = u then y is B is at least $\mu_A(u)$-true, where the truth-qualification is understood in a crisp sense. The reader is referred to (Dubois and Prade, 1990b) for a discussion of this kind of rule.

A fourth type of fuzzy rules : The inequality (30) looks like (27) when exchanging $\mu_B(v)$ and $\pi_{y|x}(v,u)$, while (31), which is equivalent to (30), is analogue to (28) in the sense that in both cases $\pi_{y|x}$ is bounded from above by a multiple-valued logic implication function (in (28) it is Dienes' implication : $a \rightarrow b = \max(1 - a, b)$ which appears). It leads to consider the inequality constraint obtained from (28) by exchanging $\mu_B(v)$ and $\pi_{y|x}(v,u)$, i.e.

$$\forall u \in U, \forall v \in V, \max(\pi_{y|x}(v,u), 1 - \mu_A(u)) \geq \mu_B(v) \qquad (35)$$

This corresponds to a fourth kind of fuzzy rules, of which we now investigate the intended meaning. (35) is perhaps more easily understood by taking the complement to 1 of each side of the inequality, i.e.

$$\forall u \in U, \forall v \in V, 1 - \mu_B(v) \geq \min(\mu_A(u), 1 - \pi_{y|x}(v,u))$$

which can be interpreted as "the more x is A and the less y is related to x, the less y is B", which corresponds to a new type of gradual rule. Using the equivalence $\min(a, 1 - t) \leq 1 - b \Leftrightarrow 1 - t \leq a \rightarrow (1 - b) \Leftrightarrow t \geq 1 - (a \rightarrow (1 - b))$, where $\rightarrow$ is Gödel's implication, we can still write (35) under the form

$$\forall u \in U, \forall v \in V, \pi_{y|x}(v,u) \geq \begin{cases} 0 \text{ if } \mu_A(u) + \mu_B(v) \leq 1 \\ \mu_B(v) \text{ if } \mu_A(u) + \mu_B(v) > 1 \end{cases}$$
$$= \min(\mu_{(1-\mu_A(u),1]}(\mu_B(v)), \mu_B(v)) \quad (36)$$
$$= \mu_{(1-\mu_A(u),1] \cap T}(\mu_B(v))$$

Unsurprisingly, the lower bound of $\pi_{y|x}(v,u)$ which is obtained, is a multiple-valued logic conjunction function of $\mu_A(u)$ and $\mu_B(v)$ (indeed $f(a,b) = 0$ if $a + b \leq 1$ and $f(a,b) = b$ otherwise, is such that $f(0,0) = f(0,1) = f(1,0) = 0$ and $f(1,1) = 1$). From (36) we see that this type of gradual rules can be interpreted in the following way, using (33) : $\forall u \in U$, if $x = u$ then y is B is at least $(1 - \mu_A(u))$-true. The membership function of the corresponding fuzzy truth-value "at least $(1 - \alpha)$-true" is

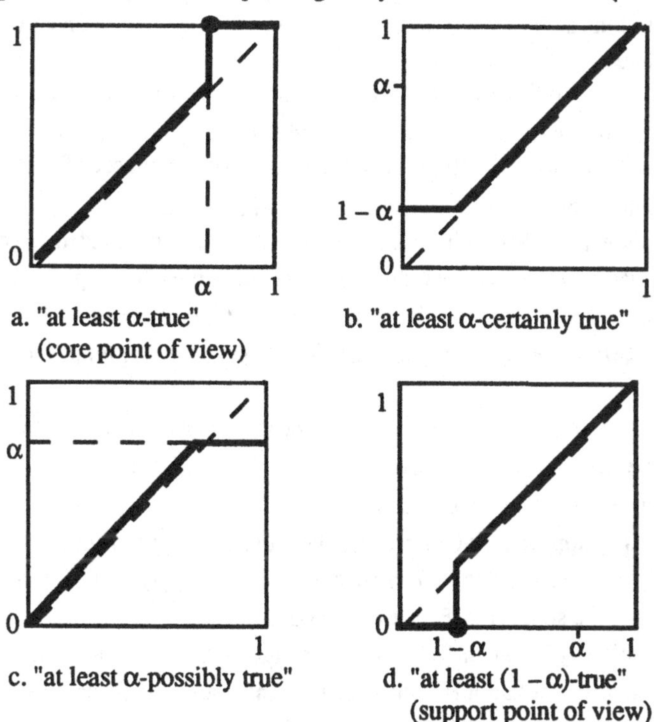

Figure 1 : Four basic types of fuzzy truth-values

pictured in Figure 1.d. As it can be observed by comparing Figures 1.a and 1.d, they correspond to two points of view in (fuzzy) truth-qualification of level α, one insisting on the complete possibility of degrees of truth greater than α (core point of view), the other insisting on the complete impossibility of degrees of truth less or equal to $1-\alpha$ (support point of view).

Figures 1.b and 1.c picture the fuzzy truth-values "at least α-certainly true" and "at least α-possibly true", whose respective membership functions are $\max(\mu_T, 1-\alpha)$ and $\min(\mu_T, \alpha)$. It can be seen on Figure 1 and formally checked that the four fuzzy truth-values we have introduced satisfy the two duality relations

'at least α-certainly true' = **comp∘ant** ('at least α-possibly true') (37)
'at least α-true (core p. of v.)' = **comp∘ant** ('at least $1-\alpha$-true (support p. of v.)) (38)

where **comp** and **ant** are two transformations reflecting the ideas of complementation and antonymy respectively, and defined by **comp**$(f(t)) = 1 - f(t)$ and **ant**$(f(t)) = f(1-t)$, $\forall\, t \in [0,1]$ and f ranging in [0,1]. Note that **comp∘ant** = **ant∘comp**. Note that when there are only two degrees of truth, 0 (false) and 1 (true), "at least α-certainly true" corresponds to the possibility distribution $\pi(\text{true}) = 1$, $\pi(\text{false}) = 1-\alpha$ and "at least α-possibly true" to $\pi(\text{false}) = 0$ and $\pi(\text{true}) = \alpha$, while the two other (fuzzy) truth-values make no sense. Dually when there are only two degrees of possibility, 0 (complete impossibility) and 1 (complete possibility), then the representations of "at least α-true" and "at least $(1-\alpha)$-true" respectively coincide with the ordinary subsets $[\alpha,1]$ and $(1-\alpha, 1]$.

The four fuzzy truth-values pictured in Figure 1 (with $\alpha = \mu_A(u)$) can be viewed as representing modifiers φ (in the sense of Zadeh (1972)) which modify the fuzzy set B into B* such that $\mu_{B*} = \varphi(\mu_B)$, in order to specify the subset of interest for y in the various rules when x = u. For summarizing, in the case of

- *possibility rules*, the possibility distribution $\pi_{y|x}(\cdot\,,u)$ is bounded from below by $\varphi(\mu_B)$ with $\varphi(t) = \min(\mu_A(u), t)$, i.e. B is truncated up to the height $\alpha = \mu_A(u)$;
- *certainty rules*, the possibility distribution $\pi_{y|x}(\cdot\,,u)$ is bounded from above by $\varphi(\mu_B)$ with $\varphi(t) = \max(t, 1-\mu_A(u))$, i.e. B is drowned in a level of indetermination $1-\alpha$;
- *gradual rules* (core point of view), the possibility distribution $\pi_{y|x}(\cdot\,,u)$ is bounded from above by $\varphi(\mu_B)$ with $\varphi(t) = \mu_A(u) \to t$ (where $\to$ denotes Gödel's implication), i.e. the core of B is enlarged ;
- *gradual rules* (support point of view), the possibility distribution $\pi_{y|x}(\cdot\,,u)$ is bounded from below by $\varphi(\mu_B)$ with $\varphi(t) = 0$ if $\mu_A(u) + t \le 1$ and $\varphi(t) = t$ otherwise, i.e. the support of B is diminished, truncated.

N.B. : The similarity between (27) and (30) suggests that "the more x is A, the more y is B", where y is in relation R with x, can be understood as meaning that $\forall\, u \in U$, B represents the statement "R(u) is a range for y which is at least possible at the degree $\mu_A(u)$", where R(u) is the fuzzy set of elements in V in relation with u.

5. INFERENCE WITH FUZZY RULES

5.1. Parallel Rules with a Precise Input

Depending on the kind of constraint induced by their representation (the possibility distribution is bounded from below or from above) the combination of several rules in parallel of the same type will be performed differently. Namely for certainty rules and for gradual rules focusing on cores, described by means of pairs (A_i,B_i), i = 1,n we have

$$\forall i = 1,n, \forall u \in U, \forall v \in V, \pi_{y|x}(v,u) \leq \mu_{A_i}(u) \ominus\!\!\rightarrow \mu_{B_i}(v)$$

where $\ominus\!\!\rightarrow$ denotes Gödel's or Dienes' implication ; then by combination we get

$$\forall u \in U, \forall v \in V, \pi_{y|x}(v,u) \leq \min_{i=1,n} (\mu_{A_i}(u) \ominus\!\!\rightarrow \mu_{B_i}(v)) \qquad (39)$$

while with possibility rules and gradual rules focusing on supports, we have

$$\forall i = 1,n, \forall u \in U, \forall v \in V, \pi_{y|x}(v,u) \geq \mu_{A_i}(u) \ \& \ \mu_{B_i}(v)$$

where & denotes the min conjunction or the non-symmetrical one introduced above ; which yields

$$\forall u \in U, \forall v \in V, \pi_{y|x}(v,u) \geq \max_{i=1,n} (\mu_{A_i}(u) \ \& \ \mu_{B_i}(v)) \qquad (40)$$

The existence of two models of combination of systems of fuzzy rules has been pointed out by several authors including Baldwin and Pilsworth (1979), Tanaka et al. (1982), Di Nola et al. (1985), when considering special cases of implication functions $\ominus\!\!\rightarrow$ in contrast with the min-conjunction for combining the fuzzy sets A_i and B_i.

The Figures 2.a and 2.b illustrate the behaviour of the four types of rules in case of two parallel rules relating A_1 and B_1 on the one hand and A_2 and B_2 on the other hand when the value x is precisely known, i.e. $\pi_x = \mu_{A'}$ with $A' = \{u_0\}$. In Figure 2.a, $x = u_0$ perfectly satisfies the requirements modelled by A_1 and A_2, i.e., $\mu_{A_1}(u_0) = \mu_{A_2}(u_0) = 1$ and we obtain as a conclusion for y, with $\pi_y = \mu_{B'}$

$$\forall v \in V, \mu_{B'}(v) \leq \min(\mu_{B_1}(v),\mu_{B_2}(v)) \qquad \text{(conjunction of the conclusions)}$$

for certainty rules and for gradual rules (focusing on cores) due to (39), and

$$\forall v \in V, \mu_{B'}(v) \geq \max(\mu_{B_1}(v),\mu_{B_2}(v)) \qquad \text{(disjunction of the conclusions)}$$

for possibility rules and for gradual rules (focusing on supports) due to (40). The fact that we obtain $B' \supseteq B_1 \cup B_2$ when $\{u_0\} \subseteq A_1 \cap A_2$ for possibility rules, for instance, should not be surprizing ; indeed we have to remember that each rule expresses that "the more x is A_i, the greater the level of possibility of B_i as a possible range for y" for i = 1,2, then since y may lie in both B_1 and B_2, $B_1 \cup B_2$ should be a possible range for y. In Figure 2.b, we have $\mu_{A_2}(u_0) = 1$ again, but now

$\mu_{A_1}(u_0) = \alpha < 1$. The difference between certainty rules and gradual rules focusing on cores appears clearly : for certainty rules, the intersection of B_2 with B_1 is pervaded with a level of uncertainty $1 - \alpha$ (i.e. $\min(\mu_{B_2}, \max(\mu_{B_1}, 1 - \alpha)))$, while the upper bound of B' for the gradual rules stays between B_1 and B_2, overlapping a little more B_2. Similarly the difference between possibility rules and gradual rules focusing on supports also appears ; for the former we obtain $\mu_{B'} \geq \max(\mu_{B_2}, \min(\mu_{B_1}, \alpha))$ which expresses that the values in B_1 are regarded as a priori

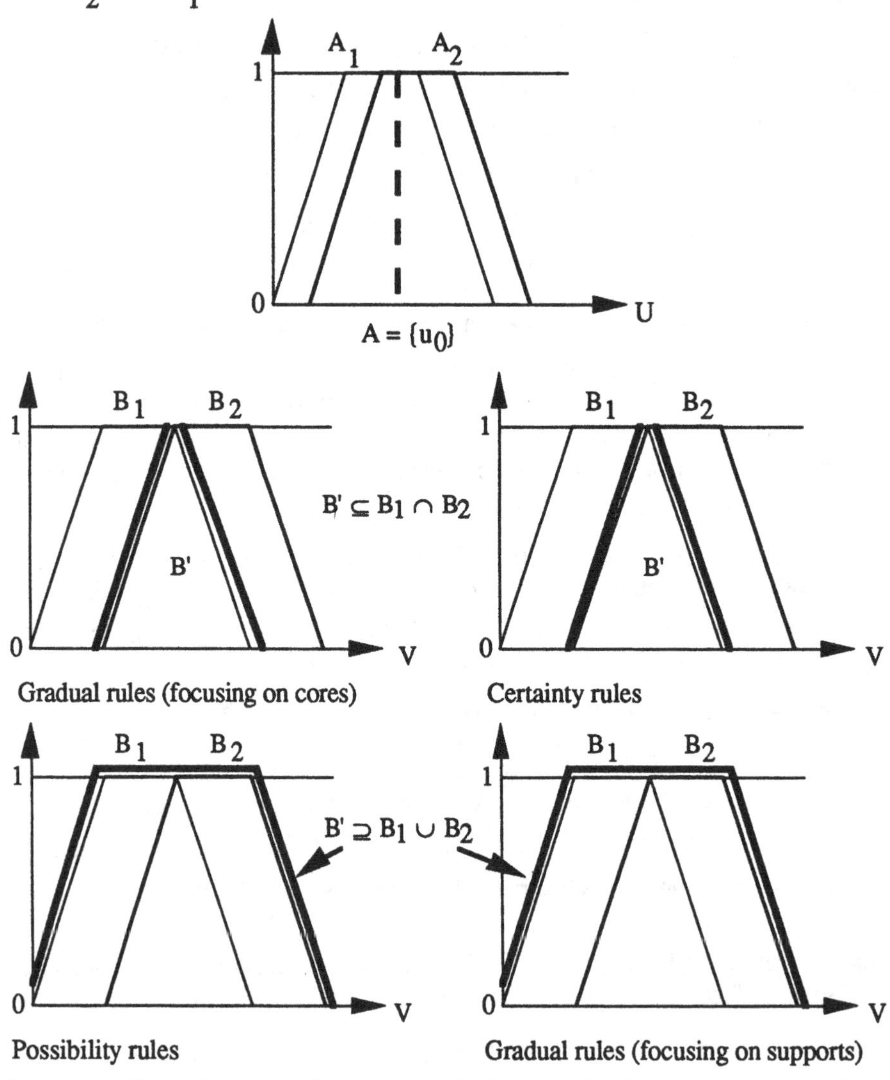

Figure 2.a: Two rules in parallel and a precise input perfectly matching the conditions

less possible than the ones in B_2 ; for the latter some values in the support of B_1 are considered as potentially impossible.

5.2. Generalized Modus Ponens with One Rule

In section 4, when studying the representation of the four types of rules considered in the paper, we have described the response B' of a rule to a precise input

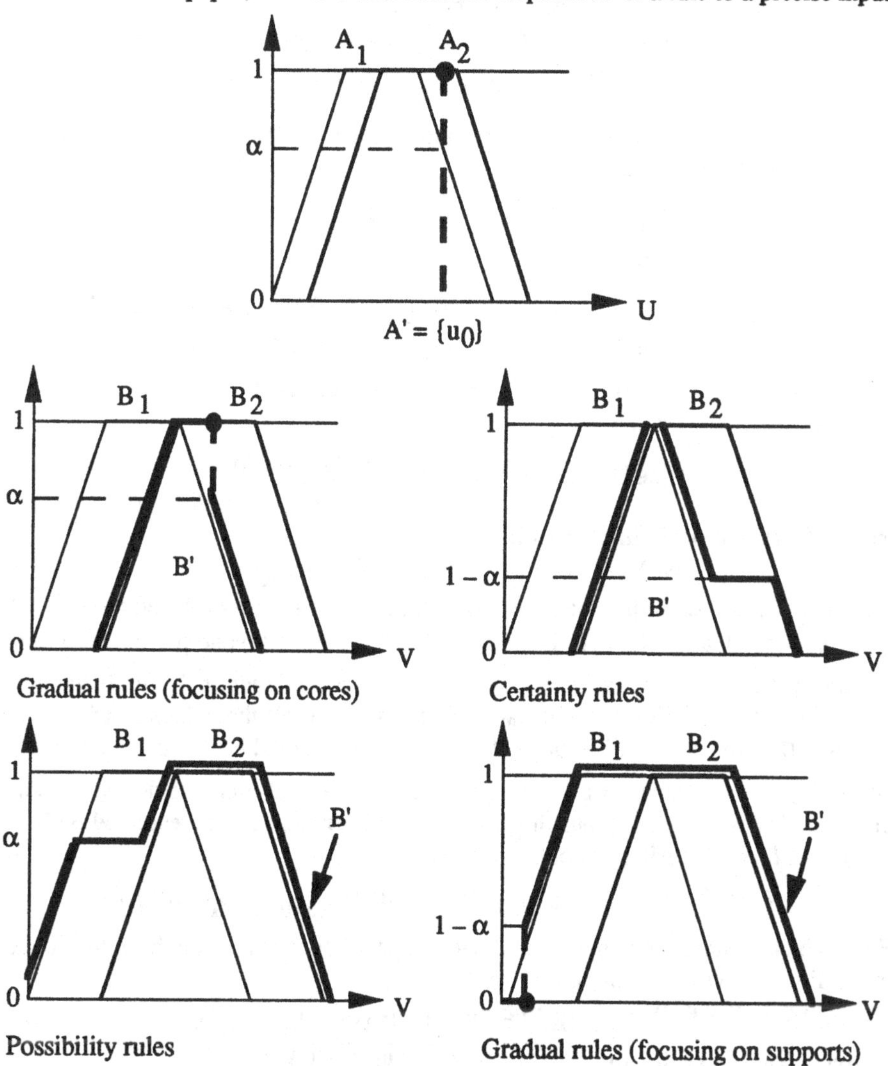

Figure 2.b: Two rules and a precise input imperfectly matching one of the conditions

$A' = \{u_0\}$ under the form $\mu_{B'} = \varphi(\mu_B)$ where φ represents a particular fuzzy truth value which acts as a modifier. We now consider the more general situation of the generalized modus ponens (Zadeh, 1979), namely

$$\frac{\begin{array}{l}\text{x is A'}\\ \text{<rule relating x is A with y is B>}\end{array}}{\text{y is B'}}$$

As usual, and in agreement with (12), "x is A' " will be understood as

$$\forall\, u \in U,\ \pi_x(u) = \mu_{A'}(u)$$

while the rule, depending on the case, is represented

by $\forall\, u \in U, \forall\, v \in V,\ \pi_{y|x}(v,u) \le \mu_A(u) \Theta\!\!\rightarrow \mu_B(v)$ (case I)

or by $\forall\, u \in U, \forall\, v \in V,\ \pi_{y|x}(v,u) \ge \mu_A(u)\ \&\ \mu_B(v)$ (case II)

Applying the combination/projection principle (Zadeh, 1979 ; see Dubois and Prade, 1990d for a discussion), i.e. here

$$\forall\, v \in V,\ \pi_y(v) = \sup_{u\in U} \min(\pi_x(u), \pi_{y|x}(v,u)) \qquad (41)$$

Thus we get, with $\forall v$, $\mu_{B'}(v) = \pi_y(v)$

$$\forall\, v \in V,\ \mu_{B'}(v) \le \sup_{u\in U} \min(\mu_{A'}(u), \mu_A(u) \Theta\!\!\rightarrow \mu_B(v)) \qquad \text{(case I)}$$

$$\forall\, v \in V,\ \mu_{B'}(v) \ge \sup_{u\in U} \min(\mu_{A'}(u), \mu_A(u)\ \&\ \mu_B(v)) \qquad \text{(case II)}$$

Let us first consider the two kinds of rules belonging to case I.
For *certainty rules* : we obtain

$$\forall\, v \in V,\ \mu_{B'}(v) \le \sup_{u\in U} \min(\mu_{A'}(u), \max(1 - \mu_A(u), \mu_B(v)) = \max(\mu_B(v), 1 - N(A\,;A')) \qquad (42)$$

provided that A' is normalized and where

$$N(A\,;A') = \inf_{u\in U} \max(\mu_A(u), 1 - \mu_{A'}(u))$$

is the dual of the possibility measure $\Pi(A\,;A')$ of the fuzzy event A defined by (23) (with $\pi_x = \mu_{A'}$), and $N(A\,;A')$ is thus equal to $1 - \Pi(\bar{A}\,;A')$ and plays a basic role in fuzzy pattern matching as briefly recalled at the end of section 3. The inequality (42) expresses the following. Our lack of certainty that all the values restricted by $\mu_{A'}$ are (highly) compatible with the requirement modelled by μ_A, induces a possibility at most equal to $1 - N(A\,;A')$ that the value of y is outside the support of B. In other words (42) means that it is $N(A\,;A')$ certain that y is restricted by B.
For *gradual rules* focusing on cores, we have

$$\forall\, v \in V,\ \mu_{B'}(v) \le \sup_{u\in U} \min(\mu_{A'}(u), \mu_{[\mu_A(u),1]\cup T}(\mu_B(v)))$$

From which it can be concluded that the least upper bounds derivable from the above inequality are given by (Dubois and Prade, 1988) :

- $\mu_{B'}(v) \le 1,\ \forall\, v \in \{v \in V,\ \mu_B(v) \ge \inf_{u\in U} \{\mu_A(u) \mid \mu_{A'}(u) = 1\}\}$
- $\mu_{B'}(v) \le \sup_{u\in U} \{\mu_{A'}(u) \mid \mu_A(u) = 0\} = \Pi(\overline{\mathrm{support}(A)}\,;A'),$
  $\forall\, v \in \{v \in V,\ \mu_B(v) = 0\}$

This shows that when A' is no longer a singleton, the enlarging effect of the core of B may be increased and a non-zero possibility $\Pi(\overline{\mathrm{support}(A)}\,;A')$ may be obtained for values outside the support of B, for the possibility distribution restricting y. The level of possibility $\Pi(\overline{\mathrm{support}(A)}\,;A')$ acknowledges the fact that some possible values of x (in A') are not consistent at all with A.

We now consider the rules belonging to case II.

For *gradual rules* focusing on support, we have

$$\forall v \in V, \mu_{B'}(v) \geq \sup_{u \in U} \min(\mu_{A'}(u), \mu_{(1-\mu_A(u),1] \cap T]}(\mu_B(v)))$$

Noticeable greatest lower bounds derivable from the above inequality are

- $\mu_{B'}(v) \geq 0, \forall v \in \{v \in V, \mu_B(v) \leq 1 - \sup\{\mu_A(u) \mid \mu_{A'}(u) > 0\}\}$
- $\mu_{B'}(v) \geq \sup_{u \in U} \{\mu_{A'}(u) \mid \mu_A(u) > 0\} = \Pi(\text{support}(A) ; A'),$ $\forall v \in \{v \in V, \mu_B(v) = 1\}$

As it can be seen, the truncation effect of the support of B may decrease, while the height of the possibility distribution attached to y, equal to Π(support(A) ; A'), may be less than 1 (without being 0), when A' is no longer a singleton. When some values compatible with A' do not belong at all to A, the lower bound on the level of possibility for values in B to be the value of y decreases.

For *possibility rules*, we have

$$\forall v \in V, \mu_{B'}(v) \geq \sup_{u \in U} \min(\mu_{A'}(u), \mu_A(u), \mu_B(v)) = \min(\mu_B(v), \Pi(A ; A')) \quad (43)$$

It expresses that as soon as there is no value in A' fully consistent with A, B is only considered as an α-possible range for y (in the sense of (13)), with $\alpha = \Pi(A ; A')$.

5.3 - Parallel Rules with a Fuzzy Input

Lastly, we consider the general problem of the generalized modus ponens in the face of a collection of n rules *of the same type*, i.e. the pattern

x is A'
<rule relating x is A_i with y is B_i> i = 1,n
———————————————
y is B'

At the theoretical level, the solution is straightforward, namely

$$\forall v \in V, \mu_{B'}(v) \leq \sup_{u \in U} \min(\mu_{A'}(u), \min_{i=1,n} \mu_{A_i}(u) \rightarrow \mu_{B_i}(v)) \quad \text{(case I)} \quad (44)$$

$$\forall v \in V, \mu_{B'}(v) \geq \sup_{u \in U} \min(\mu_{A'}(u), \max_{i=1,n} \mu_{A_i}(u) \,\&\, \mu_{B_i}(v)) \quad \text{(case II)} \quad (45)$$

However at the practical level, the computation of these expressions raises problems for some types of rules. There is no difficult problem for (45). Indeed it can be checked that (45) is equivalent to

$$\forall v \in V, \mu_{B'}(v) \geq \max_{i=1,n} \sup_{u \in U} \min(\mu_{A'}(u), \mu_{A_i}(u) \,\&\, \mu_{B_i}(v))$$

i.e., for possibility rules we get

$$\forall v \in V, \mu_{B'}(v) \geq \max_{i=1,n} \min(\mu_{B_i}(v), \Pi(A_i ; A')) \quad (46)$$

For certainty rules the following upper bound which can be derived from (44) (assuming that A' is normalized) is no longer the best one,

$$\forall v \in V, \mu_{B'}(v) \leq \min_{i=1,n} \max(\mu_{B_i}(v), 1 - N(A_i ; A')) \quad (47)$$

Indeed assume n = 2, and that A_1, A_2, A' are distinct *non-fuzzy* subsets, and

$A' = A_1 \cup A_2$, with $A' \not\subseteq A_1$, $A' \not\subseteq A_2$, then $N(A_i ; A') = 0$, $i = 1,2$, and we get the trivial result $\forall v$, $\mu_{B'}(v) \leq 1$ by (47), although using (44) it would be possible to conclude that $\forall v$, $\mu_{B'}(v) \leq \max(\mu_{B_1}(v), \mu_{B_2}(v))$, which is a satisfying result. It emphasizes the fact that the rules should be jointly processed as in (44) in order to get the more accurate result : it is not the case in (47) where the *conclusions* obtained from x is A' and from each rule i are combined. Note that (46) and (47) are respectively a weighted max and a weighted min combination ; when $\forall i$, $\Pi(A_i ; A') = 1$, $N(A_i ; A') = 1$, (46) and (47) yield the union and the intersection of the B_i's respectively, see Dubois, Prade and Testemale (1988) for instance, for more details. The case of implication-based gradual rules raises similar problems. The processing of a collection of parallel gradual rules (focusing on cores) has been investigated in (Dubois, Martin-Clouaire and Prade, 1988) and in Martin-Clouaire (1988) to which the reader is referred. It is possible from the collection of rules to build a new rule which, when applied to A', yields the optimal result, i.e. the value of the upper bound expressed by (44) ; this new rule summarizes the knowledge useful in the collection of rules for dealing with the fact "x is A' ".

Generally speaking, we have to define the consistency, the non-redundancy of the set of fuzzy rules, and this leads to put some constraints on the coverage of U by the A_i's (see the first of the two above-mentioned references for definitions of these notions). Clearly further research is needed for a complete investigation of the practical processing of a collection of rules of a given type, also including the problem of compound condition parts in the rules which have not been considered here. Figure 2.c exhibits different behaviors of the four types of rules in the case n = 2 where $A' = A_1 \cap A_2$. We notice that we obtain $B' \subseteq B_1 \cap B_2$ with gradual rules focusing on cores, which confirms the "interpolation" flavor of this behavior : if A' is between A_1 and A_2 (in the sense of the intersection), then the possible values of y are restricted by a fuzzy set in between B_1 and B_2 ; a level of uncertainty equal to 0.5 appears for certainty rules, this is due to the fact that with continuous membership functions $N(A ; A) = 0.5$ as soon as A is a fuzzy set (when $A' = A_1 \cap A_2$, we are not completely sure x that belongs to the core of A_1 and to the core of A_2. For the two types of rules corresponding to case II, we obtain $B_1 \cup B_2 \subseteq B'$ as expected (since here $\Pi(A_i ; A') = 1$ and $\sup\{\mu_{A_i}(u) \mid \mu_{A'}(u) > 0\} = 1$, for i = 1,2).

As a final remark in this section, note that we may think of using two types of rules simultaneously, especially for certainty and possibility rules, since it can be checked that the two corresponding inequalities constraining π_y are consistent, namely using both (42) and (43) we get

$$\forall v \in V, \min(\mu_B(v) ; \Pi(A ; A')) \leq \pi_y(v) \leq \max(\mu_B(v), 1 - N(A ; A')) \quad (48)$$

It corresponds to the case of a piece of knowledge saying both that "the more x is A,

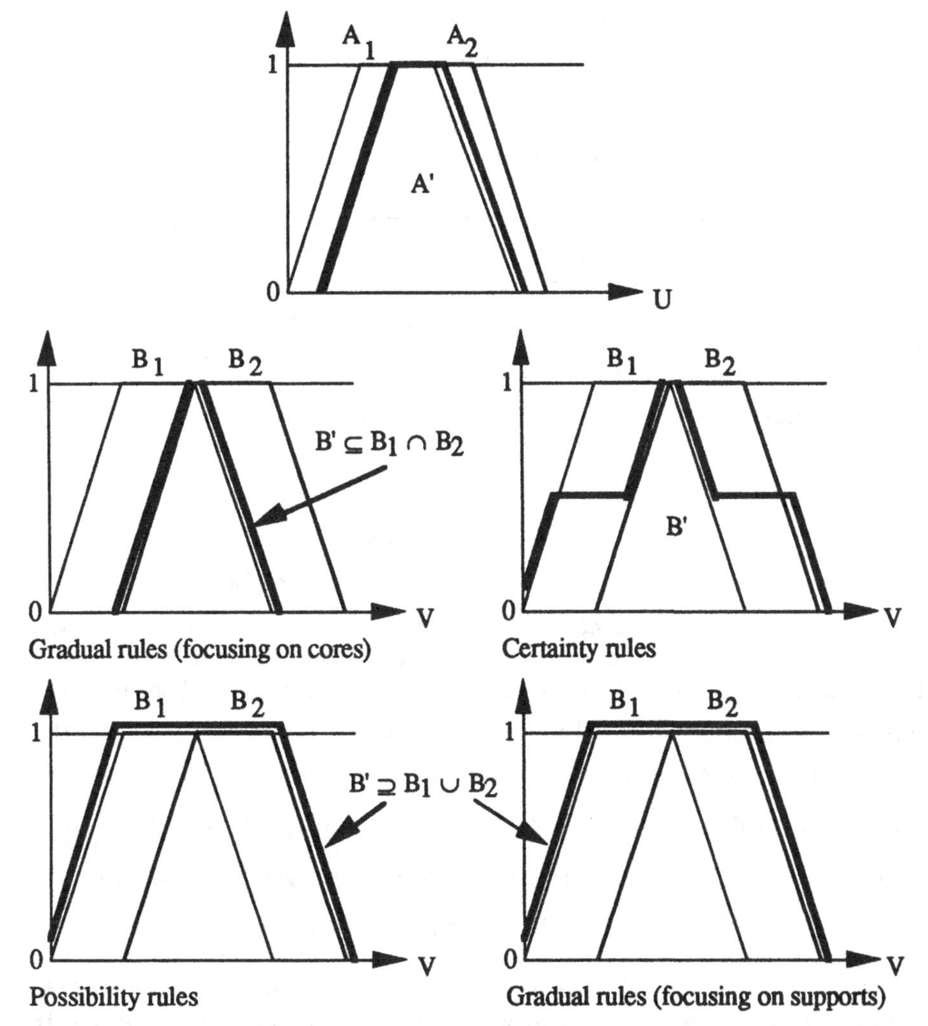

Figure 2.c: Two rules in parallel and a fuzzy input

the more possible B as a range for y and the more certain y is in B".

6 - RULES EXPRESSING PREFERENCE

So far, we have focused on "reasoning rules" whose aim is to describe the relationship between relevant parameters in some problem, e.g. a diagnostic problem. There is another class of "if... then... rules" which express preference and whose aim is to help in making choices rather than to make implicit knowledge explicit. Their format is in the crisp case :

if <situation> then <decision>.

<situation> is the description of the states of the world where some decision can be recommended. <decision> can be some action to perform, some procedure to trigger

or even an assignment statement (like "choose for y the value v_0"). When there are many possible states of the world, it is difficult to partition them into rigid classes where specific decisions can be totally recommended. As a result, the description of the states of the world where a decision is relevant is often fuzzy, because decisions can be more or less recommended. Hence the "if" part contains fuzzy predicates, and the preference rule means

"the more the state of the world corresponds to <situation>,
the more recommended is <decision>"

Let x be a vector that contains the precise description of the world, S be a fuzzy set of values of x corresponding to the description of a range of situations, and u(d) the preference degree for decision d. By definition, u(d) = 0 means that d should be rejected, u(d) = 1 means that d can be applied without any doubt. The fuzzy preference rule just indicates that u(d) can be quantified by $\mu_S(x)$.

In fuzzy control (e.g. Mamdani (1977), Sugeno (1985)), fuzzy rules can be viewed as preference rules of the form

$$\text{if } x \text{ is } A_i \text{ then } (\pi_y = \mu_{B_i})$$

where B_i is viewed as a fuzzy set of recommended (possible) actions, and an action is the selection of a value for y, the control parameter. Hence, it is a more general kind of preference rules than the one when only one decision is involved in the conclusion part. Instead of proposing a single decision in situation A_i, a weighted ordered set is proposed, as described by π_y. The preference $u_i(d)$ of assignment y = d, in the presence of $x = x_0$ can be evaluated as a function of $\mu_{B_i}(d)$ and $\mu_{A_i}(x_0)$, for a single rule i. Among natural conditions to be fulfilled is that $u_i(d) \leq \min(\mu_{B_i}(d), \mu_{A_i}(x_0))$, which claims that $\mu_{A_i}(x_0)$ stands as a upper bound on the degree of preference for y = d, induced by rule i. When the equality is taken for granted, we get the fuzzy control approach.

Usually a preference rule does not stand alone. The set of states of the world is partitioned into a family of situations, where in each situation, a decision is recommended. It corresponds to a decision table of the form

	if <situation 1> then <decision 1>
else	if <situation 2> then <decision 2>
else...	
else	if <situation n> then <decision n>
else	<decision n + 1>

where <decision n + 1> may suggest to refrain from deciding in the case of a n + 1th situation that is defined by complementarity. If <decision i> corresponds to a single decision, then, when <situation i> is fuzzy, the output is a fuzzy set of recommended decisions $\{d_i, u_i(d_i)), i=1,n+1\}$. This is what happens in the OPAL system for instance (Bensana et al., 1988) where a decision table corresponds, to a priority rule in a job shop, a decision is of the form "operation O_1 precedes operation O_2", and d_{n+1} recommends not to sequence the operations for the moment.

In fuzzy control, any elementary decision d receives an evaluation from several preference rules, since the B_i's may overlap. Hence the preference weight for d, u(d), is a function of $u_i(d)$, i = 1,n (since there is no refraining from decision in fuzzy control, i.e. d_{n+1} does not exist). A natural condition in a decision table is that a decision is selected as soon as one rule recommending the decision in triggered. Hence we should have

$$u(d) \geq \max_{i=1,n} u_i(d) \qquad (49)$$

The case of equality, again, corresponds to the choice of fuzzy control people. Strictly speaking the above scheme works provided that the actual situation is precisely described, i.e. $x = x_0$. But, since the behavior of fuzzy decision rules is based on fuzzy pattern matching between the description of the actual situation and the prototypical situations in the fuzzy rules, extension to the case of ill-described inputs is easy to envisage. If what is known about x is that x is A' where A' is a fuzzy set, u(d) should become a fuzzy number induced by the degrees of compatibility between A' and A_i (Zadeh, 1978b), i.e. $\mu_{A_i}(A')$, defined by extending the function μ_{A_i} to the fuzzy value A'. Hence we should replace max in (49) by the extended max in fuzzy arithmetic (e.g. Dubois and Prade, 1988).

In fuzzy pattern matching (Dubois, Prade and Testemale, 1988), we often use the degrees of possibility $\Pi(A_i ; A')$ and necessity $N(A_i ; A')$, instead of the fuzzy truth-value $\mu_{A_i}(A')$. Letting an equality stand in (49), and approximating $\mu_{A_i}(A')$ by the interval $[N(A_i ; A'),\Pi(A_i ; A')]$, we get the following results

$$\min(N(A_i ; A'), \mu_{B_i}(d)) \leq u_i(d) \leq \min(\Pi(A_i ; A'), \mu_{B_i}(d))$$

and

$$\max_{i=1,n} \min(N(A_i ; A'), \mu_{B_i}(d)) \leq u(d) \leq \max_{i=1,n} \min(\Pi(A_i ; A'), \mu_{B_i}(d))$$

Note that in the fuzzy control literature only the upper bound of u(d) is adopted, while the degree of necessity $N(A_i ; A')$ should be used to describe imprecision due to a fuzzy input. Also it is clear from above that preference rules do not behave like reasoning rules. Particularly, the inequality $u_i(d) \leq \min(\mu_{A_i}(x_0),\mu_{B_i}(d))$ is not in accordance with the possibility rules in the previous sections since the semantics of the latter leads to the opposite inequality.

A last issue with preference rules is that there may be more than one decision tables involved in the selection of a decision d. Indeed several points of view, objectives, etc... may be simultaneously present, and there is some strategy to adopt in the presence of conflicting decision tables. This problem is absent from the fuzzy control literature where a fuzzy controller generally involves one decision table only. However, in planning, decisions may be motivated by several conflicting criteria.

Again this is what happens in the OPAL system (Bensana et al., 1988) for instance. The problem of cooperation between decision tables has been modelled in terms of social choice (see Bel et al., 1989) and a software architecture for implementing fuzzy decision tables and cooperation strategies has been devised (Dubois, Koning and Bel, 1989). It is based on a social choice interpretation of fuzzy set aggregation connectives that is described elsewhere (Dubois and Koning, 1989).

The problem of preference rules and their implementation in rule-based systems is certainly one of the important topics in Artificial Intelligence, for the forthcoming years, as witnessed by some current activity in this area, from the standpoint of utility theory (Keeney, 1988 ; Klein and Shortliffe, 1990), or cognitive psychology (Pinson, 1987).

7 - CONCLUDING REMARKS

The semantic contents of rules in fuzzy expert systems has received little attention until now in spite of the enormous quantity of existing literature about approximate reasoning and fuzzy controllers. The paper has tried to formally derive different kinds of fuzzy rules based on very simple semantical considerations. Four types of rules have emerged corresponding to very standard alterations of a possibility distribution : enlarging its core, shrinking its support, truncating its height or, drowning it in a uniform level of uncertainty. The paper has also pointed out that fuzzy decision rules do not behave like fuzzy rules describing relationships.

Besides, rule-based expert systems have always been associated with an efficient local computation strategy where a partial conclusion obtained from a (compound) fact and a rule have to be combined with other conclusions pertaining to the same matter and derived from other facts and rules. This kind of strategy can be especially dangerous in presence of vague and uncertain pieces of knowledge since it may yield conclusions which are not as accurate as it can be expected from the available knowledge. Such conclusions may be even incorrect (see Pearl (1988), Heckerman and Horvitz (1988), Dubois and Prade (1989b) for instance). This is due to the fact that each rule, each variable to evaluate, cannot always be considered independently in the evaluation process. A possibilistic hypergraph technique coping with this problem has been recently developed by Kruse and Schwecke (1990), and by Dubois and Prade (1990c).

REFERENCES

Baldwin J.F., Pilsworth B.W. (1979) A model of fuzzy reasoning through multi-valued logic and set theory. Int. J. of Man-Machine Studies, 11, 351-380.

Bel G., Bensana E., Dubois D., Koning J.L. (1989) Handling fuzzy priority rules in a jobshop-scheduling system. Proc. of the 3rd. Inter. Fuzzy Systems Assoc. (IFSA) Congress, Seattle, Wa., Aug. 6-11, 200-203.

Bensana E., Bel G., Dubois D. (1988) OPAL : a multi-knowledge-based system for industrial job-shop scheduling. Int. J. Prod. Res., 26(5), 795-819.

Bonissone P.P., Gans S.S., Decker K.S. (1987) RUM : a layered architecture for reasoning with uncertainty. Proc. of the 10th Inter. Joint Conf. on Artificial Intelligence (IJCAI-87), Milano, Italy, 891-898.

Bouchon B. (1987) Fuzzy inferences and conditional possibility distributions. Fuzzy

Sets and Systems, 23, 33-41.
Bouchon B. (1988) Stability of linguistic modifiers compatible with a fuzzy logic. In : Uncertainty and Intelligent Systems (2nd Inter. Conf. on Information Processing and Management of Uncertainty in Knowledge-Based Systems (IPMU'88), Urbino, Italy, July 1988) (B. Bouchon, L. Saitta, R.R. Yager, eds.), Springer-Verlag, Berlin, 63-70.
Buchanan B.G., Shortliffe E.H. (1984) Rule-Based Expert Systems – The MYCIN Experiment of the Stanford Heuristic Programming Project. Addison-Wesley, Reading, Mass..
Després S. (1989) GRIF : a guide for representing fuzzy inferences. Proc. of the 3rd Inter. Fuzzy Systems Assoc. (IFSA) Congress, Seattle, Wa., Aug. 6-11, 353-356.
Di Nola A., Pedrycz W., Sessa S. (1985) Fuzzy relation equations and algorithms of inference mechanism in expert systems. In : Approximate Reasoning in Expert Systems (M.M. Gupta, A. Kandel, W. Bandler, J.B. Kiszka, eds.), North-Holland, Amsterdam, 355-367.
Dubois D., Koning J.L. (1989) Social choice axioms for fuzzy set aggregation. Workshop on Aggregation and Best Choices of Imprecise Opinions, Bruxelles, Jan. 1989. Available in Tech. Report IRIT/90-5/R, Univ. P. Sabatier, Toulouse, France. To appear in Fuzzy Sets and Systems.
Dubois D., Koning J.L., Bel G. (1989) Antagonistic decision rules in knowledge-based systems (in French). Proc. 7th AFCET Congress on Artificial Intelligence and Pattern Recognition, Paris.
Dubois D., Lang J., Prade H. (1990) Fuzzy sets in approximate reasoning – Part 2 : Logical approaches. Fuzzy Sets and Systems, 25th Anniversary Memorial Volume, to appear.
Dubois D., Martin-Clouaire R., Prade H. (1988) Practical computing in fuzzy logic. In: Fuzzy Computing – Theory, Hardware and Applications (M.M. Gupta, T. Yamakawa, eds.), North-Holland, Amsterdam, 11-34.
Dubois D., Prade H. (1982) On several representations of an uncertain body of evidence. In : Fuzzy Information and Decision Processes (M.M. Gupta, E. Sanchez, eds.), North-Holland, Amsterdam, 167-181.
Dubois D., Prade H. (1984) Fuzzy logics and the generalized modus ponens revisited. Cybernetics and Systems, 15, 293-331.
Dubois D., Prade H. (with the collaboration of Farreny H., Martin-Clouaire R., Testemale C.) (1988) Possibility Theory – An Approach to Computerized Processing of Uncertainty. Plenum Press, New York.
Dubois D., Prade H. (1989a) A typology of fuzzy "If... then..." rules. Proc. of the 3rd Inter. Fuzzy Systems Assoc. (IFSA) Congress, Seattle, Wa., Aug. 6-11, 782-785.
Dubois D., Prade H. (1989b) Handling uncertainty in expert systems : pitfalls, difficulties, remedies. In : Reliability of Expert Systems (E. Hollnagel, ed.), Ellis-Horwood, Chichester, U.K., 64-118.
Dubois D., Prade H. (1990a) Resolution principles in possibilistic logic. Int. J. of Approximate Reasoning, 3, 1-21.
Dubois D., Prade H. (1990b) Gradual inference rules in approximate reasoning. In : Tech Report IRIT/90-6/R, IRIT, Univ. P. Sabatier, Toulouse, France. Information Sciences, to appear.
Dubois D., Prade H. (1990c) Inference in possibilistic hypergraphs. In : Tech. Report IRIT/90-6/R, IRIT, Univ. P. Sabatier, Toulouse, France. Extended abstracts of the 3rd Conf. on Information Processing and Management of Uncertainty in Knowledge-Based Systems (IPMU'90), Paris, July 2-6, 228-230.
Dubois D., Prade H. (1990d) Fuzzy sets in approximate reasoning – Part 1 : Inference with possibility distributions. Fuzzy Sets and Systems, 25th Anniversary Memorial Volume, to appear.
Dubois D., Prade H., Testemale C. (1988) Weighted fuzzy pattern matching. Fuzzy Sets and Systems, 28, 313-331.
Fonck P. (1990) Representation of vague and uncertain facts. Proc. of the 3rd Inter. Conf. on Information Processing and Management of Uncertainty in Knowledge-Based Systems (IPMU'90), Paris, July 2-6, 284-288.
Godo L., López de Mántaras R., Sierra C., Verdaguer A. (1988) Managing linguistically expressed uncertainty in MILORD : application to medical diagnosis. Artificial

Intelligence Communications, 1(1), 14-31.
Heckerman D.E., Horvitz E.J. (1988) The myth of modularity in rule-based systems for reasoning with uncertainty. In : Uncertainty in Artificial Intelligence 2 (J.F. Lemmer, L.N. Kanal, eds.), North-Holland, Amsterdam, 23-34.
Keeney R. (1988) Value-driven expert systems for decision support. Decision Support Systems (4), 405-412.
Klein D.A., Shortliffe E.H. (1990) Integrating artificial intelligence & decision theory in heuristic process control systems. Proc. of the 10th Inter. Workshop on Expert Systems & their Applications, Avignon, France, May 28-June 1st, 2nd Generation Expert Systems Volume, 165-177.
Kruse R., Schwecke E. (1990) Fuzzy reasoning in a multidimentional space of hypotheses. Int. J. of Approximate Reasoning, 4, 47-68.
Magrez P., Smets P. (1989) Epistemic necessity, possibility, and truth – Tools for dealing with imprecision and uncertainty in fuzzy knowledge-based systems. Int. J. of Approximate Reasoning, 3,35-57.
Mamdani E.H. (1977) Application of fuzzy logic to approximate reasoning using linguistic systems. IEEE Trans. Comput., 26, 1182-1191.
Martin-Clouaire R. (1989) Semantics and computation of the generalized modus ponens : the long paper. Int. J. of Approximate Reasoning, 3, 195-217.
Mizumoto M., Zimmermann H.J. (1982) Comparison of fuzzy reasoning methods. Fuzzy Sets and Systems, 8, 253-283.
Pearl J. (1988) Probabilistic Reasoning in Intelligent Systems – Networks of Plausible Inference. Morgan Kaufmann Pub., San Mateo, Ca..
Pinson S. (1987) A multi-attribute approach to knowledge representation for loan granting. Proc. of the 10th Inter. Joint Conf. on Artificial Intelligence (IJCAI-87), Milano, Italy, 588-591.
Prade H. (1985) A computational approach to approximate and plausible reasoning with applications to expert systems. IEEE Trans. on Pattern Analysis and Machine Intelligence, 7(3), 260-283. Corrections, 7(6), 747-748.
Sanchez E. (1978) On possibility-qualification in natural languages. Information Sciences, 15, 45-76.
Shafer G. (1976) A Mathematical Theory of Evidence. Princeton University Press, Princeton.
Smets P., Magrez P. (1987) Implication in fuzzy logic. Int. J. of Approximate Reasoning, 1, 327-347.
Sugeno M. (ed.) (1985) Industrial Applications of Fuzzy Control. North-Holland, Amsterdam.
Tanaka H., Tsukiyama T., Asai K. (1982) A fuzzy system model based on the logical structure. In : Fuzzy Sets and Possibility Theory (R.R. Yager, ed.), Pergamon Press, New York, 257-325.
Trillas E., Valverde L. (1985) On mode and implication in approximate reasoning. In : Approximate Reasoning in Expert Systems (M.M. Gupta, A. Kandel, W. Bandler, J.B. Kiszka, eds.), North-Holland, Amsterdam, 157-166.
Yager R.R. (1984) Approximate reasoning as a basis for rule-based expert systems. IEEE Trans. on Systems, Man and Cybernetics, 14(4), 636-643.
Yager R.R. (1990) On considerations of credibility of evidence. Tech. Report #MII-1006, Machine Intelligence Institute, Iona College, New Rochelle, N.Y..
Zadeh L.A. (1971) Similarity relations and fuzzy orderings. Information Sciences, 3, 177-200.
Zadeh L.A. (1972) A fuzzy set-theoretic interpretation of linguistic hedges. J. of Cybernetics, 2(3), 4-34.
Zadeh L.A. (1978a) Fuzzy sets as a basis for a theory of possibility. Fuzzy Sets and Systems, 1, 3-28.
Zadeh L.A. (1978b) PRUF – a meaning representation language for natural languages. Int. J. of Man-Machine Studies, 10, 395-460.
Zadeh L.A. (1979) A theory of approximate reasoning. In : Machine Intelligence (J.E. Hayes, D. Michie, L.I. Mikulich, eds.), Vol. 9, Elsevier, New York, 149-194.

4

Fuzzy Logic Controllers

Hamid R. Berenji
Sterling Software
Artificial Intelligence Research Branch
NASA Ames Research Center
Mountain View, CA 94035

1 Introduction

Fuzzy Set Theory, introduced by Zadeh in 1965 [77], has been the subject of much controversy and debate. In recent years, it has found many applications in a variety of fields. Among the most successful applications of this theory has been the area of Fuzzy Logic Control (FLC) initiated by the work of Mamdani and Assilian [36]. FLC has had considerable success in Japan, where many commercial products using this technology, have been built.

In this paper, we will review the basic architecture of fuzzy logic controllers and discuss why this technology often provides controllers with performance similar to the performance of an expert human operator for ill-defined and complex systems. In section 2, an introductory survey of the basics of fuzzy set theory is presented. Next, the basic architecture of a FLC is described, followed by a brief review of the application of this theory. Finally, we discuss how a fuzzy logic based control system can learn from experience to fine-tune its performance.

2 Fuzzy sets and Fuzzy logic: The basis for Fuzzy Control

A fuzzy set is an extension of a crisp set. Crisp sets only allow full membership or no membership at all, where fuzzy sets allow partial membership. In other words, an element may partially belong to a set. In a crisp set, the membership or non-membership of an element x in set

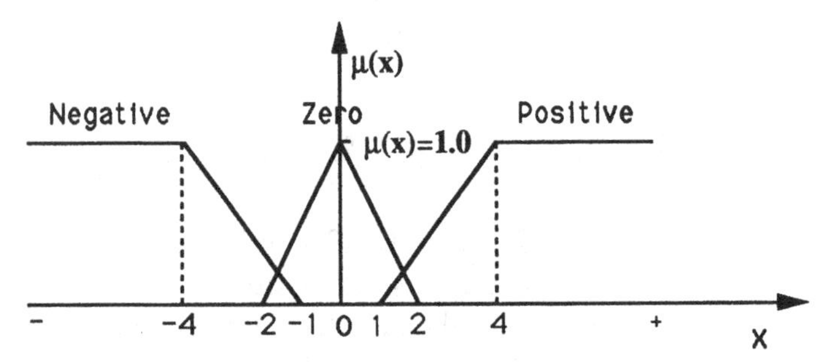

Figure 1: Examples of Fuzzy membership functions

A is described by a characteristic function $\mu_A(x)$, where:

$$\mu_A(x) = \begin{cases} 1 & \text{if } x \in A \\ 0 & \text{if } x \notin A. \end{cases}$$

Fuzzy set theory extends this concept by defining *partial memberships* which can take values ranging from 0 to 1:

$$\mu_A : X \to [0,1]$$

where X refers to the universal set defined in a specific problem. If this universal set is countable and finite, then a fuzzy set A in this universe can be defined by listing each member and its degree of membership in the set A:

$$A = \sum_{i=1}^{n} \mu_A(x_i)/x_i.$$

Similarly, if X is continuous, then a fuzzy set A can be defined by

$$A = \int_X \mu_A(x)/x.$$

Note that in the above definitions, "/" does not refer to a division and is used as a notation to separate the membership of an element from the element itself. For example, in $A = .2/element1 + .6/element2$, *element*1 has membership value of .2 and *element*2 has a membership value of .6 in the fuzzy set A. As another example, the linguistic term *Positive* as shown in Figure 1 may be defined to take the following membership function:

$$\mu_{positive}(x) = \begin{cases} 1 & \text{if } x \geq 4 \\ \frac{x-1}{3} & \text{if } 1 \geq x \leq 4 \\ 0 & \text{otherwise.} \end{cases}$$

The *support* of a fuzzy set A in the universal set X is a crisp set that contains all the elements of X which have degree of membership greater than zero[1]. In the above example, the support set includes all the real numbers for which $\mu(x) \geq 0$.

The *α-cut* of a fuzzy set A is defined as the crisp set of all the elements of the universe X which have memberships in A greater than or equal to α, where

$$A_\alpha = \{x \in X | \mu_A(x) \geq \alpha\}.$$

For example, if the fuzzy set A is described by its membership function:

$$A = \{.2/2 + .4/3 + .6/4 + .8/5 + 1/6\}$$

and $\alpha = .3$ then the *α-cut* of A is

$$A_{.3} = \{3, 4, 5, 6\}.$$

The *height* of a fuzzy set is defined as the highest membership value of the set. If $height(A) = 1$, then set A is called a *normalized* fuzzy set.

2.1 Fuzzy Set Operations

Assuming that A and B are two fuzzy sets with membership functions of μ_A and μ_B, then the following operations can be defined on these sets. The *complement* of a fuzzy set A is a fuzzy set $\bar{A}$ with a membership function

$$\mu_{\bar{A}} = 1 - \mu_A(x).$$

The *Union* of A and B is a fuzzy set with the following membership function

$$\mu_{A \cup B} = \max\{\mu_A, \mu_B\}.$$

The *Intersection* of A and B is a fuzzy set

$$\mu_{A \cap B} = \min\{\mu_A, \mu_B\}.$$

By definition, *Concentration* is a unary operation which, when applied to a fuzzy set A, results in a fuzzy subset of A in such a way that reduction in higher grades of membership is much less than the reduction in lower grades of membership. In other words, by concentrating a fuzzy set, members with low grades of membership will have even lower grades of memberships and hence the fuzzy set becomes more concentrated. A common concentration operator is to square the membership function:

$$\mu_{CON(A)}(x) = \mu_A^2(x)$$

[1]Mabuchi [35] has used this concept in comparing fuzzy subsets.

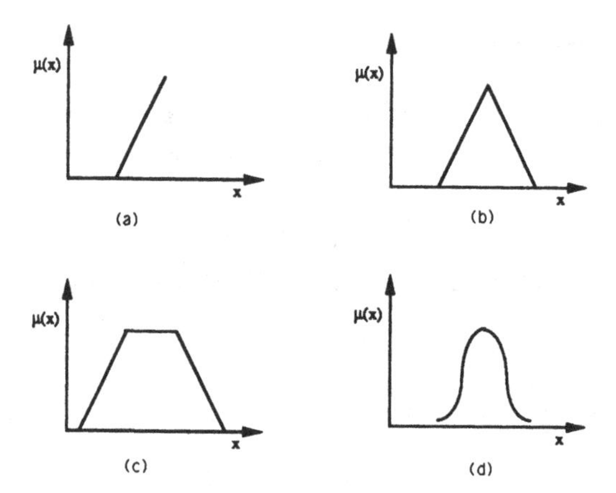

Figure 2: (a)- Monotonic, (b)- Triangular, (c)- Trapezoidal, (d)- Bell-shaped membership functions

and a typical concentration operator is the term *Very* which is also a *Linguistic Hedge* [78]. For example, the result of applying the operator *Very* on a fuzzy label *Small* is a new fuzzy label *Very Small.*

The *Dilution* operator is the converse of the concentration operator described above:

$$\mu_{DIL(A)}(x) = \sqrt{\mu_A(x)}.$$

3 The Basic Architecture Of FLC

Different methods for developing fuzzy logic controllers have been suggested over the past 15 years. In the design of a fuzzy controller, one must identify the main control parameters and determine a term set which is at the right level of granularity for describing the values of each linguistic variable[2]. For example, a term set including linguistic values such as { *Small, Medium, Large*} may not be satisfactory in some domains, and may instead require the use of a five term set such as { *Very Small, Small, Medium, Large, and Very Large*}.

Different type of fuzzy membership functions have been used in fuzzy logic control. However, four types are most common. The first type assumes monotonic membership function such as those shown in Figure 2(a). This type is simple and has been used in studies such as [8, 64]. Other types using triangular, trapezoidal, and bell-shaped functions have also been used as shown in Figure 2(b), (c), and (d) respectively.

[2] A linguistic variable is a variable which can only take linguistic values.

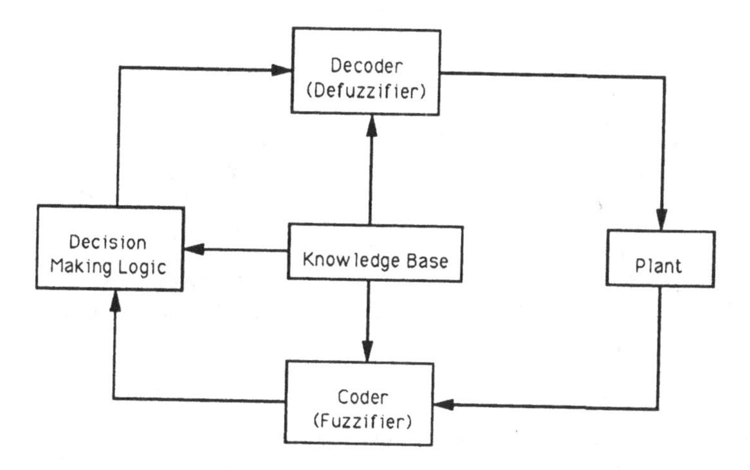

Figure 3: A simple architecture of a fuzzy logic controller

The selection of the types of fuzzy variable directly affects the type of reasoning to be performed by the rules using these variables. This is described later in 3.3. After the values of the main control parameters are determined, a knowledge base is developed using the above control variables and the values that they may take. If the knowledge base is a rule base, more than one rule may fire requiring the selection of a *conflict resolution* method for decision making, as will be described later.

Figure 3 illustrates a simple architecture for a fuzzy logic controller. The system dynamics of the plant in this architecture is measured by a set of sensors. This architecture consists of four modules whose functions are described next.

3.1 Coding the Inputs: Fuzzification

In coding the values from the sensors, one transforms the values of the sensor measurements in terms of the linguistic labels used in the preconditions of the rules.

If the sensor reading has a crisp value, then the fuzzification stage requires matching the sensor measurement against the membership function of the linguistic label as shown in Figure 4(a). If the sensor reading contains noise, it may be modeled by using a triangular membership function where the vertex of the triangle refers to the mean value of the data set of sensor measurements and the base refers to a function of the standard deviation (e.g., twice the standard deviation as used in [69]). Then in this case, fuzzification refers to finding out the intersection of the label's membership function and the distribution for the sensed data

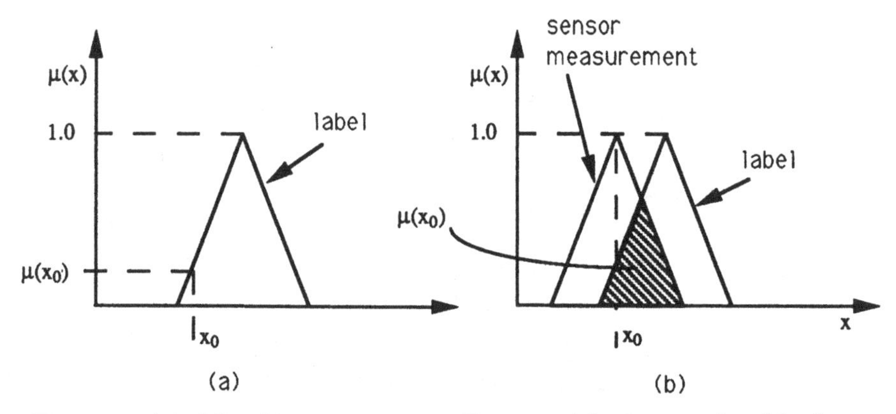

Figure 4: (a)- Matching a sensor reading x_0 with the membership function $\mu(x)$ to get $\mu(x_0)$; (a)- crisp sensor reading (b)-fuzzy sensor reading.

as shown in Figure 4(b). However, the most widely used fuzzification method is the former case when the sensor reading is crisp.

3.2 Setting up the Control Knowledge Base

There are two main tasks in designing the control knowledge base. First, a set of linguistic variables must be selected which describe the values of the main control parameters of the process. Both the main input parameters and the main output parameters must be linguistically defined in this stage using proper term sets. The selection of the level of granularity of a term set for an input variable or an output variable plays an important role in the smoothness of control. Secondly, a control knowledge base must be developed which uses the above linguistic description of the main parameters. Sugeno [49] has suggested four methods for doing this:

1. Expert's Experience and Knowledge
2. Modelling the Operator's Control Actions
3. Modeling a process
4. Self Organization

Among the above methods, the first method is the most widely used [36]. In modeling the human expert operator's knowledge, fuzzy control rules of the form:

IF Error is small and Change-in-error is small then force is small have been used in studies such as [51, 53]. This method is effective when expert human operators can express the heuristics or the knowledge that they use in controlling a process in terms of rules of the above form. Applications have been developed in process control (e.g., cement kiln operations [23]). Beside the ordinary fuzzy control rules which have been used by Mamdani and others, where the conclusion of a rule is another fuzzy variable, a rule can be developed whereby its conclusion is a function of the input parameters. For example, the following implication can be written:

$$\text{IF } X \text{ is } A_1 \text{ and } Y \text{ is } B_1 \text{ THEN Z} = f_1(X, Y)$$

where the output Z is a function of the values that X and Y may take.

The second method above, directly models the control actions of the operator. Instead of interviewing the operator, the types of control actions taken by the operators are modelled. Takagi and Sugeno [55] and Sugeno and Murakami [51] have used this method for modeling the control actions of a driver in parking a car.

The third method deals with fuzzy modeling of a process where an approximate model of the plant is configured by using implications which describe the possible states of the system. In this method a model is developed and a fuzzy controller is constructed to control the fuzzy model, making this approach similar to the traditional approach taken in control theory. Hence, structure identification and parameter identification processes are needed. For example, a rule discussed by Sugeno [49] is of the form:

$$\text{If } x_1 \text{ is } A_1^i,\ x_2 \text{ is } A_2^i, \ldots, \text{ then } y = p_0^i + p_1^i x_1 + p_2^i x_2 + \ldots + p_m^i x_m,$$

for $i = 1, \ldots, n$ where n is the number of such implications and the consequence is a linear function of the m input variables.

Finally, the fourth method refers to the research of Mamdani and his students in developing self-organizing controllers [44]. The main idea in this method is the development of rules which can be adjusted over time to improve the controllers' performance. This method is very similar to recent work in the use of neural networks in designing the knowledge base of a fuzzy logic controller which will be discussed later in this chapter.

3.3 Conflict Resolution and Decision Making

As mentioned earlier, because of the partial matching attribute of fuzzy control rules and the fact that the preconditions of the rules do overlap,

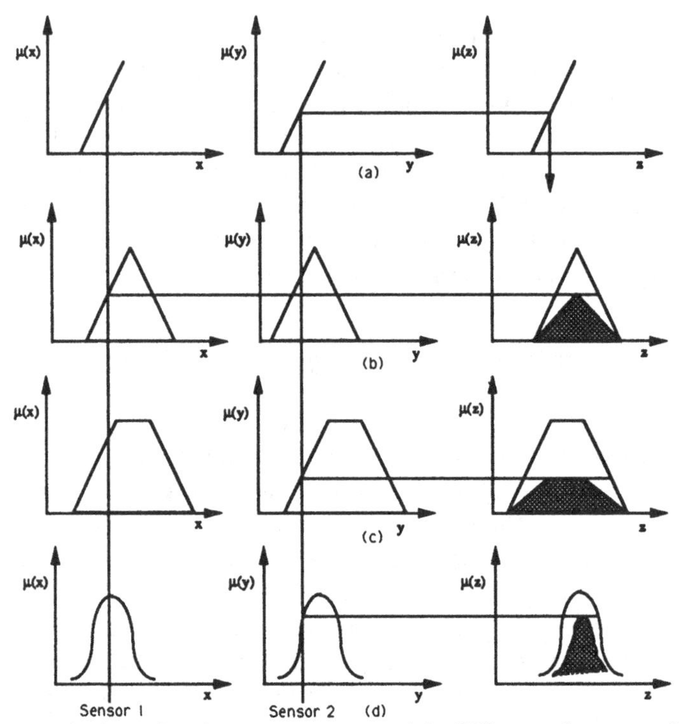

Figure 5: Examples of inference process with different fuzzy variable membership functions: (a)- monotonic, (b)-triangular, (c)- trapezoidal, (d)- bell-shaped

usually more than one fuzzy control rule can fire at one time. The methodology which is used in deciding what control action should be taken as the result of the firing of several rules can be referred to as the process of *conflict resolution*. The following example, using two rules, illustrates this process. Assume that we have the following:

Rule 1: IF X is A_1 and Y is B_1 THEN Z is C_1
Rule 2: IF X is A_2 and Y is B_2 THEN Z is C_2

Now, if we have x_0 and y_0 as the sensor readings for fuzzy variables X and Y, then their *truth values* are represented by $\mu_{A_1}(x_0)$ and $\mu_{B_1}(y_0)$ respectively for Rule 1, where μ_{A_1} represents the membership function for A_1. Similarly for Rule 2, we have $\mu_{A_2}(x_0)$ and $\mu_{B_2}(y_0)$ as the truth values of the preconditions. The *strength* of Rule 1 can be calculated by:

$$\alpha_1 = \mu_{A_1}(x_0) \ \wedge \ \mu_{B_1}(y_0)$$

where $\wedge$ refers to the *conjunction operator* which was defined earlier to be equal to *Minimum operator*. Similarly for Rule 2:

$$\alpha_2 = \mu_{A_2}(x_0) \wedge \mu_{B_2}(y_0).$$

The control output of rule 1 is calculated by applying the matching strength of its preconditions on its conclusion:

$$\mu_{C_1'}(\omega) = \alpha_1 \wedge \mu_{C_1}(\omega),$$

and for Rule 2:

$$\mu_{C_2'}(\omega) = \alpha_2 \wedge \mu_{C_2}(\omega)$$

where ω ranges over the values that the rule conclusions can take. This means that as a result of reading sensor values x_0 and y_0, Rule 1 is recommending a control action with $\mu_{C_1'}(w)$ as its membership function and Rule 2 is recommending a control action with $\mu_{C_2'}(w)$ as its membership function. The conflict-resolution process then produces

$$\mu_C(\omega) = \mu_{C_1'}(\omega) \vee \mu_{C_2'}(\omega) = [\alpha_1 \wedge \mu_{C_1}(\omega)] \vee [\alpha_2 \wedge \mu_{C_2}(\omega)]$$

where $\mu_C(\omega)$ is a pointwise membership function for the combined conclusion of Rule 1 and Rule 2. The $\wedge$ and $\vee$ operators in above are defined to be the *Min* and *Max* functions respectively [36]. The result of this last operation (i.e., $\mu_C(\omega)$) is a membership function and has to be translated (*defuzzified*) to a single value as discussed next.

3.4 Decoding the Outputs: Defuzzification

This necessary operation produces a nonfuzzy control action that best represents the membership function of an inferred fuzzy control action. Several defuzzification strategies have been suggested in literature. Among them, four methods which have been applied more often are described here.

3.4.1 Tsukamoto's defuzzification method

As shown in Figure 6(a), if monotonic membership functions are used, then a crisp control action can be calculated by:

$$Z^* = \frac{\sum_{i=1}^{n} \omega_i x_i}{\sum_{i=1}^{n} \omega_i}$$

where n is the number of rules with firing strength (ω_i) greater than 0 and x_i is the amount of control action recommended by rule i.

3.4.2 The Center Of Area (COA) method

Assuming that a control action with a pointwise membership function μ_C has been produced, the Center of Area method calculates the center of gravity of the distribution for the control action. Assuming a discrete universe of discourse, we have

$$Z^* = \frac{\sum_{j=1}^{q} z_j \mu_C(z_j)}{\sum_{j=1}^{q} \mu_C(z_j)}$$

where q is the number of quantization levels of the output, z_j is the amount of control output at the quantization level j and $\mu_C(z_j)$ represents its membership value in C.

3.4.3 The Mean of Maximum (MOM) Method

The Mean of Maximum Method (MOM) generates a crisp control action by averaging the support values which their membership values reach the maximum. For a discrete universe, this is calculated by

$$Z^* = \sum_{j=1}^{l} \frac{z_j}{l}$$

where l is the number of quantized z values which reach their maximum memberships.

3.4.4 Defuzification when the output of the rules are functions of their inputs

As mentioned earlier, fuzzy control rules may be written as a function of their inputs. For example,

Rule i: IF X is A_i and Y is B_i THEN Z is $f_i(X,Y)$

assuming that α_i is the firing strengths of the rule i, then

$$Z^* = \frac{\sum_{i=1}^{n} \alpha_i f_i(x_i, y_i)}{\sum_{j=1}^{n} \alpha_i}$$

where n is the number of firing rules.

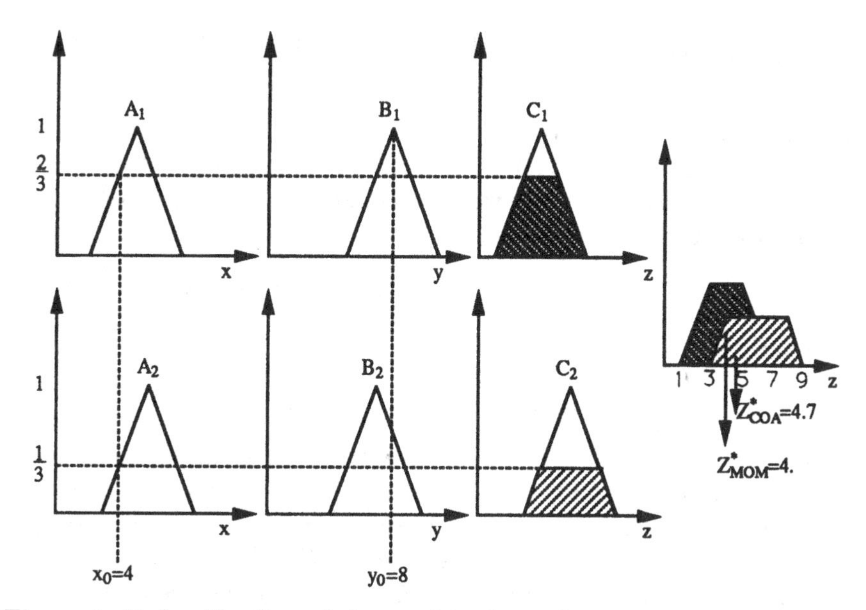

Figure 6: Defuzzification of the combined conclusion of rules described in the example.

3.4.5 An example

Assume that we have the following two rules:

Rule 1: IF X is A_1 and Y is B_1 THEN Z is C_1
Rule 2: IF X is A_2 and Y is B_2 THEN Z is C_2

Suppose x_0 and y_0 are the sensor readings for fuzzy variables X and Y, and the following are membership functions:

$$\mu_{A_1} = \begin{cases} \frac{x-2}{3} & 2 \leq x \leq 5 \\ \frac{8-x}{3} & 5 < x \leq 8 \end{cases} \qquad \mu_{A_2} = \begin{cases} \frac{x-3}{3} & 3 \leq x \leq 6 \\ \frac{9-x}{3} & 6 < x \leq 9 \end{cases}$$

$$\mu_{B_1} = \begin{cases} \frac{y-5}{3} & 5 \leq y \leq 8 \\ \frac{11-y}{3} & 8 < y \leq 11 \end{cases} \qquad \mu_{B_2} = \begin{cases} \frac{y-4}{3} & 4 \leq y \leq 7 \\ \frac{10-y}{3} & 7 < y \leq 10 \end{cases}$$

$$\mu_{C_1} = \begin{cases} \frac{z-1}{3} & 1 \leq z \leq 4 \\ \frac{7-z}{3} & 4 < z \leq 7 \end{cases} \qquad \mu_{C_2} = \begin{cases} \frac{z-3}{3} & 3 \leq z \leq 6 \\ \frac{9-z}{3} & 6 < z \leq 9 \end{cases}$$

Further assume that we are reading the sensor values $x_0 = 4$ and $y_0 = 8$. We illustrate how to calculate

1. the membership function for the control action recommended by the combination of these two rules

2. the crisp value of the control action using the COA and MOM methods.

First, the sensor readings x_0 and y_0 have to be matched against the preconditions A_1, B_1 respectively. This will produce $\mu_{A_1}(x_0) = 2/3$ and $\mu_{B_1}(y_0) = 1$. Similarly, for rule 2, we have $\mu_{A_2}(x_0) = 1/3$ and $\mu_{B_2}(y_0) = 2/3$. The strength of rule 1 is calculated by:

$$\alpha_1 = Min(\mu_{A_1}(x_0), \mu_{B_1}(y_0)) = Min(2/3, 1) = 2/3.$$

and similarly for rule 2:

$$\alpha_2 = Min(\mu_{A_2}(x_0), \mu_{B_2}(y_0)) = Min(1/3, 2/3) = 1/3.$$

Applying α_1 to the conclusion of rule 1 results in the shaded trapezoid figure shown in Figure 6 for C_1. Similarly, applying α_2 to the conclusion of rule 2 results in the dashed trapezoid shown in Figure 6 for C_2. By superimposing the resulted memberships over each other and using the *Max* operator, the membership function for the combined conclusion of these rules is found (as shown the right hand side of the Figure 6). Furthermore, using the COA method (explained earlier), the defuzzified value for the conclusion is found:

$$Z^*_{COA} = \frac{2 \cdot \frac{1}{3} + 3 \cdot \frac{2}{3} + 4 \cdot \frac{2}{3} + 5 \cdot \frac{2}{3} + 6 \cdot \frac{1}{3} + 7 \cdot \frac{1}{3} + 8 \cdot \frac{1}{3}}{\frac{1}{3} + \frac{2}{3} + \frac{2}{3} + \frac{2}{3} + \frac{1}{3} + \frac{1}{3} + \frac{1}{3}} = 4.7.$$

Using the MOM defuzzification strategy, three quantized values reach their maximum memberships in the combined membership function (i.e., 3, 4, and 5 with membership values of 2/3). Therefore,

$$Z^*_{MOM} = \frac{3 + 4 + 5}{3} = 4.0.$$

4 A Hierarchical Approach in Design of Fuzzy Controllers

Berenji et. al. [8] have proposed the following algorithm in design of fuzzy controllers with multiple goals. The algorithm has been applied in control of a cart pole balancing system.

1. Let $G = \{g_1, g_2, ... g_n\}$ be the set of goals that system should achieve and maintain. Notice that for $n = 1$ (i.e., no interacting goals), the problem becomes simpler and may be handled using the earlier methods in fuzzy control (e.g., see [36]).

2. Let $G = p(G)$ where p is a function which assigns priorities among the goals. We assume that such a function can be obtained in a particular domain. In many control problems, it is possible to specifically assign priorities to the goals. For example, in the simple problem of balancing a pole on the palm of a hand and also moving the pole to a pre-determined location, it is possible to do this by first keeping the pole as vertical as possible and then gradually moving to the desired location. Although these goals are highly interactive (i.e., as soon as we notice that the pole is falling, we may temporarily set aside the other goal of moving to the desired location), we still can assign priorities fairly well.

3. Let $U = \{u_1, u_2, ..., u_n\}$ where u_i is the set of input control parameters related to achieving g_i.

4. Let $A = \{a_1, a_2, ..., a_n\}$ where a_i is the set of linguistic values used to describe the values of the input control parameters in u_i.

5. Let $C = \{c_1, c_2, ..., c_n\}$ where c_i is the set of linguistic values used to describe the values of the output Z.

6. Acquire the rule set R_1 of approximate control rules directly related to the highest priority goal. These rules are in the general form of

   IF u_1 is a_1 THEN Z is c_1.

7. For $i = 2$ to n, subsequently form the rule sets R_i. The format of the rules in these rule sets is similar to the ones in the previous step except that they include aspects of *approximately achieving the previous goal*:

   IF g_{i-1} *is approximately achieved* and u_i is a_i THEN Z is c_i.

The approximate achievement of a goal in step 7 of the above algorithm refers to holding the goal parameters within smaller boundaries. The interactions among the goal g_i and goal g_{i-1} are handled by forming rules which include more preconditions in the left hand side. For example, let us assume that we have acquired a set of rules R_1 for keeping a pole vertical. In writing the second rule set R_2 for moving to a pre-specified location, aspects of approximately achieving g_1 should be combined with control parameters for achieving g_2. For example, a precondition such as *the pole is almost balanced* can be added while writing the rules for moving to a specific location. A fuzzy set operation known as *concentration* [78] as described earlier can be used here to systematically obtain a more focused membership functions for the parameters which represent the achievement of previous goals. The above algorithm has been applied in cart-pole balancing and more details can be found in [8].

5 Applications of Fuzzy Logic Controllers

In recent years, there has been a very significant increase in the number of applications of fuzzy logic control. Although we will not provide a complete list of the applications here, a selective number of both the laboratory prototype systems and real commercial applications will be discussed.

As mentioned earlier, Mamdani and Assilian [36] were the first to apply the fuzzy set theory to control problems (e.g., the control of a laboratory steam engine). This experiment triggered a number of other applications such as the warm water process control [22], activated sludge wastewater treatment [63], and traffic junction control [42]. Fuzzy logic control has also been applied in a diverse set of domains such as arc welding [38], refuse incineration [40], automobile speed control [37], model cars [55, 53, 51, 52], cement kiln control [65], aircraft flight control [30], robot control [34, 59, 19, 66, 41], water purification process control [70], nuclear reactor control [9], elevator control [33], process control [13, 44], adaptive control [15], automatic tuning [39], control of a liquid level rig [14], automobile transmission control [20], gasoline refinery catalytic reformer control [3], two-dimensional ping-pong game playing [18], control of biological processes [12], activated sludge plant[76], knowledge structure [46], and comparison with classical control theory [4, 58]. Among these, the cement kiln controller [65] was the first industrial application. The celebrated Hitachi's automatic train controller is among the more recent fielded applications of fuzzy logic control. In the following, we discuss a few of these systems in more details.

5.1 Automatic train control

Yasunobu and Miyamoto at Hitachi, Ltd. [73] have designed a fuzzy controller for the Automatic Train Operation (ATO) systems. This system has been in use in the city of Sendai, Japan since July 1987. The two main operations of the system are Constant Speed Control (CSC) and Train Automatic Stop Control (TASC). The CSC operation results in maintaining a constant target speed (specified by the operator at the start of the train operation) during the train travel. The TASC operation controls the speed of the train in order to stop the train at the prespecified location. The system uses only a few rules (i.e., 12 rules for each of the CSC and the TASC operations) and the control is evaluated every 100 miliseconds. These operations use the evaluation of safety, riding comfort, traceability of target velocity, accuracy of stop gap, running time and energy consumption criteria in deciding a control strategy. The control rules are of the predictive fuzzy control rule types of the form:

$$\text{IF}(\ u \text{ is } C_i \rightarrow x \text{ is } A_i \text{ and } y \text{ is } B_i) \text{ then } u \text{ is } C_i.$$

For example, when the train is in the TASC zone, the following rule is used:

If the control notch is not changed and
the train will stop at the predetermined location, then
the control notch is not changed.

The system performs as skillfully as human experts do and superior to an ordinary PID[3] automatic train operation controller in terms of stopping precision, energy consumption, riding comfort, and running time.

5.2 Fuzzy Logic Hardwares and Fuzzy Logic Computer Chips

Yamakawa [71, 72] has pioneered using fuzzy logic at the hardware level by developing systems which achieve information processing leading toward what is referred to as the *fuzzy computer*. The systems developed by Yamakawa can accept linguistic information and perform approximate reasoning-based inference at very high speeds (e.g., more than 10 million Fuzzy Logical Inferences Per Second or FLIPS). Among the many applications of Yamakawa's fuzzy electronic circuits is the control of an inverted pendulum of a short length (e.g., 15 cm, weighting 3.5 grams). Also, Yamakawa's computer chips have been used in biomedical experiments [57] and orthodentic results evaluation [75].

Fuzzy Logic Chips were first developed by Togai and Watanabe at AT& T Bell Lab [61]. Since the original design, several extensions have been provided in [60] and [67].

[3]Proportional, Integral, and Derivative.

5.3 Sugeno's model car

Sugeno has designed a model car which can automatically park itself in a garage. The fuzzy control rules are derived by modeling a human's parking actions and developed based on the third method described earlier in 3.2. The car uses the front wall distance (x), side wall distance (y), and the heading angle of the car (θ) as its input variables. Three output control variables are used: the angle of the front wheels in moving forward (backward) and a control variable for speed control. For example, a control rule for steering control in moving forward is:

$$\text{If } x \text{ is } A,\ y \text{ is } B,\ \theta \text{ is } C \text{ then } f = p_0 + p_1 x + p_2 y + p_3 \theta.$$

Eighteen control rules are used for the steering control in moving forward and sixteen control rules are used for the steering control in moving backward. The input-output data is collected while a human parks the car and is used in parameter identification (e.g., the identification of the coefficients p_0, p_1, p_2, and p_3) of the above rules. Many successful experiments have been done using the model car which is equipped with a microprocessor and sensing devices.

5.4 Sugeno's model helicopter

Sugeno has initiated several projects on applying fuzzy logic control to the control of a model helicopter. Among these are radio control by oral instructions, automatic autorotation entry in engine failure cases, and unmanned helicopter control for sea rescue [48]. Although these projects have just started, several interesting results have already been achieved. The input variables from the helicopter include pitch, roll, and yaw, and their first and second derivatives. The control rules written for the helicopter regulate the up/down, forward/backward, left/right, and nose direction. For example, the longitudinal stick controls pitch, and therefore forward/backward movement of the rotorcraft.

An example of a fuzzy control rule for hovering is as the following:

If the body rolls, then
control the lateral in reverse.

Or as another example for hovering control:

If the body pitches, then
control the longitude in reverse.

The helicopter control problems under study in these projects are challenging control problems and are already producing results which illustrate the strength of the fuzzy logic control technology.

5.5 Other recent applications

Among the other applications, fuzzy logic control has found applications in the household appliances. Examples of these which will not further be discussed here are: the air conditioner by Mitsubishi; washing machine by Matsushita and Hitachi; VCR by Sanyo and Matsushita; vacuum cleaner by Matsushita; palmtop computer by Sony; Microwave oven by Toshiba, Sharp, Sanyo, and Hitachi; photography camera by Canon and many others.

6 Learning Fuzzy Logic Controllers

In many controller design tasks, it is important to develop a controller which can learn from experience to improve its performance. Here we discuss the research on developing Self Organizing Controllers (SOCs) and the recent work in using neural networks to provide a learning attribute for the fuzzy logic controllers.

6.1 Self Organizing Controllers

The Self-Organizing Controllers (SOCs) are among the earliest fuzzy controllers which provide an ability to change control policies with respect to the process and the operating environment. The function of the SOC is a combined system identification and control task. It infers from the error and change in error, the change in control action to apply. In addition to the error and change of error scales supplied by a control designer, SOC uses a third scale to calculate the actual change in the process input. In this sense, SOC is similar to the conventional PID controllers' three gain parameters[4].

SOCs evaluate their performance using a local measure which assesses the performance over a small set of plant states and a global criterion which measures the overall performance. The performance measure resembles a human decision maker who determines the output correction required from a knowledge of the error and the change of error. However, the decision tables required by the SOCs have to be generated and

[4]Proportional, Integral, and Derivative.

stored before hand. For more complex processes than the ones discussed by Procyk and Mamdani [44](i.e., other than single-input single-output processes), the generation of these decision tables may be difficult.

6.2 Neural Networks in Fuzzy Logic Control

Similarities exist between the neural networks and the fuzzy logic controllers. They both can handle extreme nonlinearities in the system. Both techniques allow *interpolative reasoning* which frees us from the true/false restriction of logical systems such as the ones used in symbolic AI. For example, once a neural network has been trained for a set of data, it can interpolate and produce answers for the cases not present in the training data set. Similar properties hold for a fuzzy logic controller. The weighted average scheme of fuzzy control and the sum of the products of the neural nets are similar in principle.

The main idea in integrating fuzzy logic control with neural networks is to use the strength of each one collectively in the resulting neuro-fuzzy control system. This fusion allows:

1. A human understandable expression of the knowledge used in control in terms of the fuzzy control rules. This reduces the difficulties in describing the trained neural network which is usually treated as a black box.

2. The fuzzy controller learns to adjust its performance automatically using a neural network structure and hence learns by accumulating experience.

The main emphasis in the research so far has been on automatic design and fine-tuning of the membership functions used in fuzzy control through learning by neural networks. Here, we focus on only three hybrid models but many more references are available (such as the proceedings of Iizuka-88 and Iizuka-90 conferences [1, 2]).

6.3 Fuzzy Control and AHC

Lee and Berenji [32] and Lee[31] have combined the Adaptive Heuristic Critic (AHC) model of Barto, Sutton, and Anderson with fuzzy control to learn the membership functions of the conclusions of the control rules. This work builds on Barto *et. al*'s pioneering work on applying neural networks in control. Two neural-like elements are used in this model. The Adaptive Heuristic Critic learns by updating the predictions of the

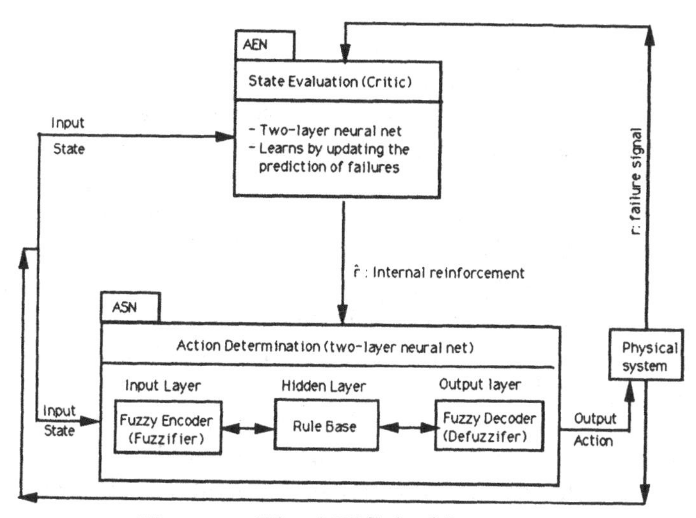

Figure 7: The ARIC Architecture

system's failure over time. The integrated fuzzy-AHC model has been tested in the domain of cart-pole balancing and the results have been consistently better when compared with the performance of the AHC model alone (e.g., in terms of speed of learning and smoothness of control). However, this model is difficult to apply for other control systems due mainly to the fact that developing the mathematical functions for the trace function and credit assignment are not trivial. The structure proposed here suffers from the lack of generality and may be difficult to apply to larger scale systems.

6.4 Fuzzy Control and two layer networks

Berenji [7, 6] has proposed a Neuro-Fuzzy Controller (ARIC) architecture which extends the Fuzzy-AHC model mentioned earlier. Figure 7 presents the architecture of the proposed ARIC model. In ARIC, two networks replace the two neural-like elements. These networks are referred to as the *Action-state Evaluation Network (AEN)* and *Action Selection Network (ASN)*. The AEN and ASN networks are multi-layered neural networks which are based on the back propagation algorithm and reinforcement learning.

6.5 Use of Clustering

Takagi and Hayashi [54] have presented an algorithm for combining neural networks with fuzzy logic which consists of three parts. The first part finds a suitable partition of the training data by using a clustering algorithm. Once the best partition of the data is found, then the second part of the algorithm identifies the membership functions of the IF parts of the rules. The last step of the algorithm is to determine the amount of control output for each rule. A neural network is used for the identification of the membership functions in the second step above. After this network is trained, it assigns the correct rule number to the combination of different sensor readings. To identify the THEN part of a rule, a backward elimination method is used. This method arbitrarily eliminates an input variable and the neural network of each THEN part is retrained. This process determines the input variable with minimal effects.

Takagi and Hayashi's model is very similar to Berenji's work in inductive learning and fuzzy control [5] where an AI clustering method (C4, a descendent of Quinlan's ID3 algorithm) is used. In Berenji's model, a decision theoretic measure is used to decide which input variable the decision tree should use to branch on next. The number of leaves of the resulting decision tree will indicate the number of control rules needed.

6.6 Other Research

Among other work in this area is Kosko's work on Fuzzy Cognitive Maps (FCM) [27]. The FCMs are graphical representations of the causal relationships between different factors, where the weights on the links represent the positive or negative causal relationships. Also, Kosko's Fuzzy Associative Memories (FAM) [28] can map fuzzy input patterns into stored fuzzy output patterns and hence are useful tools in representing the knowledge base of a fuzzy controller.

Fuzzy logic controllers and neural network controllers are complementary, and it is expected that the amount of research into hybrid approaches will grow significantly in the next few years, especially in Japan, where many applications have already been reported using a combination of these methods. In the U.S., NASA has taken an active role in integrating these two powerful techniques (e.g., sponsorship of the First and Second International Conferences on Neural Networks and Fuzzy Logic at Johnson Space Center in 1988 and 1990).

7 Discussion

Among the problems which still deserve serious attention is the problem of providing proof of stability for FLCs. In contrast to the analytical control theory, FLC lacks this necessary attribute although some theoretical work has begun producing interesting results (e.g., [29, 26, 10, 11, 45, 17, 16, 68, 24, 62, 21, 43, 74, 47, 25]).

Another area that requires attention is in what we refered to as *fuzzy modeling of systems* earlier in this paper. Here the attention should be focused on structure identification and parameter identification of the dynamics of a system in order to develop a model which could later be used to develop the fuzzy logic controller [56, 50].

Finally, as we briefly discussed in the previous section, artificial neural networks and fusion techniques are being developed in order to develop fuzzy logic controllers which can learn from experience. Despite these open issues, fuzzy logic control has achieved a huge commercial success in recent years. Because these controllers are easy to manufacture and greatly resemble human reasoning, it is expected that there will be many more applications in the near future.

References

[1] *International Conference on Fuzzy Logic & Neural Networks*, volume one and two, Iizuka, Japan, 1988.

[2] *International Conference on Fuzzy Logic & Neural Networks*, volume one and two, Iizuka, Japan, 1990.

[3] W.H. Bare, R.J. Mulholland, and S.S. Sofer. Design of a self-tuning rule based controller for a gasoline refinery catalytic reformer. *IEEE Transactions on Automatic Control*, 35(2):156–164, 1990.

[4] H. R. Berenji, Y. Y. Chen, C. C. Lee, S. Murugesan, and J. S. Jang. An experiment-based comparative study of fuzzy logic control. In *American Control Conference*, Pittsburgh, 1989.

[5] H.R. Berenji. Machine learning in fuzzy control. In *Int. Conf. on Fuzzy Logic & Neural Networks*, pages 231–234, Iizuka, Fukuoka, Japan, 1990.

[6] H.R. Berenji. A reinforcement learning based model for fuzzy logic control. *International Journal of Approximate Reasoning*, 1991 (to appear).

[7] H.R. Berenji. An architecture for designing fuzzy controllers using neural networks. In *Second Joint Technology Workshop on Neural Networks and Fuzzy Logic*, Houston, Texas, April 1990.

[8] H.R. Berenji, Y.Y. Chen, C.C. Lee, J.S. Jang, and S. Murugesan. A hierarchical approach to designing approximate reasoning-based controllers for dynamic physical systems. In *Sixth Conference on Uncertainty in Artificial Intelligence*, pages 362–369, 1990.

[9] J. A. Bernard. Use of rule-based system for process control. *IEEE Control Systems Magazine*, 8, no. 5:3–13, 1988.

[10] M. Braae and D.A. Rutherford. Theoretical and linguistic aspects of the fuzzy logic controller. *Automatica*, 15, no. 5:553–577, 1979.

[11] Y.Y. Chen. Stability analysis of fuzzy control—a lyapunov approach. In *IEEE Systems, Man, Cybernetics, Annual Conference*, volume 3, pages 1027–1031, 1987.

[12] E. Czogala and T. Rawlik. Modelling of a fuzzy controller with application to the control of biological processes. *Fuzzy Sets and Systems*, 31:13–22, 1989.

[13] J. Efstathiou. Rule-based process control using fuzzy logic. In E. Sanchez and L.A. Zadeh, editors, *Approximate Reasoning in Intelligence Systems, Decision and Control*, pages 145–148. Pergamon, New York, 1987.

[14] B. P. Graham and R. B. Newell. Fuzzy identification and control of a liquid level rig. *Fuzzy Sets and Systems*, 26:255–273, 1988.

[15] B. P. Graham and R. B. Newell. Fuzzy adaptive control of a first order process. *Fuzzy Sets and Systems*, 31:47–65, 1989.

[16] G. M. Trojan Gupta, M. M. and J. B. Kiszka. Controllability of fuzzy control systems. *IEEE Transactions on Systems, Man, and Cybernetics*, SMC-16, no. 4:576–582, 1986.

[17] W. Pedrycz Gupta, M. M. Cognitive and fuzzy logic controllers: A retrospective and perspective. In *American Control Conference*, pages 2245–2251, 1989.

[18] K. Hirota, Y. Arai, and S. Hachisu. Fuzzy controlled robot arm playing two-dimensional ping-pong game. *Fuzzy Sets and Systems*, 32:149–159, 1989.

[19] C. Isik. Identification and fuzzy rule-based control of a mobile robot motion. In *IEEE Int. Symposium on Intelligent Control*, 1987.

[20] Y. Kasai and Y. Morimoto. Electronically controlled continuously variable transmission. In *Proc. Intl. Congress on Transportation Electronics*, Dearborn, Michigan, October 1988.

[21] W. J. Kickert and E. H. Mamdani. Analysis of a fuzzy logic controller. *Fuzzy Sets and Systems*, 1(1):29–44, 1978.

[22] W. J. M. Kickert and H. R. Van Nauta Lemke. Application of a fuzzy controller in a warm water plant. *Automatica*, 12(4):301–308, 1976.

[23] P. J. King and E. H. Mamdani. The application of fuzzy control systems to industrial processes. *Automatica*, 13(3):235–242, 1977.

[24] J.B. Kiszka, M.M. Gupta, and P.N. Nikiforuk. Energistic stability of fuzzy dynamic systems. *IEEE Trans. Systems, Man and Cybernetics*, SMC-15(6), 1985.

[25] J.B. Kiszka, M.M. Gupta, and G.M. Trojan. Multivariable fuzzy controller under godel's implication. *Fuzzy Sets and Systems*, 34:301–321, 1990.

[26] S. V. Komolov, S. P. Makeev, and F. Shaknov. Optimal control of a finite automation with fuzzy constraints and a fuzzy target. *Cybernetics*, 16(6):805–810, 1979.

[27] B. Kosko. Fuzzy cognitive maps. *International Journal of Man-Machine Studies*, 24:65–75, 1986.

[28] B. Kosko. Fuzzy associative memories. In Kandel A., editor, *Fuzzy Expert Systems*. Addison-Wesley, 1987.

[29] G.R. Langari and M. Tomizuka. Stability of fuzzy linguistic control systems. In *IEEE Conference on Decision and Control*, Hawaii, December 1990.

[30] L. I. Larkin. A fuzzy logic controller for aircraft flight control. In M. Sugeno, editor, *Industrial Applications of Fuzzy Control*, pages 87–104. North-Holland, Amsterdam, 1985.

[31] C.C. Lee. Self-learning rule-based controller employing approximate reasoning and neural-net concepts. *Int. Journal of Intelligent Systems*, 1990.

[32] C.C. Lee and H.R. Berenji. An intelligent controller based on approximate reasoning and reinforcement learning. In *Proc. of IEEE Int. Symposium on Intelligent Control*, Albany, NY, 1989.

[33] Fujitec Co. Ltd. Flex-8800 series elevator group control system. Technical report, Fujitec Co. Ltd., Osaka, Japan, 1988.

[34] Scharf. E. M. and N. J. Mandic. The application of a fuzzy controller to the control of a multi-degree-freedom robot arm. In M. Sugeno, editor, *Industrial Applications of Fuzzy Control*, pages 41–62. North-Holland, Amsterdam, 1985.

[35] S. Mabuchi. An approach to the comparison of fuzzy subsets with an α-cut dependent index. *IEEE Transactions on Systems, Man, and Cybernetics*, 18(2), 1988.

[36] E. H. Mamdani and S. Assilian. An experiment in linguistic synthesis with a fuzzy logic controller. *International Journal of Man-Machine Studies*, 7(1):1–13, 1975.

[37] S. Murakami and M. Maeda. Application of fuzzy controller to automobille speed control system. In M. Sugeno, editor, *Industrial Applications of Fuzzy Control*, pages 105–124. North-Holland, Amsterdam, 1985.

[38] S. Murakami, F. Takemoto, H. Fujimura, and E. Ide. Weld-line tracking control of arc welding robot using fuzzy logic controller. *Fuzzy Sets and Systems*, 32:221–237, 1989.

[39] A. Ollero and A.J. Garcia-Cerezo. Direct digital control, autotuning and supervision using fuzzy logic. *Fuzzy Sets and Systems*, 30:135–153, 1989.

[40] H. Ono, T. Ohnishi, and Y. Terada. Combustion control of refuse incineration plant by fuzzy logic. *Fuzzy Sets and Systems*, 32:193–206, 1989.

[41] Rainer Palm. Fuzzy controller for a sensor guided manipulator. *Fuzzy Sets and Systems*, 31:133–149, 1989.

[42] C.P. Pappis and E. H. Mamdani. A fuzzy logic controller for a traffic junction. *IEEE Transactions on Systems, Man, and Cybernetics*, SMC-7(10):707–717, 1977.

[43] W. Pedrycz. An approach to the analysis of fuzzy systems. *Int. Journal of Control*, 34(3):403–421, 1981.

[44] T. J. Procyk and E. H. Mamdani. A linguistic self-organizing process controller. *Automatica*, 15(1):15–30, 1979.

[45] K. S. Ray and D. D. Majumder. Application of circle criteria for stability analysis of linear siso and mimo systems associated with fuzzy logic controller. *IEEE Transactions on Systems, Man, and Cybernetics*, SMC-14(2):345–349, 1984.

[46] Floor Van Der Rhee, Hans R. Van Nauta Lemke, and Jaap G. Dijkman. Knowledge based fuzzy control of systems. *IEEE Transactions on Automatic Control*, 35(2):148–155, 1990.

[47] William Siler and Hao Ying. Fuzzy control theory: The linear case. *Fuzzy Sets and Systems*, 33:275–290, 1989.

[48] M. Sugeno. Current projects in fuzzy control. In *Workshop on Fuzzy Control Systems and Space Station Applications*, pages 65–77, Huntington Beach, CA, 14-15 November 1990.

[49] M. Sugeno. An introductory survey of fuzzy control. *Information Science*, 36:59–83, 1985.

[50] M. Sugeno and G. T. Kang. Structure identification of fuzzy model. *Fuzzy Sets and Systems*, 28(1):15–33, 1988.

[51] M. Sugeno and K. Murakami. An experimental study on fuzzy parking control using a model car. In M. Sugeno, editor, *Industrial Applications of Fuzzy Control*, pages 125–138. North-Holland, Amsterdam, 1985.

[52] M. Sugeno, T. Murofushi, T. Mori, and Tatematsu. Fuzzy algorithmic control of a model car by oral instructions. *Fuzzy Sets and Systems*, 32:207–219, 1989.

[53] M. Sugeno and M. Nishida. Fuzzy control of model car. *Fuzzy Sets and Systems*, 16:110–113, 1985.

[54] H. Takagi and I. Hayashi. Artificial-neural-network-driven fuzzy reasoning. *Int. J. of Approximate Reasoning*, (to appear).

[55] T. Takagi and M. Sugeno. Derivation of fuzzy control rules from human operator's control actions. In *IFAC Symposium on Fuzzy Information, Knowledge Representation and Decision Analysis*, pages 55–60, Marseille, France, 1983.

[56] T. Takagi and M. Sugeno. Fuzzy identification of systems and its applications to modelling and control. *IEEE Transactions on Systems, Man, and Cybernetics*, SMC-15(1):116–132, 1985.

[57] M. Takahashi, E. Sanchez, R. Bartolin, J.P. Aurrand-Lions, E. Akaiwa, T. Yamakawa, and J.R. Monties. Biomedical applications of fuzzy logic controllers. In *Int. Conf. on Fuzzy Logic & Neural Networks*, pages 553–556, Iizuka, Fukuoka, Japan, 1990.

[58] K. L. Tang and R. J. Mulholland. Comparing fuzzy logic with classical controller designs. *IEEE Transactions on Systems, Man, and Cybernetics*, SMC-17(6):1085–1087, 1987.

[59] R. Tanscheit and E. M. Scharf. Experiments with the use of a rule-based self-organising controller for robotics applications. *Fuzzy Sets and Systems*, 26:195–214, 1988.

[60] M. Togai and S. Chiu. A fuzzy accelerator and a programming environment for real-time fuzzy control. In *Second IFSA Congress*, pages 147–151, Tokyo, Japan, 1987.

[61] M. Togai and H. Watanabe. Expert systems on a chip: an engine for real-time approximate reasoning. *IEEE Expert Systems Magazine*, 1:55–62, 1986.

[62] R. Tong. Analysis of fuzzy control algorithms using the relation matrix. *International Journal of Man-Machine Studies*, 8(6):679–686, 1976.

[63] R. Tong, M. B. Beck, and A. Latten. Fuzzy control of the activated sludge wastewater treatment process. *Automatica*, 16(6):695–701, 1980.

[64] T. Tsukamoto. An approach to fuzzy reasoning method. In M. M. Gupta, R. K. Ragade, and R. R. Yager, editors, *Advances in Fuzzy Set Theory and Applications*. North-Holland, Amsterdam, 1979.

[65] I. G. Umbers and P. J. King. An analysis of human-decision making in cement kiln control and the umplications for automation. *International Journal of Man-Machine Studies*, 12(1):11–23, 1980.

[66] M. Uragami, M. Mizumoto, and K. Tanaka. Fuzzy robot controls. *Cybernetics*, 6:39–64, 1976.

[67] H. Watanabe and W. Dettloff. Reconfigurable fuzzy logic processor: a full custom digital vlsi. In *Int. Workshop on Fuzzy System Applications*, pages 49–50, Iizuka, Japan, 1988.

[68] Chen-Wei Xu. Analysis and feedback/feedforward control of fuzzy relational systems. *Fuzzy Sets and Systems*, 35:105–113, 1990.

[69] Murayama. Y. and T. Terano. Optimising control of disel engine. In M. Sugeno, editor, *Industrial Applications of Fuzzy Control*, pages 63–72. North-Holland, Amsterdam, 1985.

[70] O. Yagishita, O. Itoh, and M. Sugeno. Application of fuzzy reasoning to the water purification process. In M. Sugeno, editor, *Industrial Applications of Fuzzy Control*, pages 19–40. North-Holland, Amsterdam, 1985.

[71] T. Yamakawa. Fuzzy microprocessors - rule chip and defuzzifier chip. In *Int. Workshop on Fuzzy System Applications*, pages 51–52, Iizuka, Japan, 1988.

[72] T. Yamakawa. Intrinsic fuzzy electronic circuits for sixth generation computer. In M.M. Gupta and T. Yamakawa, editors, *Fuzzy Computing*, pages 157–171. Elsevier Science Publishers B.V. (north Holland), north Holland, 1988.

[73] S. Yasunobu and S. Miyamoto. Automatic train operation by predictive fuzzy control. In M. Sugeno, editor, *Industrial Applications of Fuzzy Control*, pages 1–18. North-Holland, Amsterdam, 1985.

[74] Hao Ying, William Siler, and James J. Buckley. Fuzzy control theory: A nonlinear case. *Automatica*, 26(3):513–520, 1990.

[75] Y. Yoshikawa, T. Deguchi, and T. Yamakawa. Exclusive fuzzy hardware system for the appraisal of orthodentic results. In *Int. Conf. on Fuzzy Logic & Neural Networks*, pages 939–942, Iizuka, Fukuoka, Japan, 1990.

[76] C. Yu, Z. Cao, and A. Kandel. Application of fuzzy reasoning to the control of an activated sludge plant. *Fuzzy Sets and Systems*, 38:1–14, 1990.

[77] L. A. Zadeh. Fuzzy sets. *Information and Control*, 8:338–353, 1965.

[78] L.A. Zadeh. A fuzzy-set-theoretic interpretation of linguistic hedges. *Journal of Cybernetics*, 2:4–34, 1972.

5

METHODS AND APPLICATIONS OF FUZZY MATHEMATICAL PROGRAMMING

H.-J. Zimmermann
RWTH Aachen
Templergraben 55
W-5100 Aachen (Germany)

1. INTRODUCTION

In spite of other uses of the term "mathematical programming" it shall be interpreted here as it is normally done in Operations Research, i.e. an algorithmic approach to solving models of the type

$$\text{maximize } f(x)$$

$$\text{such that } g_i(x) = 0, \quad i = 1,\dots,m \tag{1}$$

Depending on the mathematical character of the objective function, f(x), and the constraints, $g_i(x)$, many types of mathematical programming algorithms exist, such as, linear programming, quadratic programming, fractional programming, convex programming etc. Exemplarily, we shall use the simplest and most commonly used type, i.e. linear programming, which focusses on the model

$$\text{maximize } f(x) = z = c^T x$$

$$\begin{aligned} \text{such that} \quad & Ax \leq b \\ & x \geq 0 \end{aligned} \tag{2}$$

with $c, x \in \mathbb{R}^n, b \in \mathbb{R}^m, A \in \mathbb{R}^{m \times n}$.

In this model it is normally assumed that all coefficients of A, b, and c are real (crisp) numbers; that "≤" is meant in a crisp sense, and that "maximize" is a strict imperative. This also implies that the violation of any single constraint renders the solution infeasable and that all constraints are of equal importance (weight). Strictly speaking, these are rather unrealistic assumptions, which are partly relaxed in "fuzzy linear programming".

If we assume that the LP-decision has to be made in fuzzy environments, quite a number of possible modifications of (2) exist. First of all, the decision maker might really not want to actually maximize or minimize the objective function. Rather he might want to reach some aspiration levels which might not even be definable crisply. Thus he might want to "improve the present cost situation considerably," and so on.

Secondly, the constraints might be vague in one of the following ways: The $\leq$ sign might not be meant in the strictly mathematical sense but smaller violations might well be acceptable. This can happen if the constraints represent aspiration levels as mentioned above or if, for instance, the constraints represent sensory requirements (taste, color, smell, etc.) which cannot adequately be approximated by a crisp constraint. Of course, the coefficients of the vectors b or c or of the matrix A itself can have a fuzzy character either because they are fuzzy in nature or because perception of them is fuzzy.

Finally the role of the constraints can be different from that in classical linear programming where the violation of any single constraint by any amount renders the solution infeasable. The decision maker might accept small violations of different constraints. Fuzzy linear programming offers a number of ways to allow for all those types of vagueness and we shall discuss some of them below.

Before we develop a specific model of linear programming in a fuzzy environment it should have become clear, that by contrast to classical linear programming "fuzzy linear programming" is not a uniquely defined type of model but that many variations are possible, depending on the assumptions or features of the real situation to be modelled.

Essentially two "families" of models can be distinguished: One interprets "fuzzy mathematical programming" as a specific decision making environment to which Bellman and Zadeh's definition of a "decision in fuzzy environments" [1970] can be applied. The other considers components of model (2) as fuzzy, makes certain assumptions, for instance, about the type of fuzzy sets which as fuzzy numbers replace the crisp coefficients in A, b, or c, and then solve the resulting mathematical problem. The former approach seems to us the more application oriented one. From experience in applications a decision maker seems to find it much easier to describe fuzzy constraints or to establish aspiration levels for the objective(s) than to specify a large number of fuzzy numbers for A, b, or c. We shall, therefore, first describe the first approach and then elaborate on the other approaches.

2. Fuzzy Mathematical Programming

2a Symmetric Fuzzy Linear Programming

As mentioned above Fuzzy LP is considered as a special case of a decision in a fuzzy environment. The basis in this case is the definition suggested by Bellman and Zadeh [1970]:

Definition 1:

Assume that we are given a fuzzy goal $\tilde{G}$ and a fuzzy constraint $\tilde{C}$ in a space of alternatives X. Then $\tilde{G}$ and $\tilde{C}$ combine to form a decision, $\tilde{D}$, which is a fuzzy set resulting from intersection of $\tilde{G}$ and $\tilde{C}$. In symbols, $\tilde{D} = \tilde{G} \cap \tilde{C}$ and correspondingly

$$\mu_{\tilde{D}} = \min \quad \{\mu_{\tilde{G}}, \mu_{\tilde{C}}\}.$$

More generally, suppose that we have n goals $\tilde{G}_1,...,\tilde{G}_n$ and m constraints $\tilde{C}_1,...,\tilde{C}_m$. Then, the resultant decision is the intersection of the given goals $\tilde{G}_1,...,\tilde{G}_n$ and the given constraints $\tilde{C}_1,...,\tilde{C}_m$. That is,

$$\tilde{D} = \tilde{G}_1 \cap \tilde{G}_2 \cap \cdots \cap \tilde{G}_n \cap \tilde{C}_1 \cap \tilde{C}_2 \cap \quad \cap \tilde{C}_m$$

and correspondingly

$$\begin{aligned} \mu_{\tilde{D}} &= \min \quad \{\mu_{\tilde{G}_1}, \mu_{\tilde{G}_2}, \cdots, \mu_{\tilde{G}_n}, \mu_{\tilde{C}_1}, \mu_{\tilde{C}_2}, \cdots, \mu_{\tilde{C}_m}\} \\ &= \min \quad \{\mu_{\tilde{G}_i}, \quad \mu_{\tilde{C}_j}\} \quad = \quad \min \quad \{\mu_i\}. \end{aligned}$$

This definition implies:

1. The "and" connecting goals and constraints in the model corresponds to the "logical and".
2. The logical "and" corresponds to the set theoretic intersection.
3. The intersection of fuzzy sets is defined in the possiblistic sense by the min-operator.

For the time being we shall accept these assumptions. An important feature of this model is also its symmetry, i.e. the fact that, eventually, it does not distinguish between constraints and objectives. This feature is not considered adequate by all authors (see, for instance, Asai et. al. [1975]). We feel, however, that this models quite well real behaviour of decision makers.

If we assume that the decision maker can establish in model (2) an aspiration level, z, of the objective function, which he wants to achieve as far as possible and if the constraints of this model can be slightly violated - without causing infeasibility of the solution - then model (2) can be written as

$$\begin{aligned} &\text{Find } x \\ &\text{such that } c^T x \gtrsim z \\ &\qquad\qquad Ax \lesssim b \\ &\qquad\qquad x \geq 0 \end{aligned} \tag{3}$$

Here $\lesssim$ denotes the fuzzified version of $\leq$ and has the linguistic interpretation "essentially smaller than or equal." $\gtrsim$ denotes the fuzzified version of $\geq$ and has the linguistic interpretation "essentially greater than or equal." The objective function in (2) might have to be written as a minimizing goal in order to consider z as an upper bound.

We see that (3) is fully symmetric with respect to objective function and constraints and we want to make that even more obvious by substituting $\binom{-c}{A} = B$ and $\binom{-z}{b} = d$. Then (3) becomes:

$$\begin{aligned} &\text{Find } x \\ &\text{such that } Bx \lesssim d \\ &\qquad\qquad x \geq 0 \end{aligned} \tag{4}$$

Each of the (m + 1) rows of (4) shall now be represented by a fuzzy set, the membership functions of which are $\mu_i(x)$. The membership function of the fuzzy set "decision" of model (4) is

$$\mu_{\tilde{D}}(x) = \min_i \{\mu_i(x)\} \tag{5}$$

$\mu_i(x)$ can be interpreted as the degree to which x fulfills (satisfies) the fuzzy inequality $B_i x \lesssim d_i$ (where B_i denotes the ith row of B).

Assuming that the decision maker is interested not in a fuzzy set but in a crisp "optimal" solution we could suggest the "maximizing solution" to (5), which is the solution to the possibly nonlinear programming problem

$$\max_{x \geq 0} \min_i \{\mu_i(x)\} = \max_{x \geq 0} \mu_{\tilde{D}}(x) \tag{6}$$

Now we have to specify the membership functions $\mu_i(x)$. $\mu_i(x)$ should be 0 if the constraints (including objective function) are strongly violated, and 1 if they are very well satisfied (i.e., satisfied in the crisp sense); and $\mu_i(x)$ should increase monotonously from 0 to 1, that is:

$$\mu_i(x) = \begin{cases} 1 & \text{if } B_i x \le d_i \\ \in[0,1] & \text{if } d_i < B_i x \le d_i + p_i \qquad i=1,\ldots m+1 \\ 0 & \text{if } B_i x > d_i + p_i \end{cases} \tag{7}$$

Using the simplest type of membership function we assume them to be linearly increasing over the "tolerance interval" p_i:

$$\mu_i(x) = \begin{cases} 1 & \text{if } B_i x \le d_i \\ 1 - \dfrac{B_i x - d_i}{p_i} & \text{if } d_i < B_i x \le d_i + p_i \qquad i=1,\ldots,m+1 \\ 0 & \text{if } B_i x > d_i + p_i \end{cases} \tag{8}$$

The p_i are subjectively chosen constants of admissable violations of the constraints and the objective function. Substituting (8) into (6) yields, after some rearrangements [Zimmermann 1976] and with some additional assumptions,

$$\max \min \left(1 - \frac{B_i x - d_i}{p_i}\right) \tag{9}$$

Introducing one new variable, λ, which corresponds essentially to (5), we arrive at

$$\begin{array}{ll} \text{maximize} & \lambda \\ \text{such that} & \lambda p_i + B_i x \le d_i + p_i \quad i=1,\ldots,m+1 \\ & x \ge 0 \end{array} \tag{10}$$

If the optimal solution to (10) is the vector (λ, x_0), the x_0 is the maximizing solution (6) of model (2) assuming membership functions as specified in (8).

The reader should realize that this maximizing solution can be found by solving one standard (crisp) LP with only one more variable and one more constraint than model (4). This makes this approach computationally very efficient.

A slightly modified version of models (9) and (10), respectively, results if the membership functions are defined as follows: A variable t_i, $i=1,\ldots,m+1$, $0 \le t_i \le p_i$, is defined which measures the degree of violation of the ith constraint: The membership function of the ith row is then

$$\mu_i(x) = 1 - \frac{t_i}{p_i} \tag{11}$$

The crisp equivalent model is then

$$\begin{array}{lll} \text{maximize} & \lambda & \\ \text{such that} & \lambda p_i + t_i \leq p_i & i=1,....,m+1 \\ & B_i x - t_i \leq d_i & \\ & t_i \leq p_i & \\ & x,t \geq 0 & \end{array} \qquad (12)$$

This model is larger than model (10) even though the set of constraints $t_i \leq p_i$ is actually redundant. Model (12) has some advantages, however, in particular when performing sensitivity analysis.

The main advantage, compared to the unfuzzy problem formulation, is the fact that the decision maker is not forced into a precise formulation because of mathematical reasons, even though he might only be able or willing to describe his problem in fuzzy terms. Linear membership functions are obviously only a very rough approximation. Membership functions which monotonically increase or decrease, respectively, in the interval of $[d_i, d_i+p_i]$ can also be handled quite easily, as will be shown later.

It should also be observed that the classical assumption of equal importance of constraints has been relaxed: the slope of the membership functions determines the "weight" or importance of the constraint. The slopes, however, are determined by the p_i's: The smaller the p_i the higher the importance of the constraint. For $p_i=0$ the constraint becomes crisp, i.e. no violation is allowed.

So far, the objective function as well as all constraints were considered fuzzy. If some of the constraints are crisp, $Dx \leq b$, then these constraints can easily be added to formulations (11) or (12), respectively. Thus (11) would, for instance, become:

$$\begin{array}{ll} \text{maximize} & \lambda \\ \text{such that} & \lambda p_i + B_i x \leq d_i + p_i \\ & Dx \leq b \\ & x,\lambda \geq 0 \end{array} \qquad (13)$$

2B LINEAR PROGRAMS WITH FUZZY CONSTRAINTS AND CRISP OBJECTIVE FUNCTIONS

So far, it has been assumed that the objective function could be calibrated by a given z and then formulated as a fuzzy set, resulting in the symmetrical model formulation. It might, however, not be possible to find in a natural way the required z. In this case the symmetry of the model can be gained by applying a specialisation of Zadeh's "maximizing set" to the objective function:

Definition 2: [Werners 1984]

Let $f: X \to \mathbb{R}^1$ be the objective function, $\tilde{R}$ = fuzzy feasible region, $S(\tilde{R})$ = support of $\tilde{R}$, and R_1 = α-level cut of $\tilde{R}$ for $\alpha = 1$. The membership function of the goal (objective function) given solution space $\tilde{R}$ is then defined as

$$\mu_{\tilde{G}}(x) = \begin{cases} 0 & \text{if } f(x) \leq \sup_{R_1} f \\ \dfrac{f(x) - \sup_{R_1} f}{\sup_{S(\tilde{R})} f - \sup_{R_1} f} & \text{if } \sup_{R_1} f < f(x) < \sup_{S(\tilde{R})} f \\ 1 & \text{if } \sup_{S(\tilde{R})} f \leq f(x) \end{cases}$$

The corresponding membership function in functional space is then

$$\mu_{\tilde{G}}(r) := \begin{cases} \sup_{x \in f^{-1}(r)} \mu_{\tilde{G}}(x) & \text{if } r \in R, f^{-1}(r) \neq 0 \\ 0 & \text{else} \end{cases} \tag{13A}$$

Adding this fuzzy set to the fuzzy sets defining the solution space gives again a symmetrical model to which (10) or (12) can be applied. Definition 2 becomes easier to understand if we apply it to a specific given LP-structure:

Let us modify (3) by adding a set of crisp constraints, $Dx \leq b$, and changing the objective function to maximize f(x). This yields model

$$\begin{array}{ll} \text{maximize} & f(x) = c^T x \\ \text{such that} & \left.\begin{array}{r} Ax \lesssim b \\ Dx \leq b' \\ x \leq 0 \end{array}\right\} \tilde{R} \end{array} \tag{14}$$

Let the membership functions of the fuzzy sets representing the fuzzy constraints be defined in analogy to (8) as

$$\mu_i(x) = \begin{cases} 1 & \text{if } A_i x \leq b_i \\ \dfrac{b_i + p_i - A_i x}{p_i} & \text{if } b_i < A_i x \leq b_i + p_i \\ 0 & \text{if } A_i x > b_i + p_i \end{cases} \tag{15}$$

On the basis of the two LP's following, the membership function of the fuzzy set defined in definition 2 can then easily be defined:

$$\begin{aligned} \text{maximize} \quad & f(x) = c^T x \\ \text{such that} \quad & Ax \le b \\ & Dx \le b' \\ & x \ge 0 \end{aligned} \tag{16}$$

The optimal solution of this model is $f_1 = \sup_{R^1} f\,(c^T x)_{opt}$.

$$\begin{aligned} \text{Maximize} \quad & f(x) = c^T x \\ \text{such that} \quad & Ax \le b + p \\ & Dx \le b' \\ & x \ge 0 \end{aligned} \tag{17}$$

The optimal solution of the model model is $f_o = \sup_{S(\tilde{R})} f = (c^T x)_{opt}$.

The membership function is therefore

$$\mu_{\tilde{G}}(x) = \begin{cases} 1 & \text{if } f_0 \le c^T x \\ \dfrac{c^T x - f_i}{f_0 - f_1} & \text{if } f_1 < c_T x < f_0 \\ 0 & \text{if } c^T x \le f_1 \end{cases} \tag{18}$$

The "equivalent" model to (14) is therefore:

$$\begin{aligned} \text{maximize} \quad & \lambda \\ \text{such that} \quad & \lambda(f_0 - f_1) - c^T x \le -f_1 \\ & \lambda p + Ax \le b + p \\ & Dx \le b' \\ & \lambda \le 1 \\ & \lambda, x \ge 0 \end{aligned} \tag{19}$$

Example:

Consider the LP-Model

$$\begin{aligned} \text{maximize} \quad & z = 2x_1 + x_2 \\ \text{such that} \quad & x_1 \le 3 \\ & x_1 + x_2 \le 4 \\ & 5x_1 + x_2 \le 3 \\ & x_1, x_2 \ge 0 \end{aligned}$$

The "tolerance intervals" of the constraints are $p_1 = 6, p_2 = 4, p_3 = 2$.
f_o and f_1 can be determined to be 7 and 16, respectively. Hence, model (19) is

$$\begin{aligned} \text{maximize} \quad & \lambda \\ \text{such that} \quad & 9\lambda - 2x_1 - x_2 \le -7 \\ & 6\lambda + x_1 \le 9 \\ & 4\lambda + x_1 + x_2 \le 8 \\ & 2\lambda + 5x_1 + x_2 \le 5 \\ & \lambda \le 1 \\ & \lambda, x_1, x_2 \ge 0 \end{aligned}$$

The solution to this problem is $x_1^o = 5.84, x_2^o = 0, \lambda_0 = .52$.

Some authors suggest not to use a "symmetrical" approach but rather to compute a fuzzy set "decision". They compute the optimal values of the objective function for all α-level-sets of the solution space. The membership function of the "decision" is then defined to be the α's corresponding to the respective optimal values of the objective function. [Orlovski 1977]

In a certain sense this philosophy is similar to that of those authors who suggest to determine to model (4) not a crisp solution (6) but the fuzzy set decision. To do that a parametric linear programming problem has to be solved [Chanas 1983]. Even though this approach leads to quite impressive results in the 2-dimensional case, it is rather questionable whether the decision maker can make use of it in a realistically sized problem.

2C EXTENSIONS

So far, two major assumptions have been made in order to arrive at "equivalent models" which can be solved efficiently by standard LP-methods:

1. Linear membership functions were assumed for all fuzzy sets involved.
2. The use of the minimum-operator for the aggregation of fuzzy sets was considered to be adequate.

The relaxation of these two assumptions leads to complication which are differently severe depending on the type of relaxation:

1. Nonlinear membership functions

The linear membership functions used so far could all be defined by fixing two points, the upper and lower aspiration levels or the two bounds of the tolerance interval. The most obvious way to handle nonlinear membership functions is probably to approximate them piece-wise by linear functions. Some authors [Hannan 1981; Nakamura 1985] have used this approach and shown that the resulting equivalent crisp problem is still a standard linear programming problem.

This problem, however, can be considerably larger than model (10) because in general one constraint will have to be added for each "linear piece" of the approximation. Quite often S-shaped membership functions have been suggested, particularly if the membership function is interpreted as a kind of utility function (representing the degree of satisfaction, acceptance etc.). Leberling [1981], for instance, suggests such a function which is also uniquely determined by two parameters. He suggests

$$\mu_H(x) = \frac{1}{2}\,\frac{\exp\left[\left(x - \frac{a+b}{2}\right)\delta\right] - \exp\left[-\left(x - \frac{a+b}{2}\right)\delta\right]}{\exp\left[\left(x - \frac{a+b}{2}\right)\delta\right] + \exp\left[-\left(x - \frac{a+b}{2}\right)\delta\right]}$$

with a, b, $\delta \geq 0$. This hyperbolic function has the following formal properties:

$\mu_H(x)$ is strictly monotonously increasing.

$$\mu_H(x) = \frac{1}{2} \text{ where } x = \frac{a+b}{2}$$

$\mu_H(x)$ is strictly convex on $[-\infty, (a+b)/2]$ and strictly concave on $[(a+b)/2, +\infty]$.

For all $x \in \mathbb{R}$: $= 0 < \mu_H(x) < 1$ and $\mu_H(x)$ approaches asymptotically $f(x) = 0$ and $f(x) = 1$, respectively.

Leberling shows that choosing as lower and upper aspiration levels for the fuzzy objective function $z = cx$ of an LP $a = \underline{c}$ (lower bound of z) and $b = \overline{c}$ (upper limit of the objective function), and representing this (fuzzy) goal by a hyperbolic function one arrives at the following crisp equivalent problem for one fuzzy goal and all crisp constraints:

$$
\begin{aligned}
\text{minimize} \quad & \lambda \\
\text{such that} \quad & \lambda - \frac{1}{2} \frac{e^{Z'(x)} - e^{-Z'(x)}}{e^{Z'(x)} + e^{-Z'(x)}} \leq \frac{1}{2} \\
& Dx \leq b' \\
& x, \lambda \geq 0
\end{aligned} \tag{20}
$$

with $Z'(x) = (\Sigma_j c_j x_j - \frac{1}{2}(\overline{c} + \underline{c})\delta$. For each additional fuzzy goal or constraint one of these exponential rows has, of course, to be added to (20).

For $x_{n+1} = \tanh^{-1}(2\lambda - 1)$, model (20) is equivalent to the following linear model:

$$
\begin{aligned}
\text{maximize} \quad & x_{n+1} \\
\text{such that} \quad & \delta \sum_j c_j x_j - x_{n+1} \geq \frac{1}{2} \delta(\overline{c} + \underline{c}) \\
& Dx \leq b' \\
& x_{n+1}, x \geq 0
\end{aligned} \tag{21}
$$

This is again a standard linear programming model which can be solved, for instance, by any available simplex code.

The above equivalence between models with nonlinear membership functions is not accidental. It has been proven that the following relationship holds [Werners 1984, p. 143].

<u>Theorem 1</u>

Let $\{f_k\}$, $k = 1,\dots,K$ be a finite family of functions $f_k: \mathbb{R}^n \to \mathbb{R}^1$, $x^0 \in X \subset \mathbb{R}^n$. $g: \mathbb{R}^1 \to \mathbb{R}^1$ strictly monotonously increasing and $\lambda, \lambda' \in \mathbb{R}$. Consider the two mathematical programming problems

$$
\begin{aligned}
\text{maximize} \quad & \lambda \\
\text{such that} \quad & \lambda \leq f_k(x) \quad k=1,\dots,K \\
& x \in X
\end{aligned}
$$

$$\text{(22)}$$

$$\begin{aligned}
&\text{maximize} \quad \lambda' \\
&\text{such that} \quad \lambda' \leq g(f_k(x)) \qquad k=1,\ldots K. \\
&\qquad\qquad\quad x \in X
\end{aligned} \tag{23}$$

If there exists a $\lambda^0 \in R'$ such that (λ^0, x^0) is the optimal solution of (22) then there exists a $\lambda'^0 \in R'$ such that (λ^0, x^0) is the optimal solution of (23).

Theorem 1 suggests that quite a number of nonlinear membership functions can be accommodated easily. Unluckily, the same optimism is not justified concerning other aggregation operators.

The computational efficiency of the approach mentioned so far has rested to a large extent on the use of the min-operator as a model for the logical "and" or the intersection of fuzzy sets, respectively. Axiomatic [Hamacher 1978] as well as empirical [Thole, Zimmermann, Zysno 1979, Zimmermann, Zysno 1980, 1983] investigations have shead some doubt on the general use of the min-operator in decision models. Quite a number of context free or context dependent operators have been suggested in the meantime [see, e.g., Zimmermann 1990b, ch. 3]. The disadvantage of these operators is, however, that the resulting crisp equivalent models are no longer linear [see, e.g., Zimmermann 1978, p.45], which reduces the computational efficiency of these approaches considerably or even renders the equivalent models unsolvable within acceptable time limits. There are, however, some exceptions to this rule, and we will present two of them in more detail.

One of the objections against the min-operator (see, for instance, Zimmermann and Zysno [1980]) is the fact that neither the logical "and" nor the min-operator is compensatory in the sense that increases in the degree of membership in the fuzzy sets "intersected" might not influence at all membership in the resulting fuzzy set (aggregated fuzzy set or intersection). There are two quite natural ways to cure this weakness:

1. Combine the (limitational) min-operator as model for the logical "and" with the fully compensatory max-operator as a model for the inclusive "or". For the former, the product operator might be used alternatively and for the latter the algebraic sum might be used. This approach departs from distinguishing between "and" and "or" aggregation as being somewhere between the "and" and the "or". (Therefore it is often called compensatory and.)
2. Stick with the distinction between "and" and "or" aggregators and introduce a certain degree of compensation into these connectives.

Compensatory "and". For some applications it seems to be important that the aggregator used maps above the max-operator and below the min-operator. The λ-operator [Zimmermann, Zysno 1980] would be such a connective. For purposes of mathematical programming it has, however, the above-mentioned disadvantage of low computational efficiency. An acceptable compromise between empirical fit and computational efficiency seems to be the convex combination of the min-operator and the max-operator:

$$\mu_c(x) = \gamma \min_{i=1}^{m} \mu_i(x) + (1-\gamma) \max_{i=1}^{m} \mu_i(x) \quad \gamma \in [0, 1] \tag{24}$$

For determining the maximizing decision the following problem has to be solved.

$$\max_{x \in X} \left(\gamma \min_{i=1}^{m} \{\mu_i(x)\} + (1-\gamma) \max_{i=1}^{m} \mu_i\{\mu_i(x)\}\right)$$

or

$$\begin{aligned} &\text{maximize} && \gamma \cdot \lambda_1 + (1-\gamma)\lambda_2 \\ &\text{such that} && \lambda_1 \leq \mu_i(x) \quad i=1,\dots, m \\ &&& \lambda_2 \leq \mu_i(x) \quad \text{for at least one } i \in \{1,\dots,m\} \\ &&& x \in X \end{aligned}$$

or

$$\begin{aligned} &\text{maximize} && \gamma\lambda_1 + (1-\gamma)\lambda_2 \\ &\text{such that} && \lambda_1 \leq \mu_i(x) && i=1,\dots, m \\ &&& \lambda_2 \leq \mu_i(x) + M\gamma_i && i=1,\dots, m \\ &&& \sum_{i=1}^{m} \gamma_i \leq m-1 \\ &\gamma \in \{0,1\}, M \text{ is a very large real number} \\ &x \in X \end{aligned} \tag{25}$$

For linear membership functions of the goals and the constraints (25) is a mixed integer linear program that can be solved by the appropriate available codes.

If one wants to distinguish between an "and"-aggregation and an "or"-aggregation (for instance, for the sake of easier modelling) one may want to use the following operators:

Definition 2 [Werners 1984]

Let $\mu_i(x)$ be the membership functions of fuzzy sets which are to be aggregrated in the sense of a fuzzy and ($\widetilde{\text{and}}$). The membership function of the resulting fuzzy set is defined to be

$$\mu_{\widetilde{and}}(x) = \gamma \cdot \min_{i=1}^{m} \mu_i(x) + (1-\gamma)\frac{1}{m}\Sigma\mu_i(x)$$

with $\gamma \in [0, 1]$.

Definition 3 [Werners 1984]

Let $\mu_i(x)$ be membership functions of fuzzy sets to be aggregated in the sense of a fuzzy or ($\widetilde{\text{or}}$). The membership function of the resulting fuzzy set is then defined as

$$\mu_{\widetilde{or}}(x) = \gamma \cdot \max_{i=1}^{m} \mu_i(x) + (1-\gamma)\frac{1}{m}\sum_{i=1}^{m}\mu_i(x)$$

with $\gamma \in [0, 1]$.

These two connectives are not inductive and associative, but they are commutative, idempotent, strictly monotonic increasing in each component, continuous, and compensatory [Werners 1984, p. 168]. These are certainly very useful and acceptable properties.

If we use the aggregation operator from definition 2 in model (4), then the "equivalent model" is:

$$\begin{aligned} &\text{maximize} && \lambda + (1-\gamma)\ \frac{1}{m}\sum_{i=1}^{m}\lambda_i \\ &\text{such that} && \lambda + \lambda_i \le \mu_i(x) \quad i=1,\dots,m \\ &&& Dx \le d \\ &&& \lambda,\ \lambda_i, x \ge 0 \\ &&& 0 \le \mu_i(x) \le 1 \end{aligned} \tag{26}$$

If $(\lambda^0, \lambda_i^0, x^0)$ is optimal solution of (26) then x^0 is a maximizing solution to (3). It is obvious that if $\mu_i(x)$ are linear (26) is again a standard linear programming problem.

So far, the reference model from which we have departed has always been the "standard LP". Depending on the type of operator chosen and the operator used the "equivalent model" turns out to be either a linear or a nonlinear programming model. Obviously, other reference models can be chosen as reference models. This has already been done, for instance, for integer programming [Zimmermann, Pollatschek 1984; Ignizio et al. 1983], Fractional Programming [Luhandjula 1984], Nonlinear Programming [Sakawa et al. 1989], etc.. The interrelationships between stochastic and fuzzy programming and their possible integration have also been investigated.[Buckley 1990; Dubois 1986]

3. FUZZY MATHEMATICAL PROGRAMMING WITH FUZZY PARAMETERS

Already for the basic approach described in section 2 a unique formulation for the "equivalent model", which eventually has to be solved, did not exist - the diversity of algorithmic approaches is even larger if other types of fuzzification of elements of mathematical programming models are considered. To demonstrate the basic idea behind most of the approaches we shall describe an easy to understand suggestion. For more general models the reader has to be referred to the literature.

Ramík and Rímánek [1985] consider the problem

$$\begin{aligned} &\text{maximize} \quad f(x) \\ &\text{such that} \quad \tilde{a}_{i1} x_1 \oplus \tilde{a}_{i2} x_2 \oplus \dots \oplus \tilde{a}_{in} x_n \leq \tilde{b}_i, \; i=1,\dots, m \\ &\qquad\qquad\quad x_j \geq 0, \; j=,\dots,n \end{aligned} \tag{27}$$

The $\tilde{a}_{ij}$ and the $\tilde{b}_i$ are supposed to be fuzzy numbers in L-R-representation. $\oplus$ denotes the extended addition. They show that for two fuzzy L-R numbers $\tilde{a} = (m, n, \alpha, \beta)_{L\text{-}R}$ and $\tilde{b} = (p, q, \gamma, \delta)_{L\text{-}R}$ $\tilde{a} \leq \tilde{b}$ holds iff the following 4 inequalities hold:

$$\begin{aligned} &\varepsilon_L (\gamma - \alpha) \leq p - m \qquad && \delta_L (\gamma - \alpha) \leq p - m \\ &\varepsilon_R (\beta - \delta) \leq q - n && \delta_R (\beta - \delta) \leq q - n \end{aligned} \tag{28}$$

where $\varepsilon_R = \sup \{u; R(u) = R(0) = 1 \}$,

$$\delta_R = \inf \{n; R(n) = \lim_{s \to \infty} R(s)\}$$

and ε_L, δ_L correspondingly for L.

For "symmetric" fuzzy numbers $\tilde{a} = (m, m, \alpha, \alpha)_{L\text{-}L}$ as shown in fig. 1 system (28) reduces to

$$\varepsilon_L \, |\alpha - \gamma| \le p - m \qquad \delta_2 \, |\alpha - \gamma| \le p - m \tag{29}$$

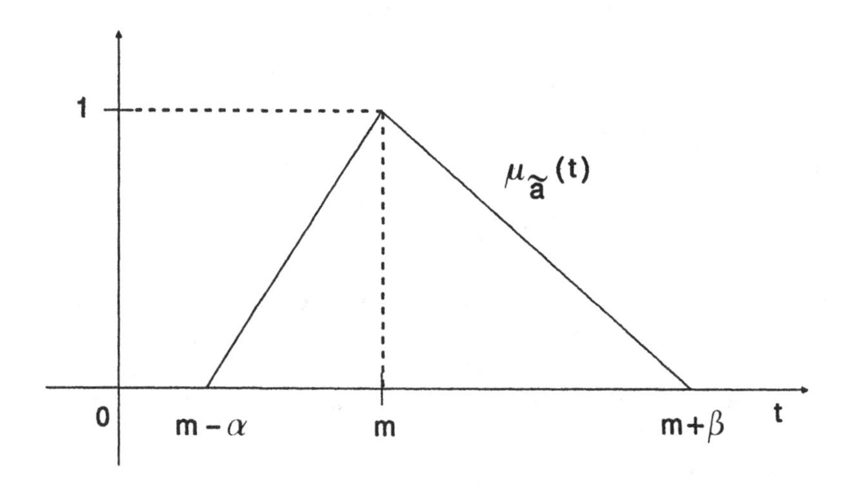

FIG. 1: Fuzzy triangular number $\tilde{a} = (m, m, \alpha, \beta)_{L\text{-}L}$

On the basis of a lemma that they proof in their paper:

$$\tilde{a}_{i1} x_1 \oplus \ldots \oplus \tilde{a}_{in} x_n = (\sum_j m_{ij} x_j, \sum_j n_{ij} x_j, \; \sum \alpha_{ij} x_j, \sum B_{ij} x_j). \tag{30}$$

Hence, the constraints of (27) can be written as

$$- \varepsilon_L \left(\sum_{j=1}^{n} \alpha_{ij} x_j - \gamma_i \right) \le p_i - \sum_{j=1}^{n} m_{ij} x_j,$$

$$- \delta_L \left(\sum_{j=1}^{n} \alpha_{ij} x_j - \gamma_i \right) \le p_i - \sum_{j=1}^{n} m_{ij} x_j,$$

$$\varepsilon_R \left(\sum_{j=1}^{n} \beta_{ij} x_j - \delta_i \right) \le q_i - \sum_{j=1}^{n} n_{ij} x_j,$$

$$\delta_R \left(\sum_{j=1}^{n} \beta_{ij} x_j - \delta_i \right) \le q_i - \sum_{j=1}^{n} n_{ij} x_j, \tag{31}$$

(31) is a system of crisp linear inequalities, which - together with the crisp objective function - can now be solved with any classical LP-method. Not counting the nonnegativity constraints, the number of rows in (31) is, however, four times as large as that of (27). It should also be noted that (28) is a specific interpretation of the fuzzy inequality relation. The authors offer two other interpretations, which lead to slightly different results.

Example 2 [Ramík, Rímánek 1985]

Consider the following linear programming problem with fuzzy constraints.

$$
\begin{aligned}
&\text{Maximize} && z = 5x_1 + 4x_2 \\
&\text{subject to} && (4, 4, 2, 1)_{L\text{-}L} x_1 \oplus (5, 5, 3, 1)_{L\text{-}L}\, x_2 \preccurlyeq (24, 24, 5, 8)_{L\text{-}L}, \\
& && (4, 4, 1, 2)_{L\text{-}L} x_1 \oplus (1, 1, 0.5, 1)_{L\text{-}L}\, x_2 \preccurlyeq (12, 12, 6, 3)_{L\text{-}L}, \\
& && x_1, x_2 \geq 0,
\end{aligned}
\tag{32}
$$

with the function L: $[0, +\infty\, [\to [o, 1]$ being defined by the formula $L(u) = \max \{0, 1 - u\}$ for $u \geq 0$.

As $\varepsilon_L = 0$, $\delta_L = 1$, applying formulae (31) the system (33) is equivalent to the system of ordinary inequalities

$$
\begin{aligned}
&4x_1 + 5x_2 \leq 24, \\
&4x_1 + x_2 \leq 12, \\
&2x_1 + 2x_2 \leq 19, \\
&3x_1 + 0.5x_2 \leq 6, \\
&5x_1 + 6x_2 \leq 32, \\
&6x_1 + 2x_2 \leq 15, \\
&x_1, x_2 \geq 0
\end{aligned}
\tag{33}
$$

In this way, the problem has been transformed to a classical linear programming problem with the optimal solution and the corresponding value of the ojective function being

$$x_1 = 1.5, \;\; x_2 = 3, \;\; z = 19.5 \tag{34}$$

With respect to section 2 complementary approaches to the one described above are Tanaka et al. [1984, 1985]. There similar assumptions concerning the fuzzy sets are made, but the objective function(s) and the nonnegativity constraints are also fuzzified. Also Rommelfanger [1989] goes into this direction.

More general treatments of this problem can be found in Delgado et al. [1989], Dubois [1987], Orlovski [1989] and others.

4. APPLICATIONS

Fuzzy Mathematical Programming has been applied to other areas of theoretical investigations as well as to practical applications.

4A METHODOLOGICAL APPLICATIONS

Due to the "symmetry" of the majority of the models in FMP the number of objective functions does not matter. In classical mathematical programming, however, normally only one objective function, which generates the order over the solution space, could be accepted. If there are more than one objective function, multi objective decision making models or "vectorial optimization" models have to be applied, which normally require a much higher computational effort. It is, therefore, quite natural that FMP has been applied extensively to the area of multi criteria analysis.

If the objective function can be calibrated by given aspiration levels either model (4) can be used directly or Tanaka et al. [1985] can be used for fuzzy parameters. If the objective functions cannot be calibrated naturally, model (19) can be used. This is even possible if the constraints are crisp rather than fuzzy.[Zimmermann 1978]

Modern systems for multi objective decision making are generally interactive. FMP has also been applied to these decision making tools [Sakawa et al. 1990; Yano et al. 1989].

An interesting application of FMP is Campos' contribution [1989] in which zero-sum matrix games with imprecise payoffs are considered.

Somewhere between methodological applications and real applications (which have been installed and used in practice) are those appplications of FMP in which solutions to functional problems have been suggested but not (yet) really been implemented. Examples for this type of application are described in Wiedey, Zimmermann [1978], (Media Selection), Ernst [1982] (Logistics), Holtz, Desonki [1981] (Maintenance), Hintz, Zimmermann [1989] (Production Planning), Nickels [1990] (Cutting Stock Problem), etc..

4B PRACTICAL APPLICATIONS

Real applications of fuzzy mathematical programming are still pretty rare. This is certainly not due to weaknesses of FLP. Experience shows that people who have been using linear programming for quite a while have become so used to "cutting problems" to LP-models that they do not see the need to allow for uncertainty. The acceptance of FLP seems to be higher amongst people who have never used LP before and who are looking for tools to solve their problems properly. Another reason for not finding interesting applications in the literature is, of course, that good applications are not published for competitive reasons and failures are not published for other obvious reasons. FLP has been applied to blending problems with sensory constraints (such as the blending of chocolate stretch, champagne cuvée, paints etc.). The paint application was published [Zimmermann et al. 1986]. Another application was in logistics by Ernst [1982], which we will sketch in the following. He suggests a fuzzy model for the determination of time schedules for containerships, which can be solved by branch and bound, and a model for the scheduling of containers on containerships, which results eventually in an LP. We shall only consider the last model (a real project).

The model contained in a realistic setting approximately 2,000 constraints and originally 21,000 variables, which could then be reduced to approximately 500 variables. Thus it could be handled adequately on a modern computer. It is obvious, however, that a description of this model in a textbook would not be possible. We shall, therefore sketch the contents of the modeling verbally and then concentrate on the aspects that included fuzziness.

The system is the core of a decision support system for the purpose of scheduling properly the inventory, movement, and availability of containers, especially empty containers, in and between 15 harbors. The containers were shipped according to known time schedules on approximately 10 big containerships worldwide on 40 routes. The demand for container space in the harbors was to a high extent stochastic. Thus the demand for empty containers in different harbors could either be satisfied by large inventories of empty containers in all harbors, causing high inventory costs, or they could be shipped from their locations to the locations where they were needed, causing high shipping costs and time delays.

Thus the system tries to control optimally primarily the movements and inventories of empty containers, the capacities of the ships, and the predetermined time schedule of the ships.

This problem was formulated as a large LP model. The objective function maximized profit (from shipping full containers) minus cost for moving empty containers minus inventory cost of empty containers. When comparing data of past periods with the model it turned out, that very often ships transported more containers than their specific maximum capacity. This, after further investigations, lead to a fuzzification of the ship's capacity constraints, which will be described in the next model.

[Ernst 1982, p. 90]

Let

$z = c^T x$	the net profit to be maximized
$Bx \leq b$	the set of crisp constraints
$Ax \lesssim d$	the set of capacity constraints for which a crisp formulation turned out to be inappropriate

Then the problem to be solved is:

$$\begin{aligned} \text{maximize} \quad & z = c^T x \\ \text{such that} \quad & Ax \lesssim d \\ & Bx \leq b \\ & x \geq 0 \end{aligned} \tag{35}$$

This corresponds to (14). Rather than using (18) to arrive at a crisp equivalent LP model the following approach was used: Basing on (11) and (12) the following membership functions were defined for those constraints that were fuzzy:

$$\mu_i(t_i) = \frac{t_i}{p_i - d_i} \quad 0 \leq t_i \leq p_i - d_i,\ i \in I,$$

I = Index set of fuzzy constraints.

As the equivalent crisp model to (35) the following LP was used:

$$\begin{aligned} \text{maximize} \quad & z' = c^T - \sum_{i \in I} s_i(p_i - b_i)\mu_i(t_i) \\ \text{such that} \quad & Ax \leq d + t \\ & Bx \leq b \\ & t \leq p - b \\ & x, t \geq 0 \end{aligned} \tag{36}$$

where the s_i are problem-dependent scaling factors with penalty character.

Formulation (36) only makes sense if problem-dependent penalty terms s_i, which also have the required scaling property, can be found and justified.

In this case the following definitions performed successfully: First the crisp constraints $Bx \leq b$ were replaced by $Bx \leq .9b$, providing a 10% leeway of capacity, which was desirable for reasons of safety. Then "tolerance" variables t were introduced:

$$Bx - t \leq .9b$$
$$t \leq .1b$$

The objective function became

$$\text{maximize } z = c^t x - s^t t$$

s was defined to be

$$s = \frac{\text{average profit of shipping a full container}}{\text{average number of time periods which elapsed between departure and arrival of a container}}$$

By the use of this definition more than 90% of the capacity of the ships was used only if and when very profitable full containers were available for shipping at the ports, a policy that seemed to be very desirable to the decision makers.

5. CONCLUSIONS

Mathematical programming is one of the areas to which fuzzy set theory has been applied extensively. Even if one considers the area of linear programming only, numerous new models - linear and nonlinear - have emerged through the application of fuzzy set theory. A good part of the models are of primarily theoretical interest. Still even from an application point of view, fuzzy mathematical programming is a valuable extension of traditional crisp optimization models. It is surprising that some areas, such as duality theory, have not yet drawn more interest. There further developments can still be expected.

REFERENCES

Behringer, F.A. [1977]. Lexicographic Quasiconcave Multiobjective Programming. In: Z. Operations Research 21, pp. 103-116.

Behringer, F.A. [1981]. A Simplex Based Agorithm for the Lexicographically Extended Linear Maxmin Problem. In: European Journal of Operational Research 7, pp. 274-283.

Bellman, R.E.; Zadeh, L.A. [1970]. Decision-making in a Fuzzy Environment. In: Management Science 17, pp. B141-164.

Buckley, J.J. [1989]. Solving Possibilistic Linear Programming Problems. In: Fuzzy Sets and Systems 31, pp. 329-341.

Buckley, J.J. [1990]. Stochastic Versus Possibilistic Programming. In: Fuzzy Sets and Systems 34, pp. 173-177.

Buckley, J.J. [1990]. Multiobjective Possibilistic Linear Programming. In: Fuzzy Sets and Systems 35, pp. 23-28.

Campos, L. [1989]. Fuzzy Linear Programming Models to Solve Fuzzy Matrix Games. In: Fuzzy Sets and Systems 32, pp. 275-289.

Chanas, S. [1983]. The Use of Parametric Programming in Fuzzy Linear Programming. In: Fuzzy Sets and Systems, pp. 243-251.

Chanas, S. [1989]. Fuzzy Programming in Multiobjective Linear Programming - A Parametric Approach. In: Fuzzy Sets and Systems 29, pp. 303-313.

Chang, L.L. [1975]. Interpretation and Execution of Fuzzy Programs. In: Zadeh et al. (eds.) 1975, pp.191-218.

Delgado, M.; Verdegay, J.L.; Vila, M.A. [1989]. A General Model for Fuzzy Linear Programming. In: Fuzzy Sets and Systems 29, pp. 21-29.

Dubois, D. [1984]. Linear Programming with Fuzzy Data. In: Bezdek, J.C.(ed.), analysis of Fuzzy Information, Vol. III - Applications in Engineering and Science, Boca Raton 1987.

Ernst, E. [1982]. Fahrplanerstellung und Umlaufdisposition im Containerschiffsverkehr (Diss. RWTH Aachen) Frankfurt/M., Bern.

Fabian, L.; Stoica, M. [1984]. Fuzzy Integer Programming. In: Zimmermann et al. (Eds.) 1984, pp. 123-132.

Fullér, R. [1989]. On Stability in Fuzzy Linear Programming Problems. In: Fuzzy Sets and Systems 30, pp. 339-344.

Hamacher, H.; Leberling, H.; Zimmermann, H.-J. [1978] Sensitivity Analysis in Fuzzy Linear Programming. In: Fuzzy Sets and Systems 1, pp. 269-281.

Hannan, E.L. [1981]. Linear Programming with Multiple Fuzzy Goals. In: Fuzzy Sets and Systems 6, pp. 235-248.

Hintz, G.-W., Zimmermann, H.-J.[1989]. A Method to Control Flexible Manufacturing Systems. In: European Journal of Operations Research 41, pp. 321-334.

Holtz, M.; Desonski, Dr. [1981]. Fuzzy-Model für Instandhaltung. In: Unscharfe Modellbildung und Steuerung IV, Karl-Marx- Stadt, pp.54-62.

Inuiguchi, M.; Ichihashi, H.; Kume, Y. [1990]. A Solution Algorithm for Fuzzy Linear Programming with Piecewise Linear Membership Functions. In: Fuzzy Sets and Systems 34, pp. 15-31.

Leberling, H. [1981]. On Finding Compromise Solutions in Multicriteria Problems Using the Fuzzy Min-Operator. In: Fuzzy Sets and Systems 6, pp. 105-118.

Luhandjula, M.K. [1984]. Fuzzy Approaches for Multiple Objective Linear Fractional Optimization. In: Fuzzy Sets and Systems 13, pp.11-24.

Nickels, W. [1990]. Ein wissensbasiertes System zur Produktionsplanung und -steuerung in der Papierindustrie (Diss. RWTH Aachen), VDI Fortschritt-Berichte Düsseldorf.

Orlovski, S.A. [1985]. Mathematical Programming Problems with Fuzzy Parameters. In: Kacprzyk, J. and Yager, R.R.(eds.), Management Decision Support Systems using Fuzzy Sets and Possibility Theory, Köln 1985, pp. 136-145.

Orlovski, S.A. [1977]. On Programming with Fuzzy Constraint Sets. In: Kybernetes 6, pp. 197-201.

Ostasiewicz, W. [1982]. A New Approach to Fuzzy Programming. In: Fuzzy Sets and Systems 7, pp. 139-152.

Ramík, J.; Rímánek, J. [1985]. Inequality Relation between Fuzzy Numbers and its Use in Fuzzy Optimization. In: Fuzzy Sets and Systems 16, pp. 123-138.

Rödder, W.; Zimmermann, H.-J. [1980]. Duality in Fuzzy Linear Programming. In: Fiacco, A.V.; Kortanek, K.O. (Eds.), Extremal Methods and Systems Analyses, New York, pp. 415-429.

Rommelfanger, H.; Hanuscheck, R.; Wolf, J. [1989]. Linear Programming with Fuzzy Objectives. In: Fuzzy Sets and Systems 29, pp. 31-48.

Rubin, P.A.; Narasimhan, R. [1984]. In: Fuzzy Sets and Systems 14, pp. 115-130.

Sakawa, M.; Yano, H. [1989]. Interactive Decision Making for Multiobjective Nonlinear Programming Problems with Fuzzy Parameters. In: Fuzzy Sets and Systems 29, pp. 315-326.

Sakawa, M. Yano, H. [1989]. An Interactive Fuzzy Satisficing Method for Multiobjective Nonlinear Programming Problems with Fuzzy Parameters. In: Fuzzy Sets and Systems 30, pp. 221-238.

Sakawa, M.; Yano, H. [1990]. An Interactive Fuzzy Satisficing Method for Generalized Multiobjective Linear Programming Problems with Fuzzy Parameters. In: Fuzzy Sets and Systems 35, pp. 125-142.

Tanaka, H.; Asai, K. [1984]. Fuzzy Linear Programming Problems with Fuzzy Numbers. In: Fuzzy Sets and Systems 13, pp. 1-10.

Tanaka, H.; Ichihashi, H.;Asai, K. [1985]. Fuzzy Decision in Linear Programming Problems with Trapezoid Fuzzy Parameters. In: Kacprzyk, J. and Yager, R.R.(eds.), Management Decision Support Systems using Fuzzy Sets and Possibility Theory, Köln 1985, pp. 146-154.

Tanaka, H.; Mizumoto, M. [1975]. Fuzzy Programs and their Execution. In: Zadeh et al. (Eds.), pp. 41-76.

Tanaka, H.; Okuda, T.; Asai, K. [1974]. On Fuzzy Mathematical Programming. In: Journal of Cybernetics 3, pp. 37-46.

Thole, U.; Zimmermann, H.-J.; Zysno, P. [1979]. On the Suitability of Minimum and Product Operators for the Intersection of Fuzzy Sets. In: Fuzzy Sets and Systems 2, pp. 167-180.

Verdegay, J.L. [1984]. A Dual Approach to Solve the Fuzzy Linear Programming Problem. In: Fuzzy Sets and Systems 14, pp.131-141.

Werners, B. [1984]. Interaktive Entscheidungsunterstützung durch ein flexibles mathematisches Programmierungssystem (Diss. RWTH Aachen), München.

Werners, B. [1987]. Interactive Multiple Objective Programming Subject to flexible Constraints. In: European Journal of Operational Research 31, pp. 324-349.

Werners, B. [1988]. Aggregation Models in Mathematical Programming. In: Mitra, G. (ed.), Mathematical Models for Decision Support, Berlin, Heidelberg, New York, pp. 295-305.

Wiedey, G.; Zimmermann, H.-J. [1978]. Media Selection and Fuzzy Linear Programming. In: Journal of the Operational Society 29, pp. 1071-1084.

Yano, H.; Sakawa, M. [1989]. Interactive Fuzzy Decision Making for Generalized Multiobjective Linear Fractional Programming Problems with Fuzzy Parameters. In: Fuzzy Sets and Systems 32, pp. 245-261.

Zadeh, L.A.; Fu, K.S.; Tanaka, K.; Simura, M. (Eds.). [1975]. Fuzzy Sets and their Applications to Cognitive and Decision Processes, New York.

Zimmermann, H.-J. [1976]. Description and Optimization of Fuzzy Systems. In: International Journal of General Systems 2, pp. 209-215.

Zimmermann, H.-J. [1978], Fuzzy Programming and Linear Programming with Several Objective Functions. In: Fuzzy Sets and Systems 1, pp. 45-55.

Zimmermann, H.-J. [1990]. Fuzzy Set Theory - and its Applications (Rev. Ed.), Boston, Dordrecht, Lancaster.

Zimmermann, H.-J. [1986]. Fuzzy Set Theory and Mathematical Programming. In: Jones, A. et al. (eds.), Fuzzy Sets Theory and Applications, pp.99-114.

Zimmermann, H.-J. [1987]. Fuzzy Sets, Decision Making and Expert Systems, Boston, Dordrecht, Lancaster.

Zimmermann, H.-J. (with Hermanns, Kaffenberger, Rödder, Selter, Stilianakis)[1986]. Lack- und Farbmischungen zu minimalen Kosten. In: Farbe & Lack 92, pp.379-382.

Zimmermann, H.-J.; Pollatschek, M.A. [1984]. Fuzzy 0-1 Programs. In: Zimmermann, Zadeh Gaines (eds.). Fuzzy Sets and Decision Analysis, Amsterdam, New York 1984. pp. 133-146.

Zimmermann, H.-J.; Zadeh, L.A.; Gaines, B.R. (Eds.) [1984]. Fuzzy Sets and Decision Analysis, New York.

Zimmermann, H.-J.; Zysno, P. [1980]. Latent Connectives in Human Decision Making. In: Fuzzy Sets and Systems, pp. 37-51.

Zimmermann, H.-J.; Zysno, P. [1983]. Decisions and Evaluations by Hierarchical Aggregation of Information. In: Fuzzy Sets and Systems 10, pp. 243-266.

6

FUZZY SET METHODS IN COMPUTER VISION

James M. Keller and Raghu Krishnapuram
Electrical and Computer Engineering
University of Missouri-Columbia
Columbia, Missouri 65211

INTRODUCTION

Computer vision is the study of theories and algorithms involving the sensing and transmission of images; preprocessing of digital images for noise removal, smoothing, or sharpening of contrast; segmentation of images to isolate objects and regions; description and recognition of the segmented regions; and finally interpretation of the scene. We normally think of images in the visible spectrum, either monochrome or color, but in fact, images can be produced by a wide range of sensing modalities including X-rays, neutrons, ultrasound, pressure sensing, laser range finding, infrared, and ultraviolet, to name a few.

Uncertainty abounds in every phase of computer vision. Some of the sources of this uncertainty include: additive and non-additive noise of various sorts and distributions in the sensing and transmission processes, questions which are often ill-posed, vagueness in class definitions, imprecisions in computations, ambiguity of representations, and general problems in the interpretation of complex scenes. The use of multiple modalities is receiving increased attention as a means of overcoming some of the limitations imposed by a single image, but the use of more than one source of information has caused new uncertainties to surface: how should the complementary and supplementary information be combined?, how should redundant information be treated?, how should conflicts be resolved?, etc.

Traditionally, probability theory was the primary mathematical model used to deal with uncertainty problems in computer vision. More recently, both Dempster-Shafer belief theory and fuzzy set theory have gained popularity in modeling and propagating uncertainty in imaging applications. While both probability theory and belief theory are important frameworks for this field, the purpose of this paper is to explore the use of fuzzy set theory in computer vision. We will consider contributions of fuzzy sets to the image model, preprocessing, segmentation, object/region recognition, and reasoning aspects of the computer vision problem. Most of the examples given in this paper are those of the authors and we apologize *a priori* to those researchers whose work we will undoubtedly (though inadvertently) omit from the references.

LOW LEVEL IMAGE PROCESSING

An image is a function $f: R^n \rightarrow R^m$ where normally n is 2 or 3 and m is 1 (intensity) or 3 (color). However, images can be constructed over numerous modalities, as well as over time, and so, the dimension of the range space can be quite large. A digital image is an image which has been discretized in both the domain and range spaces. This is commonly referred to as sampling and quantization respectively. In this paper we will restrict ourselves to two spatial coordinates, where each element $P=(x,y)$ in the domain of the image is called a pixel. If $m = 1$, then the value $f(x,y)$ is called the gray level of pixel (x,y); if $m>1$, then $f(x,y)$ is referred to a feature vector.

The first connection of fuzzy set theory to computer vision was made by Prewitt [1] who suggested that the results of image segmentation should be fuzzy subsets rather than crisp subsets of the image plane. In order to apply the rich assortment of fuzzy set theoretic operators to an image, the gray levels (or feature values) must be converted to membership values. Let X denote the domain of the digital image. Then a fuzzy subset of X is a mapping $\mu_f: X \rightarrow [0,1]$, where the value of $\mu_f(x,y)$ is dependant upon the original feature vector $f(x,y)$. The calculation of membership functions is central to the application of fuzzy set theory, just as the calculation of conditional probability density functions or basic probability assignments are crucial in the use of probabilistic or Dempster-Shafer belief models.

There are many methods of transforming pixel feature vectors into membership functions. In the case of gray scale images several authors have used S-functions when there are only two regions (object and background) and combinations of S-functions and π-functions for multiple regions (or suitable generalizations) [2-6]. These functions are defined by [7]

$$S(z; a, b, c) = \begin{cases} 0 & z \le a \\ 2\left(\frac{z-a}{c-a}\right)^2 & a < z \le b \\ 1-2\left(\frac{z-c}{c-a}\right)^2 & b < z \le c \\ 1 & z > c \end{cases}$$

$$\text{with} \quad b = \frac{a+c}{2};$$

and

$$\Pi(z; b, c) = \begin{cases} S\left(z; c-b, c-\frac{b}{2}, c\right) & z \le c \\ 1-S\left(z; c, c+\frac{b}{2}, c+b\right) & z > c, \end{cases}$$

where z is the gray level at pixel P. These functions are symmetric, but can be easily made nonsymmetric by relaxing the requirement that b be the midpoint of a and c. Intuitively, these functions correspond to the statements "z is bright" and "z is approximately c", respectively. Pal and King [2-4] used S and π functions, along with approximations of them, as the basic building blocks for both contrast enhancement and smoothing. Following Nakagawa and Rosenfeld [8], they applied min and max operations on membership values in the neighborhood of each pixel to produce smoothing or edge detection. Other approaches to edge detection using fuzzy set methods can be found in [9,10].

One problem with this approach is that the parameters which define the membership functions must be supplied, primarily in an interactive fashion by the user. Pal and Rosenfeld [11], in a two class segmentation problem, automated this process by using several choices and picking the one which optimized a certain geometric criterion which we will describe later. Recently, we have used normalized histograms of the feature values generated from training data to estimate the particular membership functions [12-14]. This has the advantages that it does not force any particular shape to the resultant distributions, can be extended to deal with multiple features instead of gray level alone, and can easily accommodate the addition of new classes.

Probably the most popular method of assigning multi-class membership values to pixels, for either segmentation or other processing, is to use the fuzzy c-means (FCM) algorithm [15,16]. Let R be the set of real numbers and R^d be the d-dimensional vector space over the reals. Let X be a finite subset of R^d, $X = \{x_1, x_2, ..., x_n\}$. In our case, each x_i is a feature vector for a pixel in the image. For an integer c, $2 \leq c \leq n$, a $c \times n$ matrix $U = [u_{ik}]$ is called a fuzzy c-partition of X whenever the entries of U satisfy three constraints:

$$\sum_{i=1}^{c} u_{ik} = 1 \quad \text{for all } k$$

$$\sum_{k=1}^{n} u_{ik} > 0 \quad \text{for all } i$$

$$u_{ik} \in [0, 1] \quad \text{for all } i, k.$$

Column j of the $c \times n$ matrix U represents membership values of x_j in the c fuzzy subsets of X. Row i of U exhibits values of a membership function u_i on X whereby $u_{ik} = u_i(x_k)$ denotes the grade of membership of x_k in the ith fuzzy subset of X.

The FCM algorithm attempts to cluster feature vectors by searching for local minima of the following objective function:

$$J_m(U, V) = \sum_{k=1}^{n} \sum_{i=1}^{c} u_{ik}^{m} \|x_k - v_i\|_A^2, \quad 1 \le m < \infty$$

where U is a fuzzy c-partition of X, $\|*\|_A$ is any inner product norm, $V = \{v_1, v_2, ..., v_c\}$ is a set of cluster centers, $v_i \in R^d$, and $m \in (1,\infty)$ is the membership weighting exponent.

Cluster center v_i is regarded as a prototypical member of class i, and the norm measures the similarity (or dissimilarity) between the feature vectors and cluster centers. When $m = 1$, J_m is the classical total within-group sum-of-squared error function; the u_i define hard clusters in X and the v_i are the centroids of the hard u_i. It is shown in [16] that for $m > 1$ under the assumption that $x_k \neq v_i$ for all i, k, (U, V) may be a local minimum of J_m only if

$$u_{ik} = \left(\sum_{j=1}^{c} \left(\frac{\|x_k - v_i\|_A}{\|x_k - v_j\|_A} \right)^{2/(m-1)} \right)^{-1} \tag{1}$$

for all i, k, and

$$v_i = \frac{\sum_{k=1}^{n} u_{ik}^{m} x_k}{\sum_{k=1}^{n} u_{ik}^{m}} \tag{2}$$

for all i.

The algorithm defined by looping iteratively through the above conditions is known to generate sequences (or subsequences) that terminate at fixed points of J_m. The FCM algorithm is comprised of the following steps

```
BEGIN
        Set c, 2 ≤ c < n
        Set ε, ε ≥ 0
        Set m, 1 ≤ m < ∞
        Initialize U^0
        Initialize j = 0
        DO UNTIL ( ||U^j - U^(j-1)|| < ε )
                Increment j
                Calculate {v_i^j} using (2) and U^(j-1)
                Compute U^j using (1) and {v_i^j}
        END DO UNTIL
END
```

The inner product norm $\| * \|_A$, or its replacement by more general distance metrics $d^2(x_k, v_i)$ (as will be used later) controls the final shape of the clusters generated by the FCM: hyperspherical, hyperellipsoidal, linear subspace, etc.

In terms of generating membership functions for later processing, the fuzzy c-means has several advantages. It is unsupervised, that is, it requires no initial set of training data; it can be used with any number of features and any number of classes; and it distributes the membership values in a normalized fashion across the various classes based on "natural" groupings in feature space. However, being unsupervised, it is not possible to predict ahead of time what type of clusters will emerge from the fuzzy c-means from a perceptual standpoint. Also, the number of classes must be specified for the algorithm to run, although as will be seen in the next section, there are modifications which avoid this problem. Finally, iteratively clustering features for a 512 × 512 resolution image can be quite time consuming. In [17, 18], approximations and simplifications were introduced to ease this computational burden.

SEGMENTATION

Image segmentation is one of the most critical components of the computer vision process. Errors made in this stage will impact all higher level activities. Therefore, methods which incorporate the uncertainty of object and region definition and the faithfulness of the features to represent various objects and regions are desirable.

The process of segmentation has been defined by Horowitz and Pavlidis [19] as follows: Given a definition of uniformity, a segmentation is a partition of the picture into connected subsets, each of which is uniform, but such that no union of adjacent subsets is uniform.

This definition is based on crisp set theory. The fuzzy c-partition introduced in the previous section can be defined as a fuzzy segmentation. Furthermore, if we define a uniformity predicate $P(\mu_{ij})$ such that assigns the value true or false to the sample point x_j based on its membership value (for example, $P(\mu_{ij}) = 1$ if $\mu_{ij} \geq \mu_{kj}$ for all k), we will have paralleled crisp segmentation. The fuzzy c-means has been successfully used as a segmentation approach by several researchers [17, 18, 20-22]. (We will see an example of this segmentation shortly).

All of the methods for converting image feature values into class membership numbers contain adjustable parameters: the cross-over point b for S and π functions, the fuzzifier m in the c-means, etc. Varying these parameters affects the final fuzzy partition, and hence the ultimate crisp segmentation of the scene. Also, the number of classes desired impacts the resultant distributions, since the memberships are required to sum to one for a fuzzy c-partition. In some cases, these problems are not serious. For example, many segmentation problems involve separating an object from its background. Here the number of classes is obviously two. However, in general situations, the choice of these parameters must be carefully considered.

The basic approach which is taken to pick the number of classes and/or the function shaping parameters iteratively varies these parameters and picks the set of values which optimizes some measure of the final fuzzy partition. The optimization criteria can be based on the geometry of the fuzzy subsets of the image or on properties of the clusters in feature space.

In a series of papers [23-25], Rosenfeld studied the geometry and topology of fuzzy subsets of the digital (i.e., image) plane. These properties were later generalized by Dubois and Jaulent [26, 27]. Many of the basic geometric properties of, and relationships among, regions can be generalized to fuzzy subsets. Rosenfeld has extended the theory of these fuzzy subsets to include the topological concepts of connectedness, adjacency and surroundedness, extent and diameter, and convexity. Rosenfeld et al. have also developed geometrical operations on fuzzy image subsets, including shrinking and expanding, and thinning. [28, 29]

Of the above-mentioned geometrical properties, we discuss here only the connectedness, area, perimeter, and compactness of a fuzzy image subset, characterized by a membership function array $\mu_f(x_{ij})$. In defining the above mentioned parameters we replace $\mu_f(x_{ij})$ by μ for simplicity.

A neighbor can be defined in several ways. In two-dimensional images, point $P = (x,y)$ of a digital image has two horizontal and two vertical neighbors, namely the points: $(x\text{-}1, y)$, $(x, y\text{-}1)$, $(x+1, y)$, and $(x, y+1)$. The four points are called the 4-connected neighbors of P, and we say that they are 4-adjacent to P. Similarly, P has four additional neighbors: $(x\text{-}1, y\text{-}1)$, $(x\text{-}1, y+1)$, $(x+1, y\text{-}1)$, and $(x+1, y+1)$. We call these eight points the 8-connected neighbors of P (8-adjacent to P) [30].

The definition of connectedness for the crisp case as defined by Rosenfeld [30] is as follows: Let P, Q be two points of an image. A path ρ of length n from P to Q in an image is a sequence of points $P = P_1, P_2, ..., P_n = Q$ such that P_i is a neighbor of P_{i-1}, $1<i<n$. There are two versions of path ρ (a 4-path or an 8-path) depending on whether "neighbor" means "4-neighbor" or "8-neighbor". If P and Q are points of an image subset S, we say that P is 4-(8-) connected to Q in S if there exists a 4- (or 8-) path from P to Q. For any P in S, the set of points which are connected to P in S is called a connected component of S.

For the fuzzy case, let μ be a mapping from X into [0,1], that is, let μ be a fuzzy subset of X. Let $P, Q \in X$. Then the degree of connectedness of P and Q with respect to μ is

$$C_\mu(P,Q) = \max_{\rho_{PQ}} \left(\min_{R \in \rho_{PQ}} \mu(R) \right)$$

where the operator max is taken over all paths ρ_{PQ} (either 4- or 8-path connected) from P to Q, and the operator min is taken over all points R on the path. P and Q are said to be connected in μ if

$$C_{\mu}(P,Q) \geq \min(\mu(P), \mu(Q)).$$

The fuzzy set μ is said to be connected if every pair of points P,Q is connected in μ.

The area of μ is defined as

$$a(\mu) = \int \mu$$

where the integral is taken over the whole image set, or for the digital case,

$$a(\mu) = \sum_{m}^{M} \sum_{n}^{N} \mu_{mn}.$$

Let us call a fuzzy subset μ of S "piecewise constant" if there exists a segmentation $\Sigma = \{S_1, ..., S_n\}$ of S such that μ has a constant value μ_i on each S_i and $\mu = 0$ on S_n (i.e., $\mu_n = 0$). Here, S_n is considered the boundary of the image. If μ is piecewise constant (for example, in a digital image) $a(\mu)$ is the weighted sum of the areas of the regions on which μ has constant values, where the areas of the regions are weighted by these values.

For the piecewise constant case, the perimeter of μ is defined as

$$p(\mu) = \sum_{ij} \sum_{k} |\mu_i - \mu_j| \, |A_{ijk}|.$$

This is just the weighted sum of the lengths of the arcs A_{ijk} along which the i-th and j-th regions having constant μ values μ_i and μ_j respectively meet, weighted by the absolute difference of these values.

Considering the 4-adjacent definition for connectedness, the above equation for $p(\mu)$ reduces to:

$$p(\mu) = \sum_{m=1}^{M} \sum_{n=1}^{N-1} |\mu_{mn} - \mu_{m,n+1}| + \sum_{n=1}^{N} \sum_{m=1}^{M-1} |\mu_{mn} - \mu_{m+1,n}|.$$

The compactness of μ is then defined as

$$\text{comp}(\mu) = \frac{a(\mu)}{p^2(\mu)}.$$

For crisp sets, the compactness is largest for a disk, where it is equal to $1/4\pi$. For a fuzzy disk where μ depends only on the distance from the origin (center), it can be shown that

$$\frac{a(\mu)}{p^2(\mu)} \geq \frac{1}{4\pi}.$$

In other words, of all possible fuzzy disks, the compactness is smallest for the crisp version.

Therefore, one approach to finding an optimal partition is to generate many candidate partitions by varying the membership generation parameters and then choosing the partition which minimizes the fuzzy compactness of the result. Pal and Rosenfeld used this technique in a two class problem (object and background) to find the best choice of S function shaping parameters [11]. In [31] Liao extended this approach to the case of several features and several classes by using the fuzzy c-means. Here the fuzzifier m was the variable. Figure 1a shows a 256×256 forward looking infrared image of a natural scene containing trees, grass areas and two vehicles. Because of the noisy nature of infrared images, this picture was smoothed using local averaging (Figure 1b). The number of classes was fixed at 4 and the fuzzifier m varied from 1.2 to 5.0. For each choice of m, the sum of the compactness values of the resultant four fuzzy subsets was computed and the value of m (m=3.0) giving minimal overall compactness was chosen. The result of the closest crisp partition segmentation is shown in Figure 1c. Note that there are still many small noise components in the segmentation. By smoothing the membership matrix, giving higher weight to the vehicle class, and performing a noise cleaning operation (shrink-and-expand) [30], the excellent segmentation shown in Figure 1d was obtained. The important point is that the initial segmentation formed a fuzzy c-partition of the image, and so, post-processing on the fuzzy subsets of the image was possible.

The *a priori* setting of the number of classes is not always possible, especially in segmentation of natural scenes. In such cases an algorithm called the Unsupervised Fuzzy Partition-Optimum Number of Clusters (UFP-ONC) algorithm [32] may be used. The UFP-ONC algorithm is derived from a combination of the fuzzy c-means algorithm and the fuzzy maximum likelihood estimation (FMLE). It attempts to obtain a satisfactory solution to the problem of large variability in cluster shapes and densities, and to the problem of unsupervised tracking of classification prototypes. There are no initial conditions on the location of cluster centroids, and classification prototypes are identified during a process of unsupervised learning [32]. The algorithm is essentially the same as the FCM algorithm described in the previous section, except that the distance measure defined by

$$d^2(x_k, v_i) = \frac{|F_i|^{1/2}}{P} \exp\left\{\frac{1}{2}(x_k - v_i)^T F_i^{-1} (x_k - v_i)\right\}, \tag{3}$$

is used instead of the inner product norm. In (3) F_i is the fuzzy covariance matrix of cluster i given by

$$F_i = \sum_k \frac{u_{ik}^m (x_k - v_i)(x_k - v_i)^T}{\sum_k u_{ik}^m}, \tag{4}$$

and P_i is the *a priori* probability of the i-th cluster defined by

$$P_i = \frac{\sum_k u_{ik}}{N}, \tag{5}$$

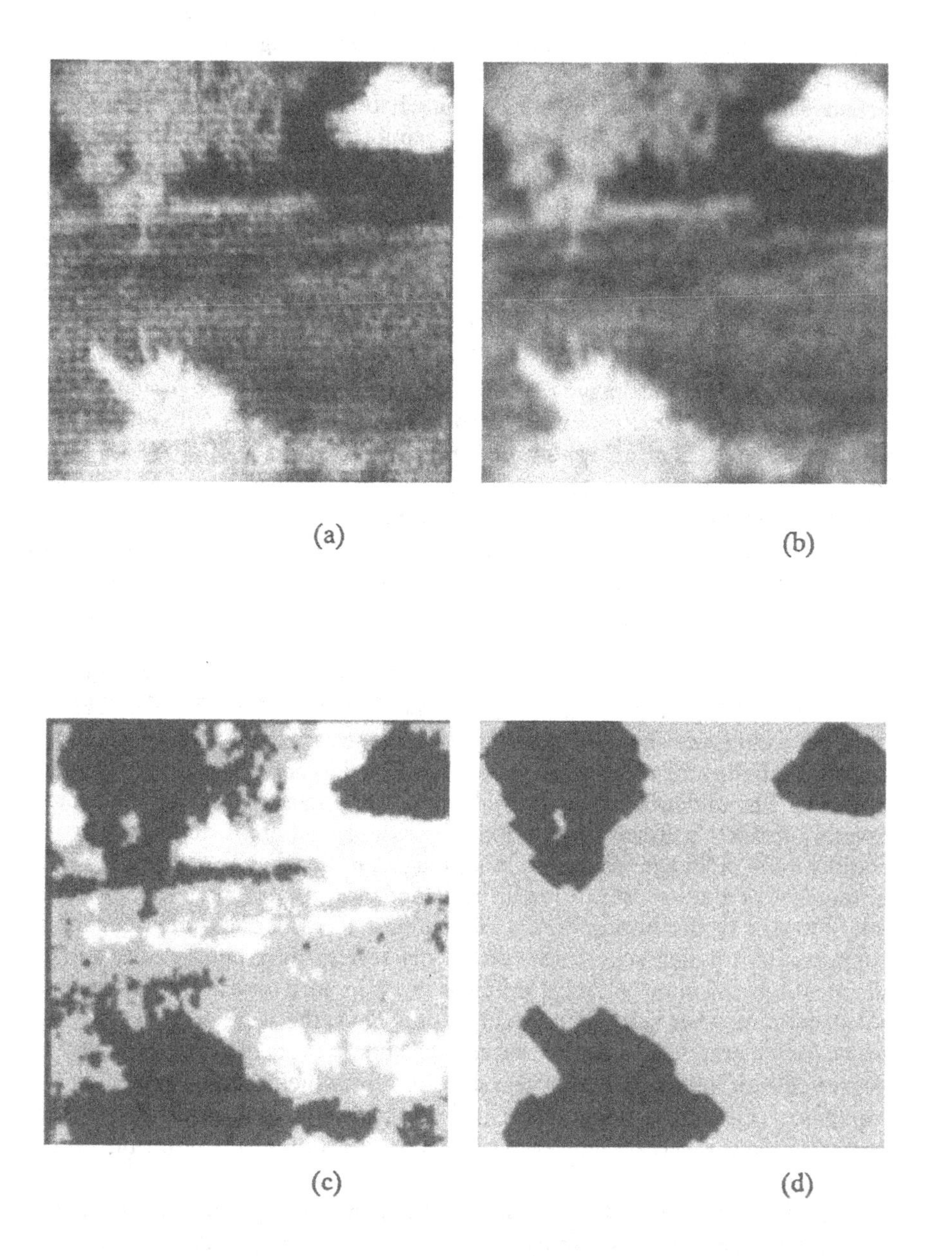

Figure 1. Segmentation using the fuzzy c-means. (a) Infrared image of scene; (b) Figure 1a smoothed by local averaging; (c) Closest crisp segmentation from fuzzy 4-means with optimal choice of m(3.0); (d) final segmentation using the fuzzy partition.

where N is the total number of feature vectors. In addition to updating the cluster prototypes using (2) and the memberships using (3), the covariance matrices F_i are also updated in every iteration. After the algorithm converges, certain performance measures are computed for the resulting fuzzy partition. This process is repeated for increasing number of clusters in the data set computing performance measures in each run, until a partition into an optimal number of subgroups is obtained.

The performance criteria in this algorithm is the minimization of the overall hypervolume of the clusters as calculated from the determinants of the fuzzy covariance matrices. Figure 2a shows an intensity image containing trees, roads, and sky regions. A set of new local features, based on fractal geometry was generated from this image [33]. Figure 2b shows the resulting segmentation when these fractal features were used with the UFP-ONC algorithm. An another example, Figure 2c shows the original range image of a block. Figure 2d shows the segmentation obtained when the mean curvature and another new differential geometric feature [34] were used as the input to the UFP-ONC algorithm. The clusters corresponding to the minimum total hypervolume in the feature space are shown in Figure 2e. As can be seen the UFP-ONC algorithm is effective in locating the ellipsoidal clusters of various sizes and orientations.

A different approach to both segmentation and object recognition is taken by Krishnapuram and Lee [35, 36] and Keller, et al. [5, 37, 38]. The two different techniques share the common idea that class labeling for segmentation or object labeling for recognition should be viewed as an aggregation of evidence problem. The evidence can be derived from several sensors (for example, color), several distinct pattern recognition algorithms, different features, or the combination of image data with non-image information (intelligence). The advantages of multi-sensor fusion lie in redundancy, complementarity, timeliness and low cost of the information. The support for a decision may depend on supports for (or degrees of satisfaction of) several different criteria, and the degree of satisfaction of each criterion may in turn depend on degrees of satisfaction of other sub-criteria, and so on. Thus, the decision process can be viewed as a hierarchical network, where each node in the network "aggregates" the degree of satisfaction of a particular criterion from the observed evidence. The inputs to each node are the degrees of satisfaction of each of the sub-criteria, and the output is the aggregated degree of satisfaction of the criterion.

For image segmentation as discussed in [35, 36], the decision making problem reduces to i) determining the structure of the aggregation network to be used, ii) determining the nature of the connectives at each node of the network, and iii) computing the input supports (degrees of satisfaction of criteria) based on observed features.

The structure of the aggregation network depends on the problem at hand [39]. The connectives used at each node of the network are based on fuzzy union, fuzzy intersection, or compensative operators (such as generalized mean

Figure 2. Fuzzy segmentation finding the optimum number of classes. (a) intensity image of natural scence and (b) range image of block; (c) & (d) optimal partition of top tow using UFP-ONC algorithm; (e) feature space clusters for the block image.

or the γ-model) [39]. The innovative aspect of this work is that a backpropagation algorithm (and convergence theory) was developed so that both the type of connective at each node, as well as the parameters associated with the connective can be learned from training data [35, 36].

As an example, consider the fusion of information from different modalities for segmentation of outdoor scenes. In particular, the modalities considered are color images of size 256 × 256 (obtained from the University of Massachusetts) with intensity components r, g and b (red, green and blue). In an initial experiment, to keep the problem tractable, the following features were used as criteria: intensity value $(r+g+b/3)$, blue-red difference b-r, excess green $2g$-r-b, and position (row number). The first three features correspond to the Ohta color space and were chosen because they have been found to correspond to meaningful colors and they have also been found to be effective for color image segmentation [40]. The position of a pixel is important for labels such as sky and road. The image was first median filtered, and the feature images were normalized so that all the values fall between 0 and 255. The following six labels were considered: sky, tree, roof, walls, grass, and road. In the example, we used one layer aggregation networks based on the generalized mean to determine the parameters of the network. About 60 training samples were taken from different parts of the image for each class. Membership values were calculated using the feature histograms of the training data. Since the histogram is very jagged, it had to be smoothed by a window of length 11 before normalizing it. After training, the network was used for segmentation of the image by assigning each pixel to the class which had the highest degree of satisfaction generated from the pixel's features.

Figure 3a shows the original intensity image and Figure 3b shows the segmented and labeled image when the γ-model was used as the aggregation operator. The labels in increasing order of grey level are: road, tree, wall, roof, grass, and sky. The results are excellent, considering the small number of features used and the simplicity of the network employed. Note that most of the misclassifications occur at areas where the true label is not any of the six labels considered. This segmented image was improved by a shrink-and-expand operator and this image is shown in Figure 3c. An important point here is that this method not only partitions the image into connected components of similar properties, but also labels these components. In other words, it produces both a segmentation and a region recognition simultaneously, while capturing an abstract model of the decision making process.

The fuzzy integral has also been used to fuse both objective information from features and (possibly subjective) information on the importance of subsets of features for segmentation in [5, 37]. This approach will be described in the section on object and region recognition.

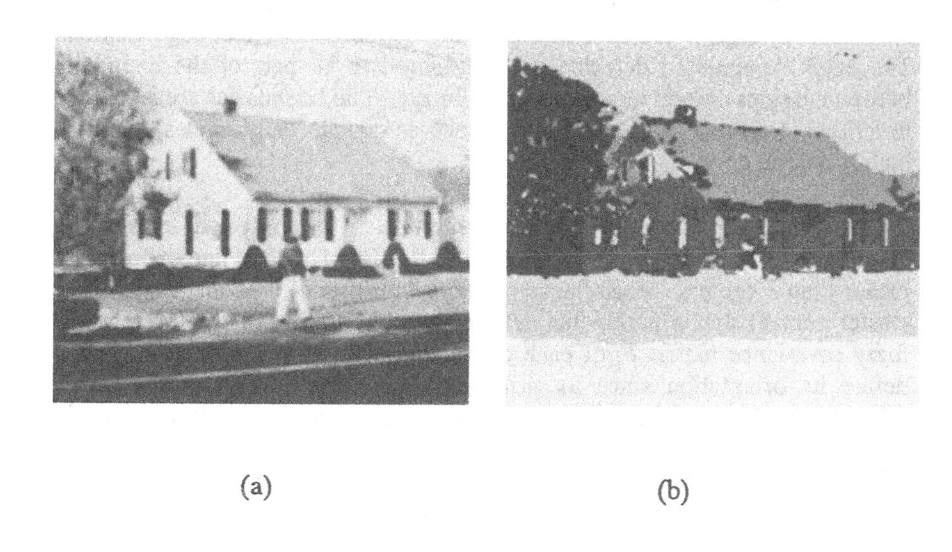

(a) (b)

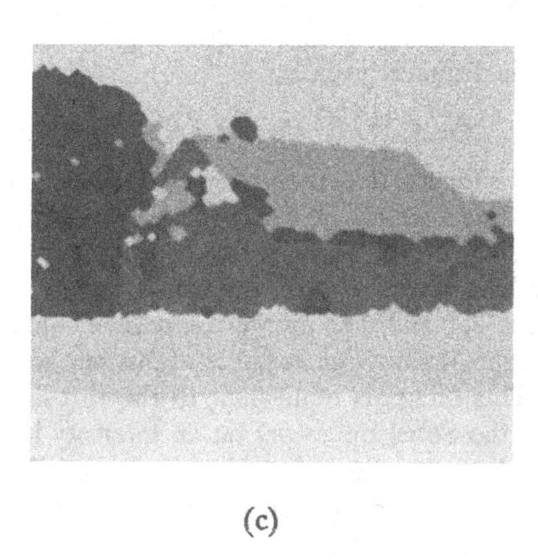

(c)

Figure 3. Multispectral segmentation by hierarchical fuzzy aggregations. (a) Intensity image of natural scene; (b) Six class segmentation and labeling; (c) Figure 3b cleaned up by shrink-and-expand operator.

BOUNDARY DETECTION

Boundary detection is another approach to segmentation. In this approach, an edge operator is first used on the image to detect edge elements. The edge elements so detected are considered to be part of the boundaries between various objects or regions in the image. The boundaries are sometimes described in terms of analytical curves such as straight lines, circles, and other higher degree curves.

The FCM algorithm can be used to detect (or fit) straight lines to edge elements. This is achieved by initializing the FCM with c linear prototypes rather than c centers. Each linear prototype consists of a point (which acts as cluster center) and a parameter defining the orientation of the cluster. The fuzzy covariance matrix F_i of each cluster (as defined in (4)) may be used to define its orientation since its principal eigenvector gives the direction of maximum variance of the cluster. The c prototypes are updated in each iteration as described in the previous section except that in each iteration the covariance matrix of each cluster is also updated. Several distance measures may be used for the detection of lines. One of them is defined by

$$d^2(x_k, v_i) = \alpha_i D_{ik}^2 + (1-\alpha_i)\,(d_{ik})^2$$

where $D_{ik}{}^2$ is the distance of the point from the line and $d_{ik}{}^2$ is the Enclidean distance between $\mathbf{x}_k$ and $\mathbf{v}_i$. α_i is chosen as $1\text{-}(\lambda_{1i}/\lambda_{2i})$, where λ_{1i} and λ_{2i} are the smaller and larger eigenvalues of cluster i [41]. We have shown that the scaled Mahalanobis distance given by

$$Z_{ik}^2 = |F_i|^{1/2}\,(x_i - v_2)^T F_i^{-1}\,(x_i - v_i) \tag{6}$$

is also very effective for the detection of lines or linear clusters. [42]. In (6) F_i is the fuzzy covariance matrix of cluster i as defined in (4). As mentioned earlier, one problem with the FCM is that the number of clusters needs to be specified. In the line detection case, one way to overcome this is to specify a relatively high value of c and then merge compatible clusters after the algorithm converges [42]. Figure 4 shows an example of this method. Figure 4a shows the original image. This image is equivalent to the threshold output of an edge operator (such as the Sobel operator) on an intensity image of the characters UMC. Figure 4b shows the clustering when c was specified to be 14. Note that the leading stroke of both the U and the M are split into two subclusters (in some examples the initial cluster organization is much worse). Figure 4c shows the clustering after compatible clusters are merged. The final optimal number of clusters was determined to be 10, which is correct in this case. In this implementation, two (or more) clusters were considered compatible if i) their orientation was the same, ii) the line joining their centers had the same orientation as the clusters and iii) the cluster centers were not more than 4 principal eigenvalues apart. The lines so found by the algorithm can then be used to describe large sections of the boundary or the linear substructures in the image.

Figure 4. Segmentation by boundary detection. (a) Simulated thresholded edge output; (b) Output of modified FCM to detect linear clusters ($c = 14$); (c) Optimal linear partitions by compatible cluster merge algorithm ($c = 10$).

The FCM algorithm with linear prototypes may be generalized to detect combinations of subspaces [43-44] and also non-linear clusters such as circles [45].

There are numerous techniques for incorporating fuzzy set theoretic operators into the segmentation process of which we have only highlighted a few. It is our belief that the benefits of producing fuzzy subsets of the image will encourage more research into the utilization of fuzzy approaches to this crucial aspect of computer vision.

OBJECT/REGION RECOGNITION AND HIGH LEVEL VISION

The area of computer vision concerned with assigning meaningful labels to regions in an image can be thought of as a subset of pattern recognition. There is a large amount of research in the use of fuzzy set theory in pattern recognition, but here we will only discuss a few approaches for object recognition in image analysis.

As was seen in the previous section, the fuzzy-connective-based hierarchical aggregation networks not only segmented an image, but also provided class labels for each pixel based on local feature evidence and training information. Normally, once segmentation has been completed, features are computed for the entire region and this data is used to classify the areas found. The aggregation networks can function well in this setting also. The reader is referred to [13] for several examples of object recognition using fuzzy aggregation networks.

The fuzzy integral is another numeric-based approach which we have used for both segmentation and object recognition [5,14,37,38]. It also uses a hierarchical network of evidence sources to arrive at a confidence value for a particular hypothesis or decision. The difference from the proceeding method is that besides this directly supplied objective evidence, the fuzzy integral utilizes information concerning the worth or importance of the sources in the decision making process.

The fuzzy integral relies on the concept of a fuzzy measure which generalizes probability measure in that it does not require additivity, replacing it with a weaker continuity condition. A particularly useful set of fuzzy measures is due to Sugeno [46]. A fuzzy measure g_λ is called a Sugeno measure if it satisfies the following additional property:

If $A \cap B = \Phi$, then $g_\lambda(A \cup B) = g_\lambda(A) + g_\lambda(B) + \lambda g_\lambda(A) g_\lambda(B)$,
for some $\lambda > -1$.

Suppose X is a finite set, $X = \{x_1, ..., x_n\}$, and let $g^i = g_\lambda(\{x_i\})$. Then the set $\{g^1, ..., g^n\}$ is called the fuzzy density function for g_λ.

Using the above definitions one can easily show that g_λ can be constructed from a fuzzy density function by

$$g_\lambda(A) = \frac{\prod_{x_i \in A}(1 + \lambda g^i) - 1}{\lambda},$$

for any subset A of X. Using the fact that $X = \bigcup_{i=1}^{n}\{x_i\}$, λ can be determined from the above equation.

Let h: $X \to [0,1]$. The fuzzy integral of h over X with respect to g_λ is defined in [46] by:

$$\int_X h(x) \circ g_\lambda = \sup_{\alpha \in [0,1]} \left[\alpha \wedge g_\lambda(F_\alpha)\right]$$

where $$F_\alpha = \{x \in X \mid h(x) \geq \alpha\}.$$

In our applications, the set X is the set of information sources (sensors, algorithms, features, etc.) and the function h supplies a confidence value for a particular hypothesis or class from the standpoint of each individual source of information. The fuzzy measure supplies the expected worth of each subset of sources from this hypothesis.

If $X = \{x_1, ..., x_n\}$, is a finite set, arranged so that $h(x_1) \geq h(x_2) \geq ... \geq h(x_n)$, then

$$\int_X h(x) \circ g_\lambda = \bigvee_{i=1}^{n} \left[h(x_i) \wedge g_\lambda(X_i)\right]$$

where $X_i = \{x_1, ..., x_i\}$. Also, given λ as calculated above, the values $g_\lambda(X_i)$ can be determined recursively from the definitions [46]. The fuzzy integral is interpreted as an evaluation of object classes where the subjectivity is embedded in the fuzzy measure. In comparison with probability theory, the fuzzy integral corresponds to the concept of expectation. In general, fuzzy integrals are nonlinear functionals (although monotone) whereas ordinary (e.g., Lebesque) integrals are linear functionals.

As an example, the fuzzy integral algorithm was tested using forward looking infrared (FLIR) images containing two tanks and an armored personnel carrier (APC) [38]. There were three sequences of 100 frames each used for training purposes. In each sequence, the vehicles appeared at a different aspect angle to the sensor (0°, 45°, 90°). In the fourth sequence the APC "circled" one of the tanks, moving in and out of a ravine and finally coming toward to sensor. This sequence was used to perform the comparison tests. The images were preprocessed to extract object of interest windows. The classification level integration was performed using four statistical features calculated from the windows. To get the partial evaluation, $h(x)$, for each feature, the fuzzy two-mean algorithm [16] was used. The fuzzy densities, the degree of importance of each feature, were assigned based on how well these features separated the two classes Tank and APC on training data [38]. The result of the fuzzy integral classifier is presented in the form of confusion matrix, in Table 1, where the count of samples listed in each row are those which belong to the corresponding class and the count of samples listed in each column are those after classification, which was made by choosing the class with the largest integral value.

The fuzzy integral outperformed a simple Bayes classifier on this data, but more importantly, the final integral values provide a different measure of certainty in the classification than posterior probabilities. The integral evaluation need not sum to one, so that lack of evidence and negative evidence can be distinguished.

This approach was also compared to a Dempster-Shafer rule-based classifier [47]. A conceptual difference between the fuzzy integral and a Dempster-Shafer classifier is in the frame of discernment [48]. For the fuzzy integral the frame of discernment contains the knowledge sources related to the hypothesis under consideration, whereas with belief theory, the frame of discernment contains all of the possible hypotheses. Thus the fuzzy integral algorithm has a means to assess the importance of all groups of knowledge sources towards answering the questions as well as the degree to which each knowledge source supports the hypothesis. With belief theory, each knowledge source would have to generate a belief function over the power set of the set of hypotheses, which are then combined using Dempster's rule. This calculation can have exponential complexity with the number of hypotheses. With the fuzzy integral, the Sugeno measure need only be calculated for n subsets (where n is the number of knowledge sources for each hypothesis). These measures are then combined with the objective evidence to produce the integral values.

TABLE 1.

FUZZY INTEGRAL CLASSIFIER FOR A TWO CLASS ATR PROBLEM

Computed densities and λ values

	g^1	g^2	g^3	g^4	λ
Tank	0.16	0.23	0.19	0.22	0.760
APC	0.15	0.24	0.18	0.23	0.764

Confusion Matrix

	Tank	APC
Tank	175	1
APC	17	49

Total correct 92.6%

Recently, Tahani has extended this information fusion approach to a large family of S-decomposable measures and generalized the definition of the fuzzy integral, thereby significantly increasing the flexibility of this powerful tool [14].

The above techniques, as well as many other fuzzy pattern recognition algorithms, are numeric feature-based procedures. On the other hand, fuzzy logic, and in general possibility theory, is inherently set-based, and so, offers the potential to manipulate higher order concepts. For example, in [49] (and refined in [50]) Keller et al used linguistic weighted averaging of possibility distributions [51] to generate object confidence from a combination of feature level results and harder-to-quantify values relating to range and motion. Rough estimates of object range and motion were used to construct trapezoidal possibility distributions which were averaged, using alpha-level set methods [51], with similar trapezoidal numbers formed from the output of fuzzy pattern recognition algorithms such as the fuzzy *k*-nearest-neighbors [52]. In [50] we developed a scaling technique to actually turn the averaging procedure into a confidence fusion methodology overcoming the spreading inherent in fuzzy arithmetic.

In [53], normalized histograms of color components of images of beef steaks were used directly in a linguistic approximation scheme to assess the degree-of-doneness of the steak. It was felt that because of the large amount of uncertainty inherent in food processing, the entire distribution of color (primarily in the red and brown regions) was important for class recognition. Note that this is conceptually distinct from those techniques described earlier which used normalized histograms of training data to calculate membership numbers for particular instances of the domain variable. Here, the object (a steak image) is represented by a group of fuzzy sets (various color histograms) and a set-based nearest prototype algorithm was used to assign class labels and confidences.

Rule-based systems have gained popularity in computer vision applications, particularly in high level vision activities. In guiding the choice of parameters for low level algorithms, a vision knowledge base may have a rule such as

IF the range is LONG, THEN
the prescreener window size is SMALL.

If LONG and SMALL are modeled by possibility distributions over appropriate domains of discourse, then fuzzy logic offers numerous approaches to translate such rules and to make inferences from the rules and facts modeled similarly. Nafarieh and Keller [54] designed a fuzzy logic rule based system for automatic target recognition which contained the above rule and approximately 40 other such rules.

Most fuzzy logic inference is based on Zadeh's composition rule. This generalizes traditional modus ponens which states that from the proposition

P_1: If X is A Then Y is B
and P_2: X is A,

we can deduce ***Y* is *B***. If proposition P_2 did not exactly match the antecedent of P_1, for example, ***X* is *A'***, then the modus ponens rule would not apply. However, in [55], Zadeh extended this rule if *A*, *B*, and *A'* are modeled by fuzzy sets. In this case, *X* and *Y* are fuzzy variables [55] defined over universes of discourse *U* and *V* respectively. As described above, the propositions ***X* is *A*** and ***Y* is *B***, where *A* and *B* are fuzzy subsets of *U* and *V* respectively, generate possibility distributions for the variables *X* and *Y*. The proposition P_1 concerns the joint fuzzy variable (*X*,*Y*) and is characterized by a fuzzy set over the cross product space *U* x *V*. Specifically, P_1 is characterized by a possibility distribution:

$$\Pi_{(X|Y)} = R \qquad \text{where}$$

$$\mu_R(u,v) = \max\{(1-\mu_A(u)),\ \mu_B(v)\}$$

It should be noted that this formula corresponds to the statement "not *A* or *B*", the logical translation of P_1. Zadeh now makes the inference *Y* is *B'* from μ_R and $\mu_{A'}$ by

$$\mu_{B'}(v) = \max_u\{\min\ \{\mu_R(u,v),\ \mu_{A'}(u)\}\}.$$

While this formulation of fuzzy inference, called the composition rule, directly extends modus ponens, it suffers from some problems. In fact, if proposition *P'* is ***X* is *A***, the resultant fuzzy set is not exactly the fuzzy set *B*.

Besides changing the way in which P_1 is translated into a possibility distribution, methods involving truth modification have been proposed. In this approach, the proposition ***X* is *A'*** is compared with ***X* is *A***, and the degree of compatibility is used to modify the membership function of *B* to get that for *B'*.

A fuzzy truth value restriction τ is a fuzzy subset of X = [0,1], and can be defined by its membership function, μ_τ, which is a mapping

$$\mu_\tau : X \rightarrow [0,1].$$

For example, we can define fuzzy truth value restrictions true, very true, false, unknown, absolutely true, absolutely false, etc.

In the truth value restriction methodology, the degree to which the actual given value *A'* of a variable *X* agrees with the antecedent value *A* in a proposition If ***X* is *A*** then ***Y* is *B*** is represented as a fuzzy subset of a truth space. This fuzzy subset of truth space is what is referred to by the phrase truth value restriction; it is used in a fuzzy deduction process to determine the corresponding restriction on the truth value of the proposition ***Y* is *B***. This latter truth value restriction is then "inverted", which means that a fuzzy proposition ***Y* is *B'*** in the *Y* universe of dis-course is found such that its agreement with ***Y* is *B*** is equal to the truth value restriction derived by the aforementioned fuzzy inference process. That is $\mu_{B'}(v) = \mu_\tau(\mu_B(v))$. The rule-based system described in [54] utilized a new inference technique based on truth value restriction which outperformed most methods of fuzzy logic inference when the inputs were exponentially defined functions of the antecedent clause (VERY LONG, MORE-OR-LESS LONG, etc.).

To ease the computational burden of performing modus ponens inferences with fuzzy sets, and to preserve the generalization capability, we introduced neural network architectures to accomplish the fuzzy logic inferences. These architectures could be trained on multiple conjunctive or disjunctive antecedent clause rules and could actually store several compatible rules in one structure, providing a natural method of conflict resolution [56-58].

CONCLUSIONS

The use of fuzzy set theory is growing in computer vision as it is in all intelligent processing. The representation capability is flexible and intuitively pleasing, the combination schemes are mathematically justifiable and can be tailored to the particular problem at hand from low level aggregation to high level inferencing, and the results of the algorithms are excellent, producing not only crisp decisions when necessary, but also corresponding degrees of support.

There is much work left to be done at all levels of computer vision. One area of particular need is the calculation and subsequent use of (fuzzy) features from the output of fuzzy segmentation algorithms. More research is also necessary in high level vision processes. Fuzzy set theory offers excellent potential for describing and manipulating object and region relationships, thereby assisting with scene interpretation. Finally, we believe that possibility distributions should be the model for the interface between (1) the human and the vision system and (2) high level vision subsystem and mid or low level vision processes.

This paper represents a short survey of fuzzy set methods in computer vision. Once again we apologize to all whose work we have inadvertently omitted from review. We strongly believe in the potential of fuzzy set theory to solve increasingly difficult computer vision problems, and hope that this survey will increase research in this area.

REFERENCES

1. J.M. Prewitt, "Object enhancement and extraction", in *Picture Processing and Psychopictorics*, B.S. Lipkin and A. Rosenfield (Eds.), Academic Press, New York, 1970, pp. 75-149.

2. S.K. Pal, and R.A. King. "Image enhancement using smoothing with fuzzy sets," *IEEE Transactions on System, Man, and Cybernetics*, Vol. SMC-11, 1981, pp. 494-501.

3. S.K. Pal, and R.A. King. "Histogram equalization with S and π functions in detecting x-ray edges", *Electronics Letters*, Vol. 17, 1981, pp. 302-304.

4. S.K. Pal, and R.A. King. "On edge detection of x-ray images using fuzzy sets," *IEEE Transactions on Pattern Analysis and Machine Intelligence*, Vol. PAMI-5, 1983, pp. 69-77.

5. J. Keller, H. Qiu, and H. Tahani, "The fuzzy integral in image segmentation," *Proceedings, NAFIPS-86*, New Orleans, June 1986, pp. 324-338.

6. R. Sankar, "Improvements in image enhancement using fuzzy sets", *Proceedings NAFIPS-86*, New Orleans, June 2-4, 1986, pp. 502-515.

7. L.A. Zadeh, "Calculus of fuzzy restrictions", in *Fuzzy Sets and Their Applications to Cognitive and Decision Processes*, L.A. Zadeh, K.S. Fu, K. Tanaka, and M. Shimura, Eds., Academic Press, London, 1975, pp. 1-26.

8. Y. Nakagowa, and A. Rosenfeld, "A note on the use of local min and max operators in digital picture processing," *IEEE Transactions on System, Man and Cybernetics*, Vol. SMC-8, 1978, pp. 632-635.

9. M.M. Gupta, G.K. Knopf, and P.N. Mikiforuk, "Edge Perception Using Fuzzy Logic", in *Fuzzy Computing: Theory Hardware and Applications*, North Holland, 1988.

10. Huntsberger, and M. Desclazi, "Color edge detection", *Pattern Recognition Letters*, 3, 1985, 205.

11. S.K. Pal and A. Rosenfeld, "Image enhances and thresholding by optimization of fuzzy compactness", *Pattern Recognition Letters*, vol. 7, 1988, pp. 77-86.

12. R. Krishnapuram and J. Lee, "Fuzzy-Compensative-Connective-Based Hierarchical Networks and their Application to Computer Vision" under review.

13. Lee, "Fuzzy-Set-Theory-Based Aggregation Networks for Information Fusion and Decision Making", Ph.D. Thesis, University of Missouri - Columbia.

14. Tahani, "The generalized fuzzy integral in computer vision," Ph.D. dissertation, University of Missouri - Columbia, 1990.

15. J.C. Dunn, A fuzzy relative of the Isodata process and its use in detecting compact well-separated clusters, *Journal Cybernet* 31(3), 1974, pp. 32-57.

16. C. Bezdek, *Pattern Recognition with Fuzzy Objective Function Algorithms*, Plenum Press, New York, 1981.

17. T. Huntsberger, C. Jacobs, and R. Cannon, "Iterative fuzzy image segmentation," *Pattern Recognition*, vol. 18, 1985, pp. 131-138.

18. R. Cannon, J. Dave and J. Bezdek, "Efficient implementation of the fuzzy c-means clustering algorithm," _IEEE Transactions on Pattern Analysis Machine Intelligence_, Vol. 8, No. 2, 1986, pp. 248-255.

19. S. Horowitz and T. Pavlidis, "Picture segmentations by a directed split and merge procedure", _Proceedings of the Second International Journal Conference Pattern Recognition_, 1974, pp. 424-433.

20. T. Huntsberger., "Representation of uncertainty in low level vision", _IEEE Transactions on Computers_, Vol. 235, No. 2, 145, 1986, p. 145.

21. R. Cannon, J. Dave, J.C. Bezdek, and M. Trivedi, "Segmentation of a thematic mapper image using the fuzzy c-means clustering algorithm," _IEEE Transactions on Geographical Science and Remote Sensing_, Vol. 24, No. 3, 1986, pp. 400-408.

22. J. Keller and C. Carpenter, "Image Segmentation in the Presence of Uncertainty," _International Journal of Intelligent Systems_, vol. 5, 1990, pp. 193-208.

23. A. Rosenfeld, "Fuzzy digital topology" _Information and Control_, 40, 1979, pp. 76-87.

24. A. Rosenfeld, "On connectivity properties of gray scale pictures", _Pattern Recognition_, 16, 1983, pp. 47-50.

25. A. Rosenfeld, "The fuzzy geometry of image subsets", _Pattern Recognition Letters_, 2, 1984, pp. 311-317.

26. D. Dubois and M.C. Jaulent, "Shape understanding via fuzzy models", _2nd IFAC/IFIP/IFORS/IEA Conference on analysis, design and evaluation of man-machine systems_, 1985, pp. 302-307.

27. D. Dubois and M.C. Jaulent, "A general approach to parameter evaluations in fuzzy digital pictures", _Pattern Recognition Letters_, to appear.

28. S. Peleg and A. Rosenfeld, "A mini-max medial axis transformation, _IEEE Transactions on Pattern Analysis and Machine Intelligence_, Vol. PAMI-3, 1981, pp. 208-210.

29. C.R. Dyer and A. Rosenfeld, "Thinning operations on grayscale pictures," _IEEE Transactions on Pattern Analysis and Machine Intelligence_, Vol. PAMI-1, 1979, pp. 88-89.

30. A. Rosenfeld and A.C. Kak, _Digital Picture Processing_, Vol. 2, Academic Press, N.Y., 1982.

31. L. Liao, "Image segmentation and enhancement by optimizing geometric parameters", M.S. Thesis, University of Missouri-Columbia, 1990.

32. I. Gath and A.B. Geva, "Unsupervised Optimal Fuzzy Clustering", _IEEE Transactions on Pattern Analysis Machine Intelligence_, vol. PAMI-11, no. 7, pp. 773-781, July 1989.

33. J. Keller and Y. Seo, "Local fractal geometric features for image segmentation", to appear _International Journal of Imaging Systems and Technology_, 1990.

34. R. Krishnapuram and A. Munshi, "Cluster-Based Segmentation of Range Images Using Differential-Geometric Features", submitted to under review.

35. R. Krishnapuram and J. Lee "Fuzzy-Connective-Based Hierarchical Aggregation Networks for Decision Making", _Fuzzy Sets and Systems_, to appear.

36. R. Krishnapuram and J. Lee "Determining the Structure of Uncertainty Management Networks", to appear in the _Proceedings of the SPIE Conference on Robotics and Computer Vision_, Philadelphia, November 1989.

37. H. Qiu and J. Keller, "Multispectral segmentation using fuzzy techniques," _Proceedings NAFIPS-87_, Purdue University, May 1987, pp. 374-387.

38. H. Tahani and J. Keller, "Information fusion in computer vision using the fuzzy integral", _IEEE Transactions on System, Man and Cybernetics_, vol. 20, no. 3, 1990, pp. 733-741.

39. H.J. Zimmermann and P. Zysno "Decisions and evaluations by hierarchical aggregation of information", _Fuzzy Sets and Systems_, vol.10, no.3, 1983 pp. 243-260.

40. Y. Ohta, _Knowledge-Based Interpretation of Outdoor National Scenes_, Pitman Advanced Publishing, Boston, 1985.

41. R. Dave, "Use of the adaptive fuzzy clustering algorithm to detect lines in digital images", _Proceedings of the Intelligent Robots and Computer Vision VIII_, vol. 1192, no. 2, 1989, pp. 600-611.

42. C.-P. Freg, "Algorithms to detect linear and planar clusters and their applications", MS Project Report, University of Missouri-Columbia, May 1990.

43. J. Bezdek, C. Cordy, R. Gunderson and J. Watson, "Detection and characterization of cluster substructure", _SIAM Journal Applied Mathematics_, Vol. 40, 1981, pp. 339-372.

44. M. Windham, "Geometrical fuzzy clustering algorithms", _Fuzzy Sets and Systems_, vol. 10, 1983, pp. 271-279.

45. R. Dave, "Fuzzy Shell-Clustering and applications to circle detection in digital images", _International Journal of General Systems_, 1990.

46. M. Sugeno, "Fuzzy measures and fuzzy integrals: A survey", in *Fuzzy Automatic and Decision Processes*, North Holland, Amsterdam, 1977, pp. 89-102.

47. J. Wootton, J. Keller, C. Carpenter, and G. Hobson, "A multiple hypothesis rule-based automatic target recognizer", in *Pattern Recognition*, Lecture Notes in Computer Science, Vol. 301, J. Kittler (ed.), Springer-Verlag, 1988, pp. 315-324.

48. G. Shafer, *A Mathematical Theory of Evidence*, Princeton University Press, Princeton, 1976.

49. J. Keller, G. Hobson, J. Wootton, A. Nafarieh, and K. Luetkemeyer, "Fuzzy confidence measures in midlevel vision," *IEEE Transactions on System, Man and Cybernetics*, Vol. SMC-17, No. 4, 1987, pp. 676-683.

50. J. Keller and D. Jeffreys, "Linguistic computations in computer vision", *Proceedings NAFIPS-90*, Vol. 2, Toronto, 1990, pp. 432-435.

51. J. Keller, H. Shah, and F. Wong, "Fuzzy Computations in risk and decision analysis", *Civil Engineering Systems*, vol. 2, 1985, pp. 201-208.

52. J. Keller, M. Gray, and J. Givens, "A fuzzy *k*-nearest neighbor algorithm," *IEEE Transactions on System, Man, and Cybernetics*, vol. 15, 1985, pp. 580-585.

53. J. Keller, D. Subhanghasen, K. Unklesbay, and N. Unklesbay, "An approximate reasoning technique for recogntion in color images of beef steaks", *International Journal General Systems*, to appear, 1990.

54. A. Nafarieh and J. Keller, "A fuzzy logic rule-based automatic target recognizer", *International Journal of Intelligent Systems* to appear, 1990.

55. L. Zadeh, "The concept of a linguistic variable and its application to approximate reasoning", *Information Sciences*, Part 1, Vol. 8, pp. 199-249; Part 2, Vol. 8, pp. 301-357; Part 3, Vol. 9, pp. 43-80, 1975.

56. J. Keller and H. Tahani, "Backpropagation neural networks for fuzzy logic", *Information Sciences*, to appear 1990.

57. J. Keller and R. Yager, "Fuzzy logic inference neural networks", *Proceedings of the SPIE Symposium on Intelligent Robots and Computer Vision VIII*, 1989, pp. 582-591.

58. J. Keller and H. Tahani, "Implementation of conjunctive and disjunctive fuzzy logic rules with neural networks", *International Journal of Approximate Reasoning*, to appear.

7

FUZZINESS, IMAGE INFORMATION AND SCENE ANALYSIS

Sankar K. Pal*
Software Technology Branch/PT4
National Aeronautics and Space Administration
Lyndon B. Johnson Space Center
Houston, Texas 77058, U.S.A.

INTRODUCTION

An application of the theory of fuzzy subsets to image processing and scene analysis problems has been described here. The problems considered are (pre)processing of 2-dimensional image pattern, extraction of primitives, and recognition and interpretation of image.

A gray tone picture possesses some ambiguity within the pixels due to the possible multivalued levels of brightness. The incertitude in an image may arise from grayness ambiguity or spatial (geometrical) ambiguity or both. Grayness ambiguity means "indefiniteness" in deciding a pixel as white or black. Spatial ambiguity refers to "indefiniteness" in shape and geometry of a region e.g., where is the boundary or edge of a region? or is this contour "sharp"?

When the regions in a image are ill-defined (fuzzy), it is natural and also appropriate to avoid committing ourselves to a specific (hard) decision e.g., segmentation/thresholding and skeletonization by allowing the segments or skeletons or contours, to be fuzzy subsets of the image. Similarly, for describing and interpreting ill-defined structural information in a pattern (when the pattern in-determinary is due to inherent vagueness rather than randomness), it is natural to define primitives and relation among them using labels of fuzzy set. For example, primitives may be defined in terms of arcs with varying grades of membership from 0 to 1 and production rules of a grammar may be fuzzified to account for the fuzziness in physical relation among the primitives; thereby increasing the generative power of a grammar.

The first part of the article consists of a definition of an image in the light of fuzzy set theory, and various information measures (arising from fuzziness) and tools relevant for processing e.g., fuzzy geometrical properties, correlation, bound functions and entropy measures. The second part provides formulation of various algorithms along with management of uncertainties (ambiguities) for image enhancement, edge detection, skeletonization, filtering, segmentation and object extraction. Ambiguity in evaluation and assessment of membership function has

* Dr. Pal is on leave from the post of Professor in the Electronics and Communication Sciences Unit, Indian Statistical Institute, Calcutta 700035, India.

also been described here. The third part describes the way of extracting various fuzzy primitives in order to describe the contours of different object regions of an image. Finally the fuzzy grammars are used to demonstrate how syntactic algorithms can be formulated for identifying different region structures/ classes of patterns. The above features have been illustrated through examples and various image data.

IMAGE DEFINITION

An image X of size MxN and L levels can be considered as an array of fuzzy singletons, each having a value of membership denoting its degree of brightness relative to some brightness level ℓ, $\ell = 0, 1, 2, \ldots L-1$. In the notation of fuzzy sets, we may therefore write

$$X = \{\mu_X(x_{mn}) = \mu_{mn}/x_{mn};\ m = 1, 2 \ldots M;\ n = 1, 2, \ldots N\} \quad (1)$$

or

$$X = \bigcup_m \bigcup_n \mu_{mn}/x_{mn},\ m = 1, 2, \ldots, M;\ n = 1, 2, \ldots N$$

where $\mu_X(x_{mn})$ or μ_{mn}/x_{mn}, $(0 \leq \mu_{mn} \leq 1)$

denotes the grade of possessing some property μ_{mn} (e.g., brightness, edginess, smoothness) by the (m,n)th pixel intensity x_{mn}. In other words, a fuzzy subset of an image X is a mapping μ from X into [0, 1]. For any point $p \in X$, $\mu(p)$ is called the degree of membership of p in μ.

One may use either global or local information of an image in defining a membership function characterizing some property. For example, brightness or darkness property can be defined only in terms of gray value of a pixel x_{mn} whereas, edginess, darkness or textural property need the neighborhood information of a pixel to define their membership functions. Similarly, positional or co-ordinate information is necessary, in addition to gray level and neighborhood information to characterize a dynamic property of an image.

Again, the aforesaid information can be used in a number of ways (in their various functional forms), depending on individuals opinion and/or the problem to his hand, to define a requisite membership function for an image property.

MEASURES OF FUZZINESS AND IMAGE INFORMATION

The definitions of various measures which represent grayness ambiguity in an image (based on individual pixel as well as a collection of pixels) are listed below.

Linear Index of Fuzziness

$$\gamma_l(X) = (2/MN) \sum_m \sum_n |\mu_{mn} - \underset{\sim}{\mu}_{mn}| \quad (2)$$

$$= (2/MN) \sum_m \sum_n \min(\mu_{mn}, 1 - \mu_{mn})$$

$$m = 1, 2, \ldots M;\ n = 1, 2, \ldots N$$

Quadratic Index of Fuzziness

$$\gamma_q(X) = \left(2/\sqrt{MN}\right)\left[\sum_m\sum_n \{\mu_{mn} - \underset{\sim}{\mu}_{mn}\}^2\right]^{0.5} \tag{3}$$

$$m = 1, 2, \ldots M;\ n = 1, 2, \ldots N$$

Entropy

$$H(X) = (1/MN \ln 2)\sum_m\sum_n Sn(\mu_{mn}) \tag{4}$$

with $S_n(\mu_{mn}) = -\mu_{mn} \ln \mu_{mn} - (1 - \mu_{mn}) \ln(1 - \mu_{mn})$

$$m = 1, 2, \ldots M;\ n = 1, 2, \ldots N$$

μ_{mn} denotes the degree of possessing some property μ by the (m, n)th pixel x_{mn}. $\underset{\sim}{\mu}_{mn}$ denotes the nearest two tone version of μ_{mn}

rth Order Entropy

$$H^r(X) = (-1/k)\sum_i \left\{\mu\left(s_i^\gamma\right)\log\left\{\mu\left(s_i^\gamma\right)\right\} + \left\{1 - \mu\left(s_i^\gamma\right)\right\}\log\left\{1 - \mu\left(s_i^\gamma\right)\right\}\right\} \tag{5}$$

$$i = 1, 2, \ldots k$$

s_i^r denotes the ith combination (sequence) of r pixels in X. k is the number of such sequences. $\mu(s_i^r)$ denotes the degree to which the combination s_i^r, as a whole, possesses the property μ.

Hybrid Entropy

$$H_{hy}(X) = -P_w \log E_w - P_b \log E_b \tag{6}$$

with

$$E_w = (1/MN) \sum_m\sum_n \mu_{mn} \exp(1 - \mu_{mn})$$

$$E_b = (1/MN) \sum_m\sum_n (1 - \mu_{mn})\exp(\mu_{mn})$$

$$m = 1, 2, \ldots M;\ n = 1, 2, \ldots N$$

μ_{mn} denotes the degree of "whiteness" of (m, n)th pixel. P_w and P_b denote probability of occurrences of white ($\mu_{mn} = 1$) and black ($\mu_{mn} = 0$) pixels respectively. E_w and E_b denote the average likeliness (possibility) of interpreting a pixel as white and black respectively.

Correlation

$$C(\mu_1, \mu_2) = 1 - 4\left[\sum_m\sum_n \{\mu_{1mn} - \mu_{2mn}\}^2\right] \Big/ (X_1 + X_2) \tag{7}$$

$$C(\mu_1, \mu_2) = 1 \quad \text{if } X_1 + X_2 = 0$$

with $$X_1 = \sum_m \sum_n \{2\mu_{1mn} - 1\}^2$$

and $$X_2 = \sum_m \sum_n \{2\mu_{2mn} - 1\}^2$$

$$m = 1, 2, \ldots M; \; n = 1, 2, \ldots N$$

$C(\mu_1, \mu_2)$ denotes the correlation between two properties μ_1 and μ_2 (defined over the same domain). μ_{1mn} and μ_{2mn} denote the degree of possessing the properties μ_1 and μ_2 respectively by the (m, n)th pixel.

These expressions (equations 2-7) are the versions extended to two dimensional image plane from those defined for a fuzzy set. For example, index of fuzziness was defined by Kaufmann [1], entropy by DeLuca and Termini [2], rth order entropy and hybrid entropy by Pal and Pal [3], and correlation by Murthy, Pal and Dutta Majumdar [4].

Index of fuzziness reflects the ambiguity present in an image by measuring the distance between its fuzzy property plane and the nearest ordinary plane. The term "entropy", on the other hand, uses Shannon's function in the property plane but its meaning is quite different from the one of classical entropy because no probabilistic concept is needed to define it. $H^r(X)$ gives a measure of the average amount of difficulty in taking a decision on any subset of size r with respect to an image property. If $r = 1$, $H^r(X)$ reduces to (unnormalized) $H(X)$ of equation (4). $H_{hy}(X)$ represents an amount of difficulty in deciding whether a pixel possesses certain properties or not by making a prevision on its probability of occurrence. In absence of fuzziness (i.e.,with proper defuzzification), H_{hy} reduces to two state classical entropy of Shannon, the states being black and white. Since a fuzzy set is a generalized version of an ordinary set, the entropy of a fuzzy set deserves to be a generalized version of classical entropy by taking into account not only the fuzziness of the set but also the underlying probability structure. In that respect, H_{hy} can be regarded as a generalized entropy such that classical entropy becomes its special case when fuzziness is properly removed.

All these terms, which give an idea of 'indefiniteness' or fuzziness of an image may be regarded as the measures of average intrinsic information which is received when one has to make a decision (as in pattern analysis) in order to classify the ensembles of patterns described by a fuzzy set.

$\gamma(X)$ and $H(X)$ are normalized in the interval [0, 1] such that

$$\text{Pr1: } \gamma_{min} = H_{min} = 0 \text{ for } \mu_{mn} = 0 \text{ for all } (m,n)(X) \tag{8a}$$

$$\text{Pr2: } \gamma_{max} = H_{max} = 1 \text{ for } \mu_{mn} = 0.5 \text{ for all } (m,n) \tag{8b}$$

$$\text{Pr3: } \gamma(X) \geq \gamma(X^*) (\text{or, } H(X) \geq H(X^*)) \tag{8c}$$

and $\quad \text{Pr 4:}\ \gamma(X) = \gamma(\overline{X})\left(\text{or, } H(X) \geq H(\overline{X})\right)$ (8d)

where X* is the 'sharpened' or 'intensified' version of X such that

$$\mu_X{}^*(x_{mn}) \geq \mu_X(x_{mn}) \quad \text{if } \mu_X(x_{mn}) \geq 0.5$$

$$\text{and } \mu_{X^*}(x_{mn}) \leq \mu_X(x_{mn}) \text{ if } \mu_X(x_{mn}) \leq 0.5 \tag{9}$$

In other words, $\gamma(X)$ or H(X) increases monotonically with μ, reaches a maximum at μ = 0.5 and then decreases monotonically. This is explained in Fig. 1.

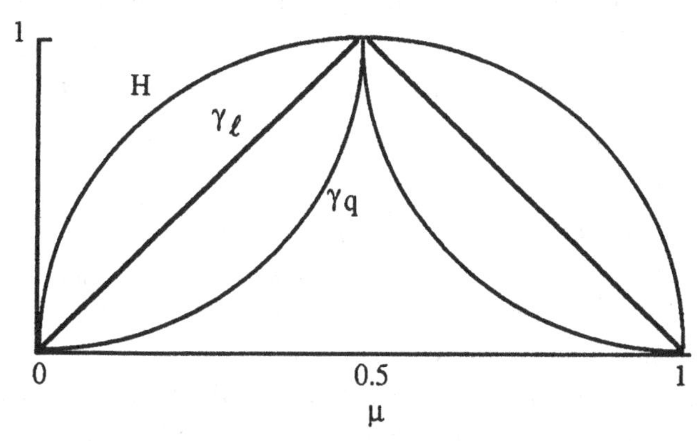

Figure 1 Variation of Fuzziness with μ.

According to property 8(c), these parameters decrease with contrast enhancement of an image. Now through processing, if we can partially remove the uncertainty on the grey levels of X, we say that we have obtained an average amount of information given by $\delta\gamma = \gamma(X) - \gamma(X^*)$ or $\delta H = H(X) - H(X^*)$ by taking a decision bright or dark on the pixels of X. The criteria $\gamma(X^*) \leq \gamma(X)$ and $H(X^*) \leq H(X)$, in order to have positive $\delta\gamma$ and δH-values, follow from Eq. (8c). If the uncertainty is completely removed, then $\gamma(X^*) = H(X^*) = 0$. In other words, $\gamma(X)$ and $H(X)$ can be regarded as measures of the average amount of information (about the grey levels of pixels) which has been lost by transforming the classical pattern (two-tone) into a fuzzy pattern X.

It is to be noted that $\gamma(X)$ or H(X) reduces to zero as long as μ_{mn} is made 0 or 1 for all (m, n), no matter whether the resulting defuzzification (or transforming process) is correct or not. In the following discussion it will be clear how H_{hy}, takes care of this situation.

$H^r(X)$ has the following properties:

Pr 1: H^r attains a maximum if $\mu_i = 0.5$ for all i.

Pr 2: H^r attains a minimum if $\mu_i = 0$ or 1 for all i.

Pr 3: $H^r \geq H^{*r}$, where H^{*r} is the rth order entropy of a sharpened version of the fuzzy set.

Pr 4: H^r is, in general, not equal to $\overline{H}^r$, where $\overline{H}^r$ is the rth order entropy of the complement set.

Pr 5: $H^r \leq H^{r+1}$ when all $\mu_i \in [0.5, 1]$.

$H^r \geq H^{r+1}$ when all $\mu_i \in [0, 0.5]$.

Note that the property P4 of equation 8(d) is not, in general, valid here. The additional property Pr 5 implies that H^r is a monotonically nonincreasing function of r for $\mu_i \in [0, 0.5]$ and a monotonically nondecreasing function of r for $\mu_i \in [0.5, 1]$ (when 'min' operator has been used to get the group membership value).

When all the μ_i values are same, $H^1(X) = H^2(X) = \ldots = H^r(X)$. This is because of the fact that the difficulty in taking a decision regarding possession of a property on an individual is same as that of a group selected therefrom. The value of H^r would, of course, be dependent on the μ_i values.

Again, the higher the similarity among singletons the quicker is the convergence to the limiting value of H^r. Based on this observation, let us define an index of similarity of supports of a fuzzy set as $S = H^1/H^2$ (when $H^2 = 0$, H^1 is also zero and S is taken as 1). Obviously, when $\mu_i \in [0.5, 1]$ and the min operator is used to assign the degree of possession of the property by a collection of supports, S will lie in [0, 1] as $H^r \leq H^{r+1}$. Similarly, when $\mu_i \in [0, 0.5]$ S may be defined as H^2/H^1 so that S lies in [0, 1]. Higher the value of S the more alike (similar) are the supports of the fuzzy set with respect to the property P. This index of similarity can therefore be regarded as a measure of the degree to which the members of a fuzzy set are alike.

Therefore, the value of conventional fuzzy entropy (H^1 or Eq. 4) can only indicate whether the fuzziness in a set is low or high. In addition to this, the value of H^r also enables one to infer whether the fuzzy set contains similar supports (or elements) or not. The similarity index thus defined can be successfully used for measuring interclass and intraclass ambiguity (i.e., class homogeneity and contrast) in pattern recognition and image processing problems.

The aforesaid features are explained in Table 1 when $\mu_i \in [0.5, 1]$, min operator is used to compute group membership and k in Eq. 5 is considered to be $^{10}C_r$, $r = 1, 2, \ldots 6$.

$H_{hy}(X)$ has the following properties. In the absence of fuzziness when MNP_b pixels become completely black $(\mu_{mn} = 0)$ and MNP_w pixels become completely

Table 1: Higher Order Entropy

Case	μ_X	H1	H2	H3	H4	H5	H6	S
1	{1,1,1,1,1,1,1,1,1,1}	0	0	0	0	0	0	1
2	{.5,.5,.5,.5,.5,.5,.5,.5,.5,.5}	1	1	1	1	1	1	1
3	{1,1,1,1,1,.5,.5,.5,.5,.5}	.5	.777	.916	.976	.996	1	.642
4	{.5,.5,.5,.5,.5,.6,.6,.6,.6,.6}	.980	.991	.996	.999	.999	1	.989
5	{.6,.6,.65,.9,.9,.9,.9,.9,.9,.915}	.538	.678	.781	.855	.905	.937	.793
6	{.8,.8,.8,.8,.8,.8,.9,.9,.9,.9}	.538	.613	.641	.649	.650	.650	.878
7	{.5,.5,.5,.5,.5,.5,.9,.9,.9,.9}	.748	.916	.979	.997	1	1	.816
8	{.7,.7,.7,.7,.7,.8,.8,.8,.8,.8}	.748	.802	.830	.841	.845	.846	.932

white $(\mu_{mn} = 1)$ then $E_w = P_w$, $E_b = P_b$ and H_{hy} boils down to two state classical entropy

$$H_c = -P_w \log P_w - P_b \log P_b, \tag{10}$$

the states being black and white. Thus, H_{hy} reduces to H_c only when a proper defuzzification process is applied to detect (restore) the pixels. $|H_{hy} - H_c|$ can therefore be acted as an objective function for enhancement and noise reduction. The lower the difference, the lesser is the fuzziness associated with the individual symbol and higher will be the accuracy in classifying them as their original value (white or black). (This property was lacking with $\gamma(X)$ and $H(X)$ measures (equations 2-4) which always reduce to zero irrespective of the defuzzification process). In other words, $|H_{hy} - H_c|$ represents an amount of information which was lost by transforming a two tone image to a gray tone.

For a given P_w and P_b $(P_w + P_b = 1,\ 0 \le P_w, P_b \le 1)$, of all possible defuzzified versions, H_{hy} is minimum for the one with properly defuzzified.

If $\quad \mu_{mn} = 0.5$ for all (m, n) then $E_w = E_b$

and $\quad H_{hy} = -\log(0.5\exp 0.5)$ (11)

i.e., H_{hy} takes a constant value and becomes independent of P_w and P_b. This is logical in the sense that the machine is unable to take decision on the pixels since all μ_{mn} values are 0.5.

Let us consider an example of a digital image in which, say, 70% pixels look white, while the remaining 30% look dark. Thus the probability of a white pixel P_w is 0.7 and that of a dark pixel P_b is 0.3. Suppose, the whiteness of the pixels is not constant, i.e., there is a variation (grayness) and similar is the case with the black pixels.

Let us now consider the effect of improper defuzzification on the pattern shown in case 1 of the Table 2. Two types of defuzzifications are considered here. In cases 2-4 all the symbols with $\mu = 0.5$ are transformed to zero when some of them were actually generated from symbol '1'. In cases 5-6 of Table 2 some of the μ values greater than 0.5 which were generated from symbol 1 (or belong to the white portion of the image) are wrongly defuzzified and brought down towards zero (instead of 1).

In both situations, it is to be noted that $|H - H_{hy}|$ does not reduce to zero. The case 7, on the other hand, has all its elements properly defuzzified. As a result, E_1 and E_0 become 0.3 and 0.7 respectively and $|H_{hy} - H_c|$ reduces to zero.

Table 2: Effect of wrong defuzzification(with $p_0 = 0.3$ and $p_1 = 0.7$)

Case	μ_X	E_1	E_0	H_{hy}	$\|H - H_{hy}\|$
1	{.9,.9,.8,.8,.7,.6,.5,.5,.4,.3}	.620	.876	.235	.375
2	{.999,.999,.9,.8,.7,.7,.3,.3,.2,.1}	.576	.776	.342	.268
3	{1,1,1,.99,.9,.9,.1,.1,0,0}	.450	.648	.542	.068
4	{1,1,1,1,1,1,0,0,0,0}	.400	.600	.632	.021
5	{.99,.99,.1,.1,.9,.8,.7,.2,.1,.1}	.630	.634	.456	.154
6	{1,1,0,0,1,1,1, 0, 0,0}	.500	.500	.693	.082
7	{1,1,1,1,1,1,1,0,0,0}	.300	.700	.611	0

$C(\mu_1, \mu_2)$ of equation (7) has the following properties.

a) If for higher values of $\mu_1(X)$, $\mu_2(X)$ takes higher values and the converse is also true then $C(\mu_1, \mu_2)$ must be very high.

b) If with increase of x, both μ_1 and μ_2 increase then $C(\mu_1, \mu_2) > 0$.

c) If with increase of x, μ_1 increases and μ_2 decreases or vice versa then $C(\mu_1, \mu_2) < 0$.

d) $C(\mu_1, \mu_1) = 1$

e) $C(\mu_1, \mu_1) \geq C(\mu_1, \mu_2)$

f) $C(\mu_1, 1-\mu_1) = -1$

g) $C(\mu_1, \mu_2) = C(\mu_2, \mu_1)$

h) $-1 \leq C(\mu_1, \mu_2) \leq 1$

i) $C(\mu_1, \mu_2) = -C(1-\mu_1, \mu_2)$

j) $C(\mu_1, \mu_2) = C(1-\mu_1, 1-\mu_2)$

IMAGE GEOMETRY

The various geometrical properties of a fuzzy image subset (characterized by $\mu_X(x_{mn})$ or simply by μ) as defined by Rosenfeld [5,6] and Pal and Ghosh [7] are given below with illustration. These provide measures of ambiguity in geometry (spatial domain) of an image.

A. *Area* The area of a fuzzy subset μ is defined as [5]

$$a(\mu) = \int \mu \tag{12}$$

where the integration is taken over a region outside which μ=0. For μ being piecewise constant (in case of digital image) the area is

$$a(\mu) = \sum \mu \tag{13}$$

where the summation is over a region outside which μ=0. Note from equation (13) that area is the weighted sum of the regions on which μ has constant value weighted by these values.

Example 1 Let μ be of the form

0.2 0.4 0.3
0.2 0.7 0.6
0.6 0.5 0.6

Area a(μ) = (0.2+0.4+0.3+0.2+0.7+0.6+0.6+0.5+0.6) = 4.1

B. *Perimeter* If μ is piecewise constant, the perimeter of μ is defined as [5]

$$p(\mu) = \sum_{i,\, j,\, k} |\mu(i) - \mu(j)| * |A(i, j, k)| \tag{14}$$

This is just the weighted sum of the lengths of the arcs A(i, j, k) along which the regions having constant μ values μ(i) and μ(j) meet, weighted by the absolute difference of these values. In case of an image if we consider the pixels as the piecewise constant regions, and the common arc length for adjacent pixels as unity then the perimeter of an image is defined by

$$p(\mu) = \sum_{i,\, j} |\mu(i) - \mu(j)| \tag{15}$$

where μ(i) and μ(j) are the membership values of two adjacent pixels.

For the fuzzy subset μ of example 1, perimeter is

$$\begin{aligned} p(\mu) = {} & |0.2 - 0.4| + |0.2 - 0.2| + |0.4 - 0.3| + |0.4 - 0.7| \\ & + |0.3 - 0.6| + |0.2 - 0.6| + |0.2 - 0.7| + |0.7 - 0.6| \\ & + |0.7 - 0.5| + |0.6 - 0.6| + |0.6 - 0.5| + |0.5 - 0.6| \\ & = 2.3 \end{aligned}$$

C. *Compactness* The compactness of a fuzzy set μ having an area of a (μ) and a perimeter of p(μ) is defined as [5]

$$\text{comp}(\mu) = \frac{a(\mu)}{(p(\mu))^2} \tag{16}$$

Physically, compactness means the fraction of maximum area (that can be encircled by the perimeter) actually occupied by the object. In non fuzzy case the value of compactness is maximum for a circle and is equal to $\pi / 4$. In case of fuzzy disc, where the membership value is only dependent on its distance from the center, this compactness value is $\geq \pi / 4$ [6]. Of all possible fuzzy discs compactness is therefore minimum for its crisp version.

For the fuzzy subset μ of example 1, comp(μ) = 4.1/(2.3*2.3) = 0.775.

D. *Height and Width* The height of a fuzzy set μ is defined as [5]

$$h(\mu) = \int \max_{m} \mu_{mn} dn \tag{17}$$

where the integration is taken over a region outside which $\mu_{mn} = 0$.

Similarly the width of the fuzzy set is defined by

$$w(\mu) = \int \max_{n} \mu_{mn} dm \tag{18}$$

with the same condition over integration as above. For digital pictures m and n can take only discrete values, and since $\mu = 0$ outside the bounded region, the max operators are taken over a finite set. In this case the definitions take the form

$$h(\mu) = \sum_{n} \max_{m} \mu_{mn} \tag{19}$$

and

$$w(\mu) = \sum_{m} \max_{n} \mu_{mn} \tag{20}$$

$$m = 1, 2, \ldots M; \; n = 1, 2, \ldots N$$

So physically, in case of a digital picture, height is the sum of the maximum membership values of each row. Similarly, by width we mean the sum of the maximum membership values of each column.

For the fuzzy subset μ of example 1, height is $h(\mu) = 0.4+0.7+0.6 = 1.7$ and width is $w(\mu) = 0.6+0.7+0.6 = 1.9$.

E. *Length and Breadth* The length of a fuzzy set μ is defined as [7]

$$l(\mu) = \max_{m} \left(\int \mu_{mn} dn \right) \tag{21}$$

where the integration is taken over the region outside which $\mu_{mn} = 0$. In case of a digital picture where m and n can take only discrete values the expression takes the form

$$l(\mu) = \max_{m} \left(\sum_{n} \mu_{mn} \right) \tag{22}$$

Physically speaking, the length of an image fuzzy subset gives its longest expansion in the column direction. If μ is crisp, $\mu_{mn} = 0$ or 1; in this case length is the maximum number of pixels in a column. Comparing equation (22) with (19) we notice that the length is different from height in the sense, the former takes the summation of the entries in a column first and then maximizes over different columns whereas, the later maximizes the entries in a column and then sums over different columns.

The breadth of a fuzzy set μ is defined as

$$b(\mu) = \max_{n} \left(\int \mu_{mn} dm \right) \tag{23}$$

where the integration is taken over the region outside which $\mu_{mn} = 0$. In case of a digital picture the expression takes the form

$$b(\mu) = \max_n \left(\sum_m \mu_{mn} \right) \tag{24}$$

Physically speaking, the breadth of an image fuzzy subset gives its longest expansion in the row direction. If μ is crisp, μ_{mn} = 0 or 1; in this case breadth is the maximum number of pixels in a row. The difference between width and breadth is same as that between height and length.

For the fuzzy subset μ in example 1, length is $l(\mu) = 0.4 + 0.7 + 0.5 = 1.6$ and breadth is $b(\mu) = 0.6 + 0.5 + 0.6 = 1.7$.

F. *Index of Area Coverage* (IOAC) The index of area coverage of a fuzzy set may be defined as [7]

$$\mathrm{IOAC}(\mu) = \frac{\mathrm{area}(\mu)}{l(\mu) * b(\mu)} \tag{25}$$

In nonfuzzy case, the IOAC has value of 1 for a rectangle (placed along the axes of measurement). For a circle this value is $\pi r^2 / (2r * 2r) = \pi / 4$. Physically by IOAC of a fuzzy image we mean the fraction (which may be improper also) of the maximum area (that can be covered by the length and breadth of the image) actually covered by the image.

For the fuzzy subset μ of example 1, the maximum area that can be covered by its length and breadth is 1.6*1.7 = 2.72 whereas, the actual area is 4.1, so the IOAC = 4.1 / 2.72 = 1.51.

It is to be noted that $l(X)/h(X) \le 1$ (26)

$b(X)/w(X) \le 1$ (27)

When equality holds for (26) or (27) the object is either vertically or horizontally oriented.

G. *Degree of Adjacency* The degree to which two regions S and T of an image are adjacent is defined as

$$a(S,T) = \sum_{p \in BP(S)} \frac{1}{1 + |\mu(p) - r(q)|} * \frac{1}{1 + d(p)} \tag{28}$$

Here d(p) is the shortest distance between p and q, q is a border pixel (BP) of T and p is a border pixel of S. The other symbols are having their same meaning as in the previous discussion.

The degree of adjacency of two regions is maximum (=1) only when they are physically adjacent i.e., d(p)=0 and their membership values are also equal i.e., $\mu(p) = r(q)$. If two regions are physically adjacent then their degree of adjacency is determined only by the difference of their membership values. Similarly, if the membership values of two regions are equal their degree of adjacency is determined by their physical distance only.

IMAGE PROCESSING OPERATIONS

In this section we will be explaining how the various grayness and geometrical ambiguity measures can be used for image enhancement, segmentation, edge detection and skeleton extraction problems. The algorithms which will be described here provide both fuzzy and nonfuzzy (as a special case) outputs.

Segmentation and Object Extraction

The problem of grey level thresholding plays an important role in image processing. For example, in enhancing contrast in a image we need to select proper threshold levels from its histogram so that some suitable non-linear transformation can highlight a desirable set of pixel intensities compared to others. Similarly, in image segmentation one needs proper histogram thresholding whose objective is to establish boundaries in order to partition the image spaces into meaningful regions. This Section illustrates an application of theory of fuzzy sets to make this task automatic so that an optimum threshold (or set of thresholds) may be estimated without the need to refer directly to the histogram.

Criteria for Threshold Selection

Let us consider, first of all, the parameters $\gamma(X)$ or $H(X)$ to explain the criterion of thresholding.

Consider the standard S-function [8]

$$P_{mn} = \mu_X(x_{mn}) = S(x_{mn};\ a,\ b,\ c) = 0,\ x_{mn} \le a \qquad (29a)$$

$$= 2\left[(x_{mn} - a)/(c-a)\right]^2, a \le x_{mn} \le b \qquad (29b)$$

$$= 1 - 2\left[(x_{mn} - c)/(c-a)\right]^2, b \le x_{mn} \le c \qquad (29c)$$

$$= 1,\ x_{mn} \ge c \qquad (29d)$$

with $b = (a+c)/2$, $b-a = c-b = \Delta b$,
for obtaining the μ_{mn} plane from the spatial x_{mn} plane of the image X and for computing $\gamma(X)$ and $H(X)$ values from Eqs. (2), (3) and (4). The parameter b is the cross-over point, i.e., $S(b;\ a,\ b,\ c) = 0.5$. Δb is the bandwidth. This is explained in Fig. 2 for an L-level image. Such a μ plane may be viewed to represent a fuzzy set "bright image" so that the degree of brightness of a pixel increases with its gray value.

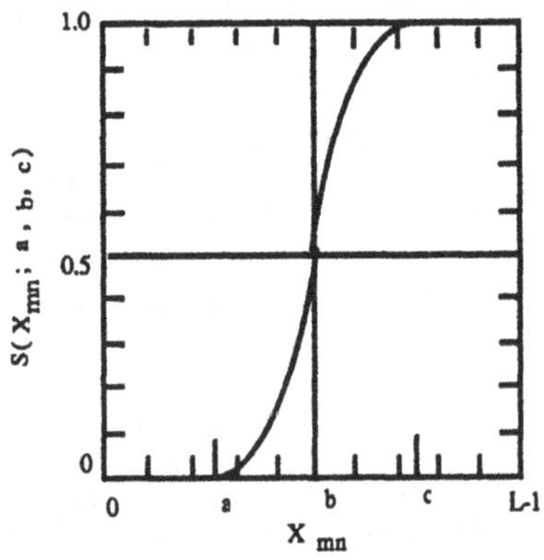

Figure 2 Standard S function for an L-level image.

For a particular cross-over point, say, $b = l_c$ we have $\mu_x(l_c) = 0.5$ and the μ_{mn} plane would contain values > 0.5 or < 0.5 corresponding to $x_{mn} > l_c$ or $< l_c$. The terms $\gamma(X)$ and H(X) then measure the average ambiguity in X by computing $\mu_{X \cap \bar{X}}(x_{mn})$ or $Sn(\mu_X(x_{mn}))$ which is 0 if $\mu_X(x_{mn}) = 0$ or 1 and is maximum for $\mu_X(x_{mn}) = 0.5$.

The selection of a cross-over point at $b = l_c$ implies the allocation of grey levels $< l_c$ and $> l_c$ within the two clusters namely, background and object of a binodal image. The contribution of the levels towards $\gamma(X)$ and H(X) is mostly from those around l_c and would decrease as we move away from l_c. Again, since the nearest ordinary plane $\underset{\sim}{X}$ (which gives the two-tone version of X) is dependent on the position of cross-over point, a proper selection of b may therefore be obtained which will result in appropriate segmentation of object and background. In other words, if the grey level of image X has binodal distribution, then the above criteria for different values of b would result in a minimum γ or H value only when b corresponds to the appropriate boundary between the two clusters.

For such a position of the threshold (cross-over point), there will be minimum number of pixel intensities in X having $\mu_{mn} \simeq 0.5$ (resulting in γ or H $\simeq$ 1) and maximum number of pixel intensities having $\mu_{mn} \simeq 0$ or 1(resulting in γ or H $\simeq$ 0) thus contributing least towards $\gamma(X)$ or H(X). This optimum (minimum) value would be greater for any other selection of the cross-over point.

This suggests that modification of the cross-over point will result in variation of the parameters $\gamma(X)$ and H(X) and so an optimum threshold may be estimated for automatic histogram-thresholding problems without the need to refer directly to the histogram of X. The above concept can also be extended to an image having multimodal distribution in grey levels in which one would have several minima in γ and H values corresponding to different threshold points in the histogram.

Let us now consider the geometrical parameters comp(X) and IOAC(X) (equations 16 and 25). It has been noticed that for crisp sets the value of index of area coverage (IOAC) is maximum for a rectangle. Again, of all possible fuzzy rectangles IOAC is minimum for its crisp version. Similarly, in a nonfuzzy case the compactness is maximum for a circle and of all possible fuzzy discs compactness is minimum for its crisp version [6]. For this reason, we will use minimization (rather than maximization) of fuzzy compactness/IOAC as a criterion for image segmentation [9].

Suppose we use equation (29) for obtaining the 'bright image' μ(X) of an image X. Then for a particular cross over point of S function, compactness (μ) and IOAC(μ) reflect the average amount of ambiguity in the geometry (i.e., in spatial domain) of X. Therefore, modification of the cross over point will result in different μ(X) planes (and hence different segmented versions), with varying amount of compactness or IOAC denoting fuzziness in the spatial domain. The μ(X) plane having minimum IOAC or compactness value can be regarded as an optimum fuzzy segmented version of X.

For obtaining the nonfuzzy threshold one may take the cross over point (which is considered to be the maximum ambiguous level) as the threshold between object and background. For images having multiple regions, one would have a set of such optimum $\mu(X)$ planes. The algorithm developed using these criteria is given below.

Algorithm 1

Given an L level image X of dimension MxN with minimum and maximum gray vales ℓ_{min} and ℓ_{max} respectively,

Step 1: Construct the membership plane using equation (29) as

$$\mu_{mn} = \mu(\ell) = S(\ell;\ a,\ b,\ c)$$

(called bright image plane if the object regions possess higher gray values)

or $$\mu_{mn} = \mu(\ell) = 1 - S(\ell;\ a,\ b,\ c)$$

(called dark image plane if the object regions possess lower gray values) with cross-over point b and a bandwidth Δ b.

Step 2: Compute $\gamma(X)$,H(X), Comp(X) and IOAC(X)

Step 3: Vary b between ℓ_{min} and ℓ_{max} and select those b for which I(X) (where I(X)) denotes one of the aforesaid measures or a combination of them) has local minima. Among the local minima let the global one have a cross over point s.

The level s, therefore, denotes the cross over point of the fuzzy image plane μ_{mn}, which has minimum grayness and/or geometrical ambiguity. The μ_{mn} plane then can be viewed as a fuzzy segmented version of the image X. For the purpose of nonfuzzy segmentation, we can take s as the threshold or boundary for classifying or segmenting image into object and background.

The measure I(X) in Step 3 can represent either grayness ambiguity (i.e., $\gamma(X)$ or H(X)) or geometrical ambiguity (i.e., comp(X) or IOAC(X) or a(S,T)) or both (i.e., product of grayness and geometrical ambiguities).

Faster Method of Computation

From the algorithm 1 it appears that one needs to scan an L level image L times (corresponding to L cross over points of the membership function) for computing the parameters for detecting its threshold. The time of computation can be reduced significantly by scanning it only once for computing its co-occurrence matrix, row histogram and column histogram, and by computing $\mu(l), l = 1, 2, \ldots L$ every time with the membership function of a particular cross over point.

The computations of $\gamma(X)$ (or H(X)), a(X), p(X), 1(X) and b(X) can be made faster in the following way. Let h(i), i=1,2..L be the number of occurrences of the level i, c[i,j], i = 1, 2 . . L, j = 1, 2 . . L the co-occurrence matrix and $\mu(i)$, i = 1, 2.. L the membership vector for a fixed cross over point of an L level image X.

Determine $\gamma(X)$, area and perimeter as

$$\gamma(X) = \frac{2}{MN} \sum_{i=1}^{L} T(i)\, h(i) \tag{30a}$$

$$T(i) = \min\{\mu(i), 1 - \mu(i)\} \tag{30b}$$

$$a(X) = \sum_{i=1}^{L} h(i) \, . \mu(i) \tag{31}$$

$$p(X) = \sum_{i=1}^{L} \sum_{j=1}^{L} C[i, j] . \; |\mu(i) - \mu(j)| \tag{32}$$

For calculating length and breadth following steps can be used. Compute the row histogram R[m, 1], m = 1, . . . M, 1 = 1 . . L, where R[m, 1] represents the number of occurrences of the gray level 1 in the mth row of the image. Find the column histogram C[n, 1], n = 1. . N, 1 = 1 . . L, where C[n, 1] represents the number of occurrences of the gray level 1 in the nth column of the image. Calculate length and breadth as

$$l(X) = \max_{n} \sum_{l=1}^{L} C[n, 1] . \; \mu(1) \tag{33}$$

$$b(X) = \max_{m} \sum_{l=1}^{L} R[m, 1] . \; \mu(1) \tag{34}$$

Some Remarks

The grayness ambiguity measure e.g., $\gamma(X)$ or H(X) basically sharpens the histogram of X using its global information only and it detects a single threshold in its valley region. Therefore, if the histogram does not have a valley, the above measures will not be able to select a threshold for partitioning the histogram. This can readily be seen from Equation (30) which shows that the minima of $\gamma(X)$ measure will only correspond to those regions of gray level which has minimum occurrences (i.e., valley region). Comp (X) or IOAC(X), on the other hand, uses local information to determine the fuzziness in spatial domain of an image. As a result, these are expected to result better segmentation by detecting thresholds even in the absence of a valley in the histogram.

Again, comp(X) measure attempts to make a circular approximation of the object region for its extraction, whereas, the IOAC(X) goes by the rectangular approximation. Their suitability to an image should therefore be guided by this criterion.

Choice of Membership Function

In the aforesaid algorithm $w = 2\Delta b$ is the length of the interval which is shifted over the entire dynamic range of gray scale. As w decreases, the $\mu(x_{mn})$ plane would have more intensified contrast around the cross-over point resulting in decrease of ambiguity in X. As a result, the possibility of detecting some undesirable thresholds (spurious minima) increases because of the smaller value of Δ b. On the other hand, increase of w results in a higher value of fuzziness and thus leads towards the possibility of losing some of the weak minima.

The criteria regarding the selection of membership function and the length of window (i.e., w) have been reported recently by Murthy and Pal [10] assuming continuous function for both histogram and membership function. For a fuzzy set "bright image plane", the membership function μ: $[0,w] \rightarrow [0,1]$ should be such that

i) μ is continuous, $\mu(0) = 0$, $\mu(w) = 1$
ii) μ is montominally non-decreasing, and
iii) $\mu(x) = 1 - \mu(w-x)$ for all $x \in [0,w]$ where $w>0$ is the length of the window.

Furthermore, μ should satisfy the bound criteria derived based on the correlation measure (equation 7). The main properties on which correlation was formulated are

P_1: If for higher values of μ_1, μ_2 takes higher values and for lower values of μ_1, μ_2 also takes lower values then $C(\mu_1, \mu_2) > 0$

P_2: If $\mu_1 \uparrow$ and $\mu_2 \uparrow$ then $C(\mu_1, \mu_2) > 0$

P_3: If $\mu_1 \uparrow$ and $\mu_2 \downarrow$ then $C(\mu_1, \mu_2) < 0$

[$\uparrow$ denotes increases and $\downarrow$ denotes decreases].

It is to be mentioned that P2 and P3 should not be considered in isolation of P_1. Had this been the case, one can cite several examples when $\mu_1 \uparrow$ and $\mu_2 \uparrow$ but $C(\mu_1, \mu_2) < 0$ and $\mu_1 \uparrow$ and $\mu_2 \downarrow$ but $C(\mu_1, \mu_2) > 0$. Subsequently, the type of membership functions which should not be considered in fuzzy set theory are categorized with the help of correlation. Bound functions h_1 and h_2 are accordingly derived [11]. They are

$$h_1(x) = 0, \quad 0 \le x \le \epsilon \qquad (35)$$
$$= x - \epsilon, \quad \epsilon \le x \le 1$$

$$h_2(x) = x + \epsilon, \quad 0 \le x \le 1 - \epsilon \qquad (36)$$
$$= 1, \quad 1 - \epsilon \le x \le 1$$

where $\epsilon = 0.25$. The bounds for membership function μ are such that

$$h_1(x) \le \mu(x) \le h_2(x) \text{ for } x \in [0,1].$$

For x belonging to any arbitrary interval, the bound functions will be changed proportionately. For $h_1 \le \mu \le h_2$, $C(h_1, h_2) \ge 0$, $C(h_1, \mu) \ge 0$ and $C(h_2, \mu) \ge 0$. The function μ lying in between h_1 and h_2 does not have most of its variation concentrated (i) in a very small interval, (ii) towards one of the end points of the interval under consideration and (iii) towards both the end points of the interval under consideration.

Figure 3 shows such bound functions. It is to be noted that Zadeh's standard S function (equation 29) satisfies these bounds.

It has been shown [10] that for detecting a minimum in the valley region of a histogram, the window length w of the μ function should be less than the distance between two peaks around that valley region.

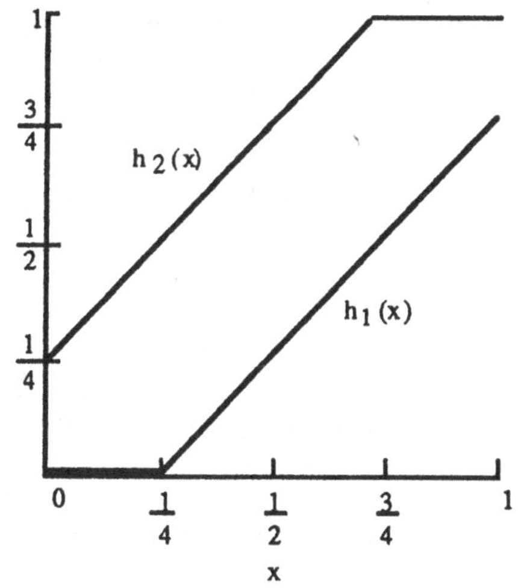

Figure 3 Bound Functions for $\mu(x)$.

H^r as an Objective Criterion

Let us now explain another way of extracting object by minimizing higher order fuzzy entropy (equation 5) of both object and background regions. Before explaining the algorithm, let us describe the membership function and its selection procedure.

Let s be an assumed threshold which partitions the image X into two parts namely, object and background. Suppose the gray level ranges [1 - s] and [s + 1 - L] denote, respectively, the object and background of the image X. An inverse π-type function as shown by the solid line in the Figure 4 is used here to obtain μ_{mn} values of X. The inverse π-type function is seen (from Fig. 4) to be generated by taking union of S(x ; (s - (L - s)), s, L) and 1 - S(x; 1, s, (s + s - 1)), where S denotes the standard S function defined by Zadeh (equation 29).

The resulting function as shown by the solid line, makes μ lie in [0.5,1]. Since the ambiguity (difficulty) in deciding a level as a member of the object or the background is maximum for the boundary level S, it has been assigned a membership value of 0.5 (i.e., cross-over point). Ambiguity decreases (i.e., degree of belongingness to either object or background increases) as the gray value moves away from s on either side. The μ_{mn} thus obtained denotes the degree of belongingness of a pixel x_{mn} to either object or background.

Since s is not necessarily the mid point of the entire gray scale, the membership function (solid line if Fig. 4) may not be a symmmetric one. It is further to be noted that one may use any linear or nonlinear equation (instead of Zadeh's standard S function) to represent the membership function in Fig. 4. Unlike the Algorithm-1, the membership function does not need any parameter selection to control the output.

Algorithm 2

Assume a threshold s, $1 \leq s \leq L$ and execute the following steps.

Step 1: Apply an inverse π - type function [Fig. 4] to get the fuzzy μ_{mn} plane,

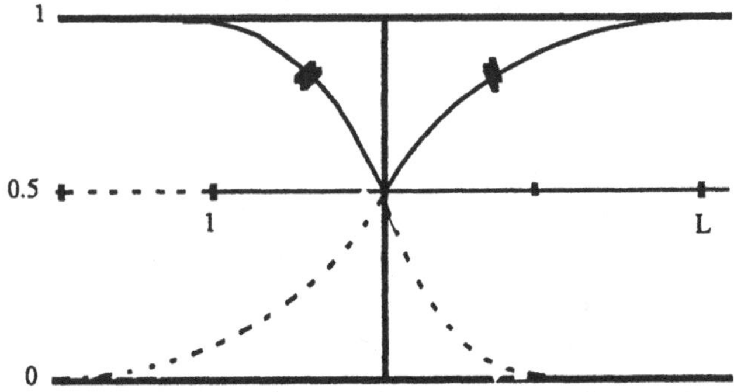

Figure 4 Inverse π function (solid line) for computing object and background entropy.

with $\mu_{mn} \in [0.5, 1]$. (The membership function is in general asymmetric).

Step 2: Compute the rth order fuzzy entropy of the object H_O^r and the background H_B^r considering only the spatially adjacent sequences of pixels present within the object and background respectively. Use the 'min' operator to get the membership value of a sequence of pixels.
Step 3: Compute the total rth order fuzzy entropy of the partitioned image as

$$H_s^r = H_O^r + H_B^r.$$

Step 4: Minimize H_s^r with respect to s to get the threshold for object background classification.

Referring back to the Table 1, we have seen that H^2 reflects the homogeneity among the supports in a set, in a better way than H^1 does. Higher the value of r, the stronger is the validity of this fact. Thus, considering the problem of object-background classification, H^r seems to be more sensitive (as r increases) to the selection of appropriate threshold; i.e., the improper selection of the threshold is more strongly reflected by H^r than H^{r-1} For example, the thresholds obtained by H^2 measure has more validity than those by H^1 (which only takes into account the histogram information). Similar arguments hold good for even higher order $(r > 2)$ entropy.

Example 2

Figures 5 and 6 show the images of Lincoln and blurred chromosome along with the histogram. Table 3 shows the thresholds obtained by comp (X) and IOAC (X) measures for various window sizes w when Zadeh's S function is used as membership function. Lincoln image is of 64x64 with 32 gray levels whereas, chromosome image is of 64x64 with 64 gray levels.

FIgure 5(a) Input.

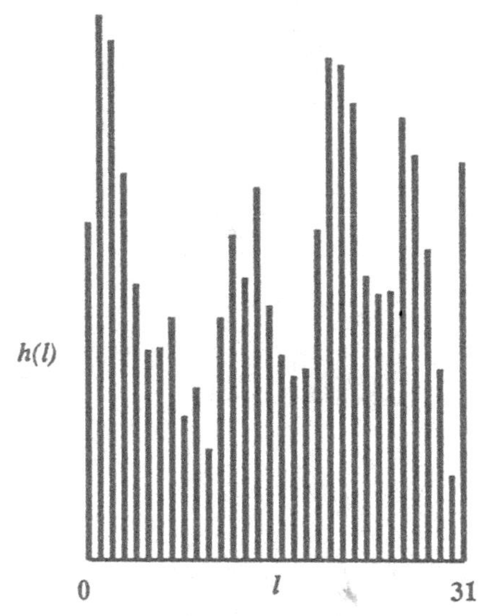

Figure 5(b) Histogram

Figure 6(a) Input.

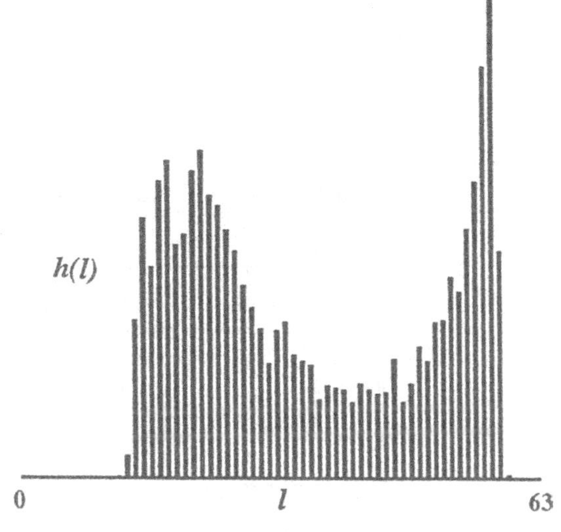

Figure 6(b) Histogram.

Figure 7(a) Threshold = 10.

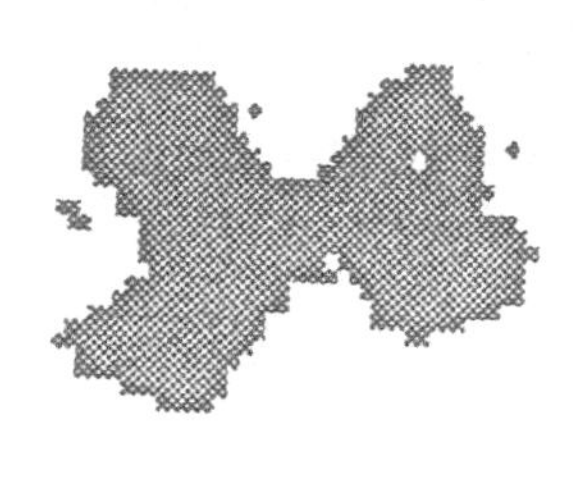

Figure 7(b) Threshold = 32.

Figure 7(c) Threshold = 56.

Table 3 Various Thresholds (* denotes global minimum)

W	Lincoln	
	Comp	IOAC
8	10	11 * 23
10	10	11 * 23
12	10	11 * 23
16	9	11

W	Chromosome	
	Comp	IOAC
12	33 56	* 30 * 51
16	55	31 * 49
20	54	32 * 46
24	52	34

Threshold produced by H^2 measure (Algorithm 2) is 8 for Lincoln image. Some typical nonfuzzy thresholded outputs of these images are shown in Figure 7. Recently, transitional correlation and within class correlation have been defined [12] based on equation (7) for image segmentation which takes both local and global information into account.

Image Enhancement

The object of enhancement technique is to process a given image so that the result is more suitable than the original for a specific application. The term 'specific' is of course, problem oriented. The techniques used here are based on the modification of pixels in the fuzzy property domain of an image. Three kinds of enhancement operations namely, contrast enhancement, smoothing and edge detection will be discussed here.

Enhancement in Property Domain [13-16]

The contrast intensification operator on a fuzzy set A generates another fuzzy set A' = INT (A) in which the fuzziness is reduced by increasing the values of $\mu_A(x)$ which are above 0.5 and decreasing those which are below it. Define this INT operator by a transformation T_1 of the membership function μ_{mn} or P_{mn} as

$$T_1(P_{mn}) = T_1'(P_{mn}) = 2P_{mn}^2, \; 0 \le P_{mn} \le 0.5 \quad (37a)$$

$$= T_1''(P_{mn}) = 1 - 2(1 - P_{mn})^2, \; 0.5 \le P_{mn} \le 1 \quad (37b)$$

$$m = 1, 2, \ldots M, \; n = 1, 2, \ldots N$$

In general, each P_{mn} or μ_{mn} in X (Eq. 1) may be modified to P'_{mn} to enhance the image X in the property domain by a transformation function T_r where

$$P'_{mn} = T_r(P_{mn}) = T_r'(P_{mn}), 0 \le P_{mn} \le 0.5 \quad (38a)$$

$$= T_1''(P_{mn}) = \; 0.5 \le P_{mn} \le 1 \quad (38b)$$

$$r = 1, 2, \ldots$$

The transformation function T_r is defined as successive applications of T_1 by the recursive relationship

$$T_s(P_{mn}) = T_1 \{T_{s-1}(P_{mn})\}, \; s = 1, 2, \ldots \quad (39)$$

and $T_{s-1}(P_{mn})$ represents the operator INT defined in (37).

This is shown graphically in Figure 8. As r increases, the curve tends to be steeper because of the successive application of INT. In the limiting case, as $r \to \infty, T_r$ produces a two-level (binary) image. It is to be noted here that, corresponding to a particular operation of T', one can use any of the multiple operations of T'', and vice versa, to attain a desired amount of enhancement. Now it is up to the user how he will interpret and exploit this flexibility depending on the problems to hand. It is further to be noted from equation 8(c) that H(X) or $\gamma(X)$ of an image decreases with its contrast enhancement [16].

The membership plane μ_{mn} for enhancing contrast around a cross-over point may be obtained from [13]

$$\mu_{mn} = G(x_{mn}) = \left[1 + \left(|\hat{x} - x_{mn}|/F_d\right)^{F_e}\right]^{-1} \tag{40}$$

where the position of cross-over points, bandwidth and hence the symmetry of the curves are determined by the fuzzifiers F_e and F_d. When $\hat{x} = x_{max}$ (maximum level in X), μ_{mn} represents an S type function. When $\hat{x}$ = any arbitrary level ℓ, μ_{mn} represents a π type function. Zadeh's standard functions do not have the provision for controlling its cross-over point. The parameters F_e and F_d of equation (40) are determined from the cross-over point across which contrast enhancement is desired.

After enhancement in the fuzzy property domain, the enhanced spatial domain x'_{mn} may be obtained from

$$x'_{mn} = G^{-1}(\mu'_{mn}),\ \alpha \le \mu'_{mn} \le 1 \tag{41}$$

where α is the value of μ_{mn} when $x_{mn}=0$.

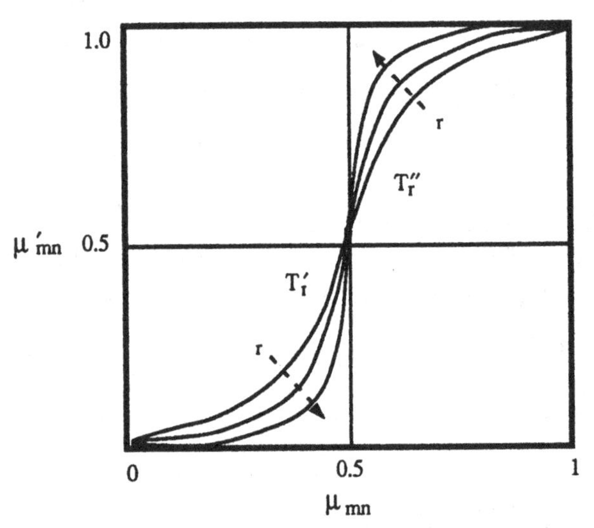

Figure 8 INT Transformation function for contract enhancement in property plane.

Smoothing Algorithm

The idea of smoothing is based on the property that image points which are spatially close to each other tend to possess nearly equal grey levels. Smoothing of an image X may be obtained by using q successive applications of 'min' and then 'max' operators within neighbors such that the smoothed grey level value of (m, n)th pixel is [13,14]

$$x'_{mn} = \max_{Q_1}{}^{q} \min_{Q_1}{}^{q} \{x_{ij}\}, \tag{42}$$

$$(i, j) \ne (m, n), (i, j) \in Q_1,\ q = 1, 2 \ldots$$

Smoothing operation blurs the image by attenuating the high spatial frequency components associated with edges and other abrupt changes in grey levels. The higher the values of Q_1 and q, the greater is the degree of blurring.

Edge Detection

If x'_{mn} denotes the edge intensity corresponding to a pixel x_{mn} then edges of the image are defined as [13,15]

$$\text{Edges} \triangleq \bigcup_m \bigcup_n x'_{mn} \tag{43a}$$

where
$$x'_{mn} = |x_{mn} - \min_Q \{x_{ij}\}| \tag{43b}$$

or,
$$x'_{mn} = |x_{mn} - \max_Q \{x_{ij}\}| \tag{43c}$$

or,
$$x'_{mn} = \max_Q \{x_{ij}\} - \min_Q \{x_{ij}\}, \quad (i,j) \in Q \tag{43d}$$

Q is a set of N coordinates (i, j) which are on/within a circle of radius R centered at the point (m,n). Equation (43 c) as compared with (43b) causes the boundary to be expanded by one pixel. Equation (43 d), on the other hand, results in a boundary of two-pixel width. It therefor appears from Eq. (43) that the better the contrast enhancement between the regions, the easier is the detection and the higher is the intensity of contours x'_{mn} among them.

Other operations based on max and min operators are available in [17,18]. Automatic selection of an appropriate enhancement operator based on fuzzy geometry is available in [19].

Edginess Measure

Let us now describe an edginess measure [20,21] based on H^1(Equation 5) which denotes an amount of difficulty is deciding whether a pixel can be called an edge or not. Let $N^3_{x,y}$ be a 3 x 3 neighborhood of a pixel at (x, y) such that

$$N^3_{x,y} = \{(x, y), (x-1,y), (x+1,y), (x,y-1), (x,y+1), (x-1, y-1), (x-1, y+1), (x+1,y-1), (x+1,y+1)\} \tag{44}$$

The edge-entropy, $H^E_{x,y}$ of the pixel (x, y), giving a measure of edginess at (x, y) may be computed as follows. For every pixel (x, y), compute the average, maximum and minimum values of gray levels over $N^3_{x,y}$. Let us denote the average, maximum and minimum values by Avg, Max, Min respectively. Now define the following parameters.

$$D = \max\{\,Max - Avg,\ Avg - Min\}\tag{45}$$

$$B = Avg\tag{46}$$

$$A = B - D\tag{47}$$

$$C = B + D\tag{48}$$

A π-type membership function is then used to compute μ_{xy} for all (x, y) $\in N^3_{x,y}$, such that $\mu(A) = \mu(C) = 0.5$ and $\mu(B) = 1$. It is to be noted that $\mu_{xy} \geq 0.5$. Such a μ_{xy}, therefore, gives the degree to which a gray level is close to the average value computed over $N^3_{x,y}$. In other words, it represents a fuzzy set "pixel intensity close to its average value", averaged over $N^3_{x,y}$. When all pixel values over $N^3_{x,y}$ are either equal or close to each other (i.e., they are within the same region), such a transformation will make all $\mu_{xy} = 1$ or close to 1. In other words, if there is no edge, pixel values will be close to each other and the μ values will be close to one(1); thus resulting in a low value of H^1. On the other hand, if there is an edge (dissimilarity in gray values over $N^3_{x,y}$), then the μ values will be more away from unity; thus resulting in a high value of H^1. Therefore, the entropy H^1 over $N^3_{x,y}$ can be viewed as a measure of edginess ($H^E_{x,y}$) at the point (x, y). The higher the value of $H^E_{x,y}$, the stronger is the edge intensity and the easier is its detection. As mentioned before, there are several ways in which one can define a π-type function as shown in Fig. 9.

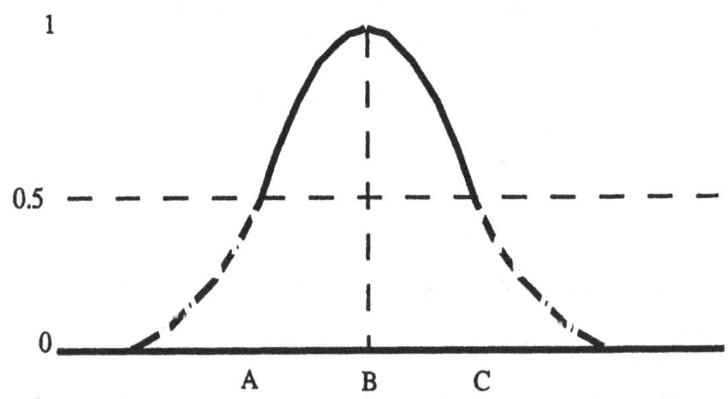

Figure 9 π function for computing edge entropy.

The proposed entropic measure is less sensitive to noise because of the use of a

dynamic membership function based on a local neighborhood. The method is also not sensitive to the direction of edges. Figure 10 shows edge output of biplane and Lincoln images. Other edginess measures are available in [13,22].

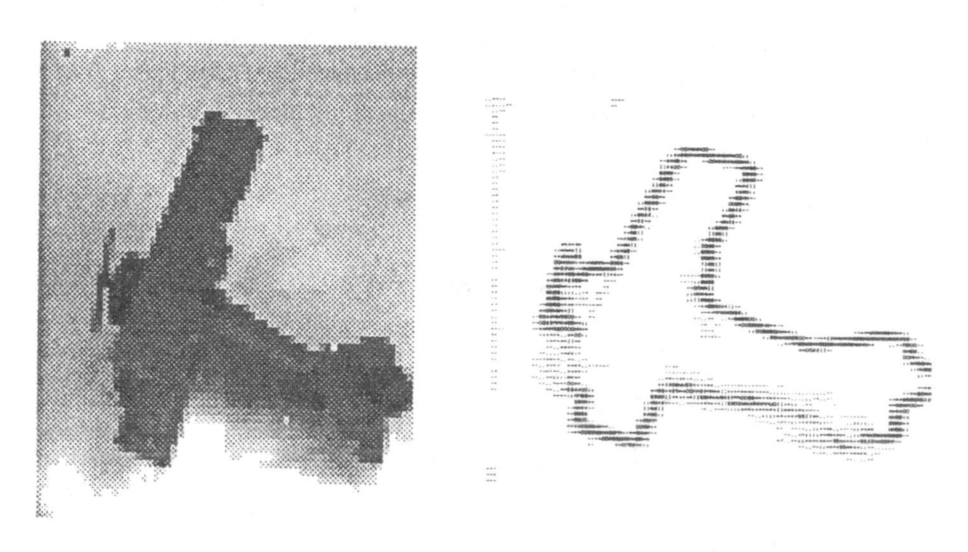

Input

Figure 10 Edge output of Lincoln and Biplane.

Fuzzy Skeletonization

The problem of skeletonization or thinning plays a key role in image analysis and recognition because of the simplicity of object representation it allows. Let us now explain a skeletonization technique [23] based on minimization of compactness property over the fuzzy core line plane. The output is fuzzy and one may obtain its nonfuzzy (crisp) single pixel width version by retaining only those pixels which have strong skeleton-membership value compared to their neighbors.

Coreline Membership Plane

After obtaining a fuzzy segmented version (as described before) of the input image X, the membership function of a pixel denoting the degree of its belonging to the subset 'Core line' (skeleton) is determined by three factors. These include the properties of possessing maximum intensity, and occupying vertically and horizontally middle positions from the edges (pixels beyond which the membership value in the fuzzy segmented image is zero) of the object.

Let x_{max} be the maximum pixel intensity in the image and $P_o(x_{mn})$ be the function which assigns the degree of possessing maximum brightness to the (m,n)th pixel. Then the simplest way to define $P_o(x_{mn})$ is

$$P_o(x_{mn}) = x_{mn}/x_{max}. \tag{49}$$

It is to be mentioned here that one may use other monotonically nondecreasing functions to define $P_o(x)$ with a flexibility of varying cross-over point. Equation 49 is the simplest one with fixed cross-over point at $x_{max}/2$.

Let x_1 and x_2 be the distances of x_{mn} from the left and right edges respectively. (The distance being measured by the number of units separating the pixel under consideration from the first background pixel along that direction). Then $P_h(x_{mn})$ denoting the degree of occupying the horizontally central position in the object is defined as

$$\begin{aligned} P_h(x_{mn}) &= \frac{x_1}{x_2} \quad \text{if } d(x_1,x_2) \le 1 \text{ and } x_1 \le x_2, \\ &= \frac{x_2}{x_1} \quad \text{if } d(x_1,x_2) \le 1 \text{ and } x_1 \ge x_2, \\ &= \frac{2x_1}{x_2(x_1+x_2)} \quad \text{if } d(x_1,x_2) > 1 \text{ and } x_1 < x_2, \\ &= \frac{2x_2}{x_1(x_1+x_2)} \quad \text{if } d(x_1,x_2) > 1 \text{ and } x_1 > x_2, \end{aligned} \tag{50}$$

where $d(x_1,x_2) = |x_1 - x_2|.$

Similarly, the vertical function is defined as

$$P_v(x_{mn}) = \frac{y_1}{y_2} \quad \text{if } d(y_1, y_2) \le 1 \text{ and } y_1 \le y_2,$$

$$= \frac{y_2}{y_1} \quad \text{if } d(y_1, y_2) \le 1 \text{ and } y_1 \ge y_2,$$

$$= \frac{2y_1}{y_2(y_1 + y_2)} \quad \text{if } d(y_1, y_2) > 1 \text{ and } y_1 < y_2,$$

$$= \frac{2y_2}{y_1(y_1 + y_2)} \quad \text{if } d(y_1, y_2) > 1 \text{ and } y_1 > y_2, \tag{51}$$

Equations (50,51) assign high values (≈ 1.0) for pixels near core and low values to pixels away from the core. The factor $(x_1 + x_2)$ or $(y_1 + y_2)$ in the denominator takes into consideration the extent of the object segment so that there is an appreciable amount of changes in the property value for the pixels not belonging to the core.

These primary membership functions P_o, P_h and P_v may be combined as either

$$\mu_C(x_{mn}) = \max\{\min(P_o, P_h), \min(P_o, P_v), \min(P_h, P_v)\} \tag{52}$$

or

$$\mu_C(x_{mn}) = W_1 P_o + W_2 P_h + W_3 P_v \tag{53a}$$

$$\text{with } w1 + w2 + w3 = 1 \tag{53b}$$

to define the grade of belonging of x_{mn} to the subset 'Core Line' of the image.

Equation (52) involves connective properties using max and min operators such that $\mu_c = 1$ when at least two of the three primary properties take values of unity. All the three primary membership values are given equal weight in computing the μ_c value. Equation (53), on the other hand, involves a weighted sum (weights being denoted by W_1, W_2 and W_3). Usually, one can consider the weight W_1 attributed to P_o (property corresponding to pixel intensity) to be higher than the other two and $W_2 = W_3$.

Equation (52) or (53) therefore extracts (using both gray level and spatial information) the subset 'Core line' such that the membership value decreases as one moves away towards the edges (boundary) of object regions.

Optimum α-cut

Given the $\mu_c(x_{mn})$ plane developed in the previous stage with the pixels having been assigned values indicating their degree of membership to 'Core line', the optimum (in the sense of minimizing ambiguity in geometry or in spatial domain) skeleton can be extracted from one of its α-cuts having minimum comp(μ) value (Eq. (16)). The α-cut of $\mu_c(x_{mn})$ is defined as

$$\mu_{C\alpha} = \{x_{mn} \in X | \ \mu_C(x_{mn}) \geq \alpha, \ 0 < \alpha < 1\} \tag{54}$$

Modification of α will therefore result in different fuzzy skeleton planes with varying comp(μ) value. As α increases, the comp(μ) value initially decreases to a certain minimum and then for a further increase in α, the comp(μ) measure increases.

The initial decrease in comp(μ) value can be explained by observing that for every value of α, the border pixels having μ-values less than α are not taken into consideration. So, both area (Eq. (13)) and perimeter (Eq. (14)) are less than those for the previous value of α. But the decrease in area is more than the decrease in its perimeter and hence the compactness (Eq. 16) decreases (initially) to a certain minimum corresponding to a value $\alpha = \alpha'$, say.

Further increase in α (i,e,. for $\alpha > \alpha'$), results in a $\mu_{C\alpha}$ plane consisting of a number of disconnected regions (because majority of core line pixels being dropped out). As a result, decrease in perimeter here is more than the decrease in area and comp(μ) increases. The $\mu_{C\alpha'}$ plane having minimum compactness value can be taken as an optimum fuzzy skeleton version of the image X. This is optimum in the sense that for any other selection of α (i,e,. $\alpha \neq \alpha'$) the comp(μ) value would be greater.

If a nonfuzzy (crisp) single-pixel width skeleton is deserved, it can be obtained by a contour tracking algorithm [24] which takes into account the direction of contour, multiple crossing pixels, lost path due to spurious wiggles etc. based on octal chain code.

Fig. 11 shows the optimum fuzzy skeleton of biplane image (Fig. 10). This corresponds to $\alpha = 0.55$. The connectivity of the skeleton in the optimum version can be preserved, if necessary, by inserting pixels having intensity equal to the minimum of those of pairs of neighbors in the object.

Figure 11 Optimum fuzzy skeleton of biplane.

PRIMITIVE EXTRACTION

In picture recognition and scene analysis problems, the structural information is very abundant and important, and the recognition process includes not only the capability to assign the input pattern to a pattern class, but also the capacity to describe the characteristics of the pattern that make it ineligible for assignment to another class. In these cases the recognition requirement can only be satisfied by a description of pattern-rather than by classification.

In such cases complex patterns are described as hierarchical or tree-like structures of simpler subpatterns and each simpler subpattern is again described in terms of even simpler subpatterns and so on. Evidently, for this approach to be advantageous, the simplest subpatterns, called pattern primitives are to be selected.

Another activity which needs attention in this connection is the subject of shape-analysis that has become an important subject in its own right. Shape analysis is of primal importance in feature/primitive selection and extraction problems. Shape analysis also has two approaches, namely, description of shape in terms of scalar measurements and through structural descriptions. In this connection, it needs to be mentioned that shape description algorithms should be information-preserving in the sense that is is possible to reconstruct the shapes with some reasonable approximation from the descriptors.

This section presents a method [24] to demonstrate an application of the theory of fuzzy sets in automatic description and primitive extraction of gray-tone edge-detected images. The ultimate aim is to recognize the pattern using syntactic approach as described in the next section.

The method described here provides a natural way of viewing the primitives in terms of arcs with varying grades of membership from 0 to 1.

Encoding

The gray tone contour of an image can be encoded into one-dimensional symbol strings using the rectangular (octal) array method. The directions of the octal codes are shown in Figure 12. An octal code is used to describe a w-pixel ($w>1$) length contour by taking the maximum of its grades of membership corresponding to 'vertical', 'horizontal' and 'oblique' lines. This approximation of using w-pixel (instead of one-pixel) length line saves computational time and storage requirement without affecting the system performance.

$\mu_V(x)$, $\mu_H(x)$ and $\mu_{ob}(x)$ representing the membership functions for vertical, horizontal and oblique lines respectively of a line segment x marking an angle θ with the horizontal line H (Figure 13) may be defined as [13,24].

$$\mu_V(x) = 1 - |1/m_x|^{F_e}, |m_x| > 1, \qquad (55)$$
$$= 0 \text{ otherwise}$$

$$\mu_H(x) = 1 - |m_x|^{F_e}, |m_x| < 1, \qquad (56)$$
$$= 0 \text{ otherwise}$$

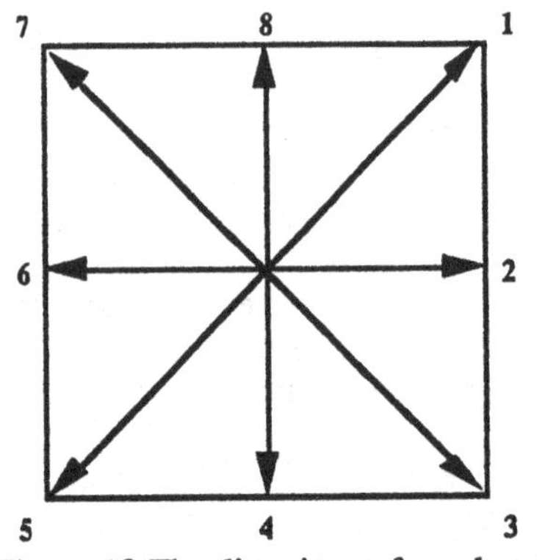

Figure 12 The directions of octal codes.

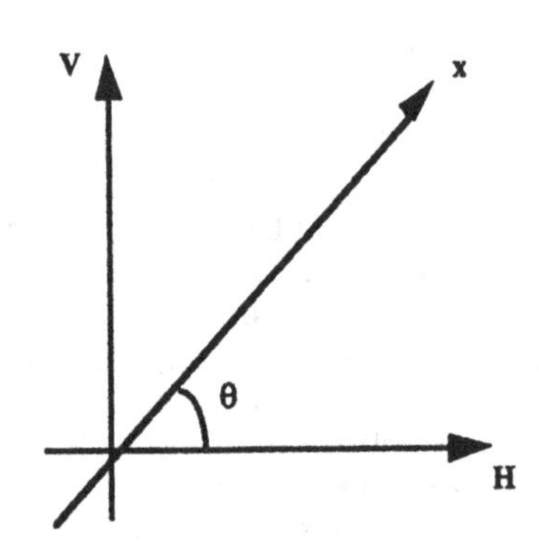

Figure 13 Membership function for vertical and horizontal lines.

$$\mu_{ob}(x) = 1 - |(\theta - 45)/45|^{F_e}, 0 < |m_X| < \infty, \quad (57)$$
$$= 0 \quad \text{otherwise}$$

F_e is a positive constant which controls the fuzziness in a set and $m_X = \tan\theta$.
The equations (55-57) are such that

$\mu_v(x) \rightarrow 1$ as $|\theta| \rightarrow 90°$,

$\mu_H(x) \rightarrow 1$ as $|\theta| \rightarrow 0°$,

$\mu_{ob}(x) \rightarrow 1$ as $|\theta| \rightarrow 45°$,

and $\mu_v(x) \lessgtr 1$ as $|\theta| \lessgtr 45°$,

The details of encoding technique are available in [13].

Segmentation and Contour Description

The next task before extraction of primitives and description of contours is the process of segmentation of the octal coded strings. Splitting up of a chain is dependent on the constant increase/decrease in code values. For extracting an arc, the string is segmented at a position whenever a decrease/increase after constant increase/decrease in values of codes is found [13]. Again, if the number of codes between two successive changes exceeds a prespecified limit, a straight line is said to exist between two curves. In the case of a closed curve, a provision may be kept for increasing the length of the chain by adding first two starting codes to the tail of the string. This enables one to take the continuity of the chain into account in order to reflect its proper segmentation [13].

After segmentation one needs to provide a measure of curvature along with direction of the different arcs and also to measure the length of lines in order to extract the primitives. The degree of 'arcness' of a line segment x is obtained using the function

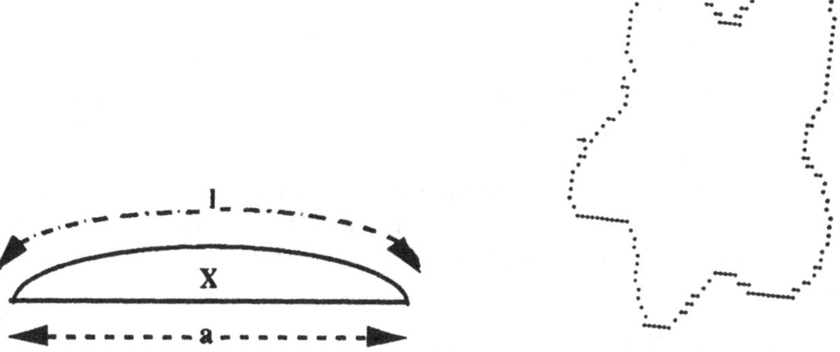

Figure 14 Membership function for arc. FIgure 15 Nuclear Pattern of brain cell.

$$\mu_{arc}(x) = (1 - a/l)^{F_e}. \tag{58}$$

a is the length of the line joining the two extreme points of an arc x (Figure 14), ℓ is the arc-length such that the lower the ratio a/ℓ is, the higher is the degree of 'arcness'.

For example, consider a sequence of codes 5 6 6 7 denoting an arc x. For computing its ℓ note that if a code represents an oblique line, the corresponding increase in arc-length would be $\sqrt{2}$, otherwise increase is by unity. Arc diameter a is computed by measuring the resulting shifts Δm and Δn of spatial coordinates (along mth and nth axes) due to those codes in question. For the aforesaid example we have

$$\Delta m = 1 + 0 + 0 + -1 = 0,$$

$$\Delta n = -1 - 1 - 1 - 1 = -4,$$

$$a = \sqrt{\Delta m^2 + \Delta n^2} = 4,$$

$$\ell = 4.828,$$

$$\mu_{arc}(x) = 0.643 \text{ (for } F_e = 0.25).$$

Since the initial code (5) is lower than the final code (7), the sense of the curve is positive (clockwise).

Similarly, for sequences 5 6 and 5 6 7 the μ_{arc} values are respectively 0.52 and 0.682. The figures thus obtained for the different sequences agree well with our intuition as far as their degree of arcness (curvature) is concerned. Also note that the sequences like 5 5 6 6 7 7 and 5 5 5 6 6 6 7 7 7 have the same μ_{arc} values as obtained with the sequence 5 6 7. Similarly, the sequences 5 5 6 6 and 5 6 have the same μ_{arc} value.

Example 3

To explain the aforesaid features, let us consider the Fig. 15 showing a two-tone contour of nuclear pattern of brain neurosecretory cells [25]. The string descriptions of Fig. 15 in terms of arcs (of different arcness) and lines are shown below.

11, 187, 7881, 112233, 321, 11, 112, 223, 334, 44, 445, 543, 3345, 55, 56, 667, 7665, 55, 5667, 78, 776, 66, 678, 11

$L\overline{V}_{.68}V_{.64}V_{.68}\overline{V}_{.64}LV_{.49}V_{.51}V_{.49}LV_{.51}\overline{V}_{.68}V_{.49}LV_{.52}V_{.51}\overline{V}_{.64}LV_{.64}V_{.52}\overline{V}_{.49}LV_{.68}L$

Here, L, V, and $\overline{V}$ denote the straight line, 'clockwise arc' and 'anticlockwise arc' respectively. The suffix of V represents the degree of arcness of the arc V. The positions of segmentation are shown by a comma (,).

It is to be mentioned here that the approach adopted here to define and to extract arcs with varying grades of membership is not the only way of doing this. One may change the procedure so as to result in segments with membership values different from those mentioned here.

FUZZY SYNTACTIC ANALYSIS

The syntactic approach to pattern recognition involves the representation of a pattern by a string of concatenated subpatterns called primitives. These primitives are considered to be the terminal alphabets of a formal grammar whose language is the set of patterns belonging to the same class. Recognition therefor involves a parsing of the string.

The syntactic approach has incorporated the concept of fuzzy sets at two levels. First, the pattern primitives are themselves considered to be labels of fuzzy sets, i.e., such subpatterns as 'almost circular arcs', 'gentle', 'fair' and 'sharp' curves are considered. Secondly, the structural relations among the subpatterns may be fuzzy, so that the formal grammar is fuzzified by the weighted production rules and the grade of membership of a string is obtained by min-max composition of the grades of the production used in the derivations. Inference of a fuzzy grammar is another interesting problem which infers from the specified fuzzy language, the productions as well as the weights of these rules. In this section we will be explaining the elementary notions of fuzzy grammar with examples.

The formal definition of a fuzzy grammar is as follows:

Definition: A fuzzy grammar FG is a 6-tuple

$$FG = (V_N, V_T, P, S, J, \mu)$$

where

V_N : a set of non-terminals (which are essentially labels for certain fuzzy subsets called fuzzy syntactic categories of V_T^*)

V_T : a set of terminals, such that $V_N \cap V_T = 0$ (null set)

P : a set of production (or rewriting) rules of the type $\alpha \rightarrow \beta$ (α is replaced by β), where $\alpha, \beta \in (V_N \cup V_T)^*$

S : a starting symbol, such that $S \in V_N$

J : $\{r_i | \ i = 1, 2 \ldots n, \ n = \text{cardinality of P}\}$, is the set of labels for the production rules

μ : a mapping $\mu: J \rightarrow [0,1]$ such that $\mu(r_i)$ denotes the membership in P of the rule labelled r_i

V_T^* : the set of finite strings obtained by the concatenation of elements of V_T

$(V_N \cup V_T)^*$: the set of finite strings obtained by the concatenation of element of $V_N \cup V_T$

A fuzzy grammar FG generates a fuzzy language L(FG) as follows:

A string $X \in V_T^*$ is said to be in L(FG) iff it is derivable from S and its grade of membership $\mu_{L(FG)}(X)$ in L(FG) is > 0, where

$$\mu_{L(FG)}(X) = \max_{1 \le k \le m} \left[\min_{1 \le i \le l_k} \mu\left(r_i^k\right) \right] \tag{59}$$

where m in the number of derivations that X has in FG; l_k is the length of the kth derivation chain, k = 1(1)m; and r_i^k is the label of the ith production used in the kth derivation chain, i = 1, 2, . . . , l_k.

Clearly, if a production $\alpha \rightarrow \beta$ be visualized as a chain link of strength $\mu(r)$, r being the label of $\alpha \rightarrow \beta$, then the strength of a derivation chain is the strength of its weakest link, and hence

$\mu_{L(FG)}(X)$ = strength of the strongest derivation chain from S to X for all $X \in V_T^*$

Example 4: Suppose FG_1=({A, B, S}, {a, b}, P, S, {1, 2, 3, 4}, μ) where J, P and μ are as follows

1:	$S \rightarrow AB$	with	$\mu(1) = 0.8$
2:	$S \rightarrow aSb$		$\mu(2) = 0.2$
3:	$A \rightarrow a$		$\mu(3) = 1$
4:	$B \rightarrow b$		$\mu(4) = 1$

Clearly, the fuzzy language generated is $FL_1 = \{X \mid X = a^n b^n, n = 1, 2, \ldots\}$

with $\mu_{FL_1}\left(a^n b^n\right) = 0.8 \quad \text{if } n = 1$

$= 0.2 \quad \text{if } n \ge 1$

Example 5: Consider the fuzzy grammar $FG_2 = (\{S, A, B\}, \{a, b, c\}, P, S, J, \mu)$ where J, P and μ are as follows

r_1:	$S \rightarrow aA$	$\mu(r_1) = \mu_H(a)$
r_2:	$A \rightarrow bB$	$\mu(r_2) = \mu_V(b)$
r_3:	$B \rightarrow c$	$\mu(r_3) = \mu_{ab}(c)$

the primitives a, b, c being 'horizontal', 'vertical' and 'oblique' directed line segments respectively; the terms 'horizontal; 'vertical' and 'oblique' are taken to be fuzzy with membership values μ_H, μ_V and μ_{ob} respectively as defined in the previous section.

Further, the concatenation considered is of the 'head-tail' type. Hence the only string generated is X = abc which is in reality a triangle having membership

$$\mu_{L(FG_2)}(abc) = \min(\mu_H(a), \mu_V(b), \mu_{ob}(c))$$

which attains its maximum value 1 when abc is an isosceles right triangle. Thus $L(FG_2)$ is the fuzzy set of isosceles right triangles.

The membership of a pattern triangle given in Fig. 16b is

$$\min(1.0, 1.0, 0.66) = 0.66$$

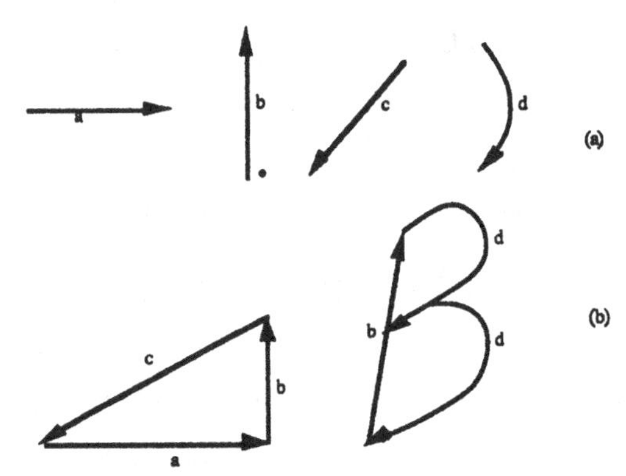

Figure 16 (a) Primitive (b) Production of Triangle and Letter B.

Example 6: Consider the following fuzzy grammar for generating the fuzzy set representing the English upper case letter B

$V_N = \{S, A, B, C, D)$

$V_T = \{a, b\}$

where the primitive a denotes a directed 'vertical' (fuzzy) line segment and b denotes a directed arc (clockwise). The concatenation considered here is again of the 'head-tail' type.

Also J, P and μ are as follows

r_1:	$S \rightarrow aB$	$\mu(r_1) = \mu_V(a)$
r_2:	$B \rightarrow aC$	$\mu(r_2) = \mu_V(a)$
r_3:	$C \rightarrow bD$	$\mu(r_3) = \mu_{Cir}(b)$
r_4:	$D \rightarrow b$	$\mu(r_4) = \mu_{Cir}(b)$

The string generated is X=aabb having the following membership in set B.

$$\mu_B(X) = \min(\mu_V(a)_l, \mu_V(a)_u, \mu_{Cir}(b)_l, \mu_{Cir}(b)_u)$$

where the suffices ℓ and u denote the locations ('lower' and 'upper') of the primitives a and b.

For the pattern given in the Fig. 16b

$$\mu_V(a)_u = \mu_V(a)_l = 0.83, \mu_{Cir}(a)_u = 0.36, \mu_{Cir}(a)_l = 0.5$$

so that $\mu_B(X) = 0.36$

DePalma and Yau [26] introduced the concept of fractionally fuzzy grammars with a view to tackling some of the drawbacks of fuzzy grammars which make them unsuitable for use in pattern recognition problems. Some of these drawbacks are as follows:

(i) Memory requirements are greatly increased when fuzzy grammars are implemented with the help of parsing algorithms that require backtracking. This is because when fuzzy grammars are being used, it is not sufficient to keep track of the current derivation tree alone. The fuzzy value at each preceding step must also be simultaneously remembered at each node, in case back-tracking in needed at some step.

(ii) All strings in the language L(FG) generated by a fuzzy grammar FG can be classified into a finite number of subsets by their membership in the language. The number of such subsets is strictly limited by the number of productions in FG.

Definition: A fractionally fuzzy grammar (FFG) is a 7-tuple

$$FFG = (V_N, V_T, P, S, J, g, h)$$

where V_N, V_T, P, S, J are as before, and g and h are mappings from J into the set of non-negative integers such that $g(r) \leq h(r)$ for all $r \in J$. Various applications of fuzzy and fractionally fuzzy grammars are available in [13,25,27].

The incorporation of the element of fuzziness in defining 'sharp', 'fair' and 'gentle' curves in the grammars enables one to work with a much smaller number of primitives. By introducing fuzziness in the physical relations among the primitives, it was also possible to use the same set of production rules and non-terminal at each stage. This is expected to reduce, to some extent, the time required for parsing in the sense that parsing needs to be done only once at each stage, unlike the case of the non-fuzzy approach [28], where each string has to be parsed more than once, in general, at each stage. However, this merit has to be balanced against the fact that the fuzzy grammars are not as simple as the corresponding nonfuzzy grammars.

Acknowledgements

This work was done while the author held an NRC-NASA research Associateship at the Johnson Space Center, Houston, Texas. The author gratefully acknowledges Dr. Robert N. Lea for his interest in this work, Ms. Dianne Rader, Ms. Kim Herhold, and Mr. Todd Carlson for typing the manuscript and Mr. Albert Leigh for his assistance in getting some results.

References

1. A Kaufmann, *Introduction to the Theory of Fuzzy Subsets-Fundamental Theoretical Elements*, vol 1, Academic Press, NY, 1975.
2. A De Luca and S. Termini, A definition of nonprobabilistic entropy in the setting of fuzzy set theory, *Inform and Control*, vol 20, pp 301-312, 1972.
3. N.R. Pal and S.K. Pal, Higher order fuzzy entropy and hybrid entropy of a set, *Information Sciences* (to appear).
4. C.A. Murthy, S.K.Pal and D. Dutta Majumder, Correlation between two fuzzy membership functions, *Fuzzy Sets and Systems*, vol. 7, no. 1, pp 23-38, 1985.
5. A. Rosenfeld, The fuzzy geometry of image subsets, *Patt. Recog. Lett.*, vol 2, pp 311-317, 1984.
6. A. Rosenfeld and S. Haber, The perimeter of a fuzzy set, *Technical Report*, University of Maryland, Center for Automation Research, TR-8, 1983.
7. S.K. Pal and A. Ghosh, Index of area coverage of fuzzy image subsets and object extraction, *Patt. Recog. Lett.*, vol. 11, pp. 831-841, 1990.
8. L.A. Zadeh, K.S. Fu, K. Tanaka and M. Shimura, *Fuzzy Sets and Their Applications to Cognitive and Decision Processes*, Academic Press, London, 1975.
9. S.K. Pal and A. Rosenfeld, Image enhancement and thresholding by optimization of fuzzy compactness, *Patt. Recog. Lett.*, vol. 7, pp 77-86, 1988.
10. C.A. Murthy and S.K. Pal, Histogram thresholding by minimizing graylevel fuzziness, *Information Sciences* (to appear).
11. C.A. Murthy and S.K. Pal, Bounds for membership functions: correlation based approach, *Information Sciences* (to appear).
12. S.K. Pal and A. Ghosh, Image segmentation using fuzzy correlation, *Information Sciences* (to appear).
13. S.K. Pal and D. Dutta Majumdar, *Fuzzy Mathematical Approach to Pattern Recognition*, John Wiley and Sons, (Halsted Press), NY, 1986.
14. S.K. Pal and R.A. King, Image enhancement using smoothing with fuzzy sets, *IEEE Trans. Syst., Man and Cyberns.*, vol. SMC-11, pp. 494-501, 1981.
15. S.K. Pal and R.A. King, On edge detection of x-ray images using fuzzy set, *IEEE Trans. Patt. Anal. and Machine Intell.*, vol. PAMI-5, pp. 69-77, 1983.
16. S.K. Pal, A note on the quantitative measure of image enhancement through fuzziness, *IEEE Trans. Patt. Anal. and Machine Intell.* vol. PAMI-4, pp. 204-208, 1982.
17. Y. Nakagowa and A. Rosenfeld, A note on the use of local min and max operations in digital picture processing, *IEEE Trans. Syst., Man and Cyberns.*, vol. SMC-8, pp. 632-635, 1978.
18. S. Peleg and A. Rosenfeld, A min-max medial axis transformation, *IEEE Trans. Patt. Anal. and Mach. Intell.* vol. PAMI-3, no. 2, 1981.
19. M.K. Kundu and S.K. Pal, Automatic selection of object enhancement operator with quantitative justification based on fuzzy set theoretic measure, *Patt. Recog. Lett.*, vol. 11, pp. 811-829, 1990.

20. N.R. Pal, *On Image Information Measure and Object Extraction*, Ph. D. Thesis, Indian Statistical Institute, Calcutta, India, March 1990.
21. S.K. Pal and N.R. Pal, Higher order entropy, hybrid entropy and their applications, *Proc. INDO-US Workshop on Spectrum analysis in one and two dimensions*, Nov 27-29, 1990, New Delhi, NBH Oxford Publishing Co. New Delhi (to appear).
22. S.K. Pal, A measure of edge ambiguity using fuzzy sets, *Patt. Recog. Lett.*, vol. 4, pp. 51-56, 1986.
23. S.K. Pal, Fuzzy skeletonization of an image, *Patt. Recog. Lett.*, vol. 10, pp. 17-23, 1989.
24. S.K. Pal, R.A. King and A.A. Hashim, Image description and primitive extraction using fuzzy set, *IEEE Trans. Syst., Man and Cyberns.*, vol. SMC-13, pp. 94-100, 1983.
25. S.K. Pal and A. Bhattacharyya, Pattern recognition technique in analyzing the effect of thiourea on brain neurosecretory cells, *Patt. Recog. Lett.*, vol. 11, pp. 443-452, 1990.
26. G.F. DePalma and S.S. Yau, Fractionally fuzzy grammars with applications to pattern recognition, in [8], pp. 329-351.
27. A. Pathak and S.K. Pal, Fuzzy grammars in syntactic recognition of skeletal maturity from X-rays, *IEEE Trans. Syst., Man and Cyberns.*, vol. SMC-16, pp. 657-667, 1986.
28. K.S. Fu, *Syntactic Pattern Recognition and Applications*, Prentice-Hall, N.J., 1982.

8

FUZZY SETS IN NATURAL LANGUAGE PROCESSING

Vilém Novák

Czechoslovak Academy of Sciences,
Mining Institute, Studentská 1768,
708 00 Ostrava-Poruba, Czechoslovakia

1 INTRODUCTION

Natural language is one of the most complicated structures a man has met with. It plays a fundamental role not only in human communication but even in human way of thinking and regarding the world. Therefore, it is extremely important to study it in all its respects. Much has been done in understanding its structure, especially the phonetic and syntactic aspects. Less, however, is understood its semantics. There are many linguistic systems, often based on set theory and logic, attempting to grasp (at least some phenomena) of the natural language. However, none them is fully acceptted and satisfactory in all respects.

A serious obstacle on the way to this goal is, besides the complexity mentioned, also the vagueness of the meaning of separate lexical units as well as of the sentences and longer text. On the other side, the capability of human mind to take vagueness into account and to handle it, which is reflected in the semantics of natural language, is the main cause of the extreme power of natural language to convey relevant and succint information. There is no way out than to cope with the vagueness in the models of natural language semantics.

Fuzzy set theory is a mathematical theory whose program is to provide us with methods and tools which may make us possible to grasp vague phenomena instrumentally. Therefore, it seems to be appropriate for using in modelling of natural language semantics.

In this paper, we provide the reader with an overview of the main results obtained so far in processing of some phenomena of the semantics of natural language using fuzzy sets.

2 FUZZY SETS

In this section, we briefly touch the notion of a fuzzy set, especially those aspects which are important for our further explanation. Words and more complex syntagms [1] of natural language can in general be considered as names of properties encountered by a man in the world.

An object is a phenomenon to which we concede its individuality keeping it together and separating it from the other phenomena. Objects are usually accompanied by properties. In general, however, the same property accompanies more objects. If all such objects are grouped together then they can be seen as one, new object of a special kind. A grouping of objects being seen as an object is called a *class*. Hence, if φ is a property then there is a class X of objects x having φ. In symbols we can write

$$X = \{x; \varphi(x)\}. \tag{1}$$

If the property φ is simple and sharp then the class X forms a set. However, most properties a man meets in the world are not of this kind. Then the class X is not separeted sharply, i.e. there is no way how to name or imagine all the objects x from X without any doubt whether a given object x has the property φ, or not. Thus, we encountered the phenomenon of vagueness. The above mentioned doubt, which probably stems from the inner, still not understood, complexity of φ is a core of the phenomenon of vagueness being encountered. Classical mathematics has no other possibility than to model the grouping X using (sharp) sets. Therefore, the result cannot be satisfactory from the very beginning. Unlike classical set theory, fuzzy set theory attempts at finding a more suitable model of the class (1).

Let us take the objects x from some sufficiently big set U called the *universe*. Note that this assumption is not restrictive since such a set always exist. For example, consider the property φ :=*to be a small number*[2]. Then there surely exists a number $z \in N$ which is not small (e.g. $z = 2^{10}$ and we may put $U = \{x \in N; x \leq z\}$).

Our doubt whether an object $x \in U$ has the given property φ can be expressed by means of a certain scale L having the smallest 0 and greatest 1 elements, respectively. Thus, 1 expresses that $\varphi(x)$ (x has the property φ) with no doubt while 0 means that $\varphi(x)$ does not hold at all. We obtain a function

$$A : U \longrightarrow L \tag{2}$$

assigning an element

$$Ax \in L$$

[1] A syntagm is any part of a sentence (even a word or a whole sentence) that is constructed according to the gramatical rules.

[2] A natural number, for simplicity.

from the scale L to each element

$$x \in U.$$

This function serves us as a certain characterization of the class X in (1) and it is called the ***fuzzy set***. We can view the fuzzy set A as a set

$$\{Ax/x; x \in U\}. \tag{3}$$

The element

$$Ax \in L$$

is called the ***membership degree*** of x and thus (2) is often called the ***membership function*** of the fuzzy set A. One can see that a fuzzy set is identified with its membership function. If A is a fuzzy set (3) in the universe U then we write $A \underset{\sim}{\subseteq} U$. The scale L is usually put to be $L =< 0,1 >$ and it is assumed to form the structure

$$\mathcal{L} =<< 0,1 >, \vee, \wedge, \otimes, \rightarrow, 0, 1 > \tag{4}$$

where $\vee$ and $\wedge$ are the operations of ***supremum*** (maximum) and ***infimum*** (minimum) respectively, $\otimes$ is the operation of ***bold product*** defined by

$$a \otimes b = 0 \wedge (a + b - 1)$$

and $\rightarrow$ is the operation of ***residuum*** defined by

$$a \rightarrow b = 1 \wedge (1 - a + b)$$

for all the $a, b \in < 0,1 >$.

There are deep reasons for the choice of this structure. The reader may find them in [13,16]. The operations with fuzzy sets form the structure 4. The basic ones are

union

$$C = A \cup B \text{ iff } Cx = Ax \vee Bx$$

intersection

$$C = A \cap B \text{ iff } Cx = Ax \wedge Bx$$

bold intersection

$$C = A \cap B \text{ iff } Cx = Ax \otimes Bx$$

residuum

$$C = A \ominus B \text{ iff } Cx = Ax \rightarrow Bx$$

On the basis of residuum, one can define the complement $\overline{A} = A \ominus \emptyset$ [3] where $\emptyset$ is the empty fuzzy set

$$\emptyset = \{0/x; x \in U\}$$

[3]This definition gives $\overline{A}x = 1 - Ax$ for all $x \in U$ which is the usual definition of the complement.

In modelling of natural language semantics it is necessary to introduce also new, additional operations. Put

$$a \leftrightarrow b = (a \rightarrow b) \wedge (b \rightarrow a)$$

(biresiduation) and

$$a^p = \underbrace{a \otimes \ldots \otimes a}_{p\text{-times}}$$

(power) for all the $a, b \in < 0, 1 >$.

When introducing a new $n - ary$ operation o on L, the following *fitting condition* must be fulfilled: there are $p_1, \ldots, p_n$ such that

$$(a_1 \leftrightarrow b_1)^{p_1} \otimes \ldots \otimes (a_n \leftrightarrow b_n)^{p_n} \leq o(a_1, \ldots, a_n) \leftrightarrow o(b_1, \ldots, b_n) \qquad (5)$$

holds for every $a_i, b_i \in L, i = 1, \ldots, n$. The justification of the fitting condition can be found in [16,15]. Note that all the basic operations fulfil the fitting condition. Moreover, the folowing holds true:

Theorem 1 *All the the operations derived from the operations fulfilling the fitting condition fulfil it, as well.*

Proof – see [16].

The following operations are fitting:

product

$$a \cdot b$$

bounded sum

$$a \oplus b = 1 \wedge (a + b)$$

concentration

$$\mathrm{CON}(a) = a^2$$

dilation [4]

$$\mathrm{DIL}(a) = 2a - a^2$$

intensification

$$\mathrm{INT}(a) = \begin{cases} 2a^2 & a \in < 0, 0.5 > \\ 1 - 2(1-a)^2 & a \in (0.5, 1 > \end{cases}$$

for all the $a, b \in < 0, 1 >$.

[4] The widely used operation of dilation $\mathrm{DIL}(a) = a^{0.5}$ is not fittig and thus it cannot be used.

The operations in L lead to the operations with fuzzy sets as follows. Let

$$o : L^n \longrightarrow L$$

and $A_1, \ldots, A_n \subsetapprox U$ be fuzzy sets. Then o is a basis of the operation O assigning a fuzzy set $C \subsetapprox U$ to $A_1, \ldots, A_n$ when we put

$$C = O(A_1, \ldots, A_n) \text{ iff } Cx = o(A_1x_1, \ldots, A_nx_n) \tag{6}$$

for every $x \in U$.

For example, we can define the operation of bounded sum of fuzzy sets by putting

$$C = A \uplus B \text{ iff } Cx = Ax \oplus Bx$$

for every $x \in U$.

A very important notion is that of a fuzzy *cardinality* of a fuzzy set. There are several kinds of them [13,22]. We will use the following ones defined for fuzzy sets with finite support: *Absolute fuzzy cardinality* of $A \subsetapprox U$:

$$\text{FCard}(A) = \{\alpha_n/n; n \in N\} \tag{7}$$

where

$$\alpha_n = \bigvee\{\beta; \text{Card}(A_\beta) = n\}$$

and A_β is a β cut of A. *Relative fuzzy cardinality* of A with respect to B where $A, B \subsetapprox U$:

$$\text{FCard}_A(B) = \{\alpha_r/r; r \in Re\} \tag{8}$$

where

$$\alpha_r = \bigvee\{\beta; \frac{\text{Card}(A_\beta \cap B_\beta)}{\text{Card}(A_\beta)} = r\}$$

The notions introduced above will be used in the sequel. For other notions and operations with fuzzy sets see e.g. [13,4].

3 THE USE OF FUZZY SETS IN MODELLING OF NATURAL LANGUAGE SEMANTICS

3.1 The general representation of the meaning

Let us turn our attention to the problem of grasping of natural language semantics. A sentence of natural language can be viewed from several points. In classical linguistic, it is usual to talk about representation of a sentence on various levels. In the system called the *functional generative description* of natural language (FGD),(see [18]) five levels are differetiated, namely *phonetic* (PH) (how a sentence is composed as a system of sounds), *phonemic* (PM) (how words of a sentence are composed), *morphemic* (MR) (how a sentence is composed of its words), *surface syntax* (SS) (the system of grammatical rules) and tectogrammatical (TR) which is the highest level corresponding to the semantics. The latter is also called the *deep structure* of the sentence and this structure is the objective of possible application of fuzzy set theory. As has already been stated, words and more complex syntagms of natural language can be understood to be names of properties enocountered by a man in the world. In the light of the previous section, fuzzy sets can be used as follows: let $\mathcal{A}$ be a syntagm of natural language and φ the corresponding property. If the class (1) determined by φ is approximated by a fuzzy set $A \underset{\sim}{\subseteq} U$ then the meaning $M(\mathcal{A})$ of $\mathcal{A}$ is

$$M(\mathcal{A}) = A. \tag{9}$$

Thus, our job consists in determining of the membership function A.

However, the situation is by no means simple as not every word of natural language coresponds to such a property and, above all, there are various relations between words. Thus, determination of the membership function which corresponds to a complex syntagm may be a very complicated task.

On the tectogrammatical level, a meaning of a sentence is represented as a complex dependency structure which can be depicted in the form of a labelled graph. For example, the sentence

Peter writes a short letter to his friend

can be depicted in the form of a graph on Fig. 1.

For the detailed explanation of this graph see [18,17]. The letters t and f mean *topic* and *focus*, respectively. A topic is a part of a sentence containing the theme which is spoken about and focus contains a new information conveyed by the sentence. Of course, one surface structure of a sentence may lead to several deep structures. Up to now, we are far from the detailed understanding to all the nuances of the sentence semantics. The present state of art makes us possible to model the meaning of only some simple syntagms, i.e. certain branches of the tectogrammatical tree such as that on Fig. 1. This will be disscused in the subsequent sections.

3.2 Fuzzy semantics of selected syntagms

First, let us stop at the modelling of the semantics of nouns. In general, if $\mathcal{S}$ is a noun, then its meaning is a fuzzy set

$$M(\mathcal{S}) = S, S \underset{\sim}{\subseteq} U.$$

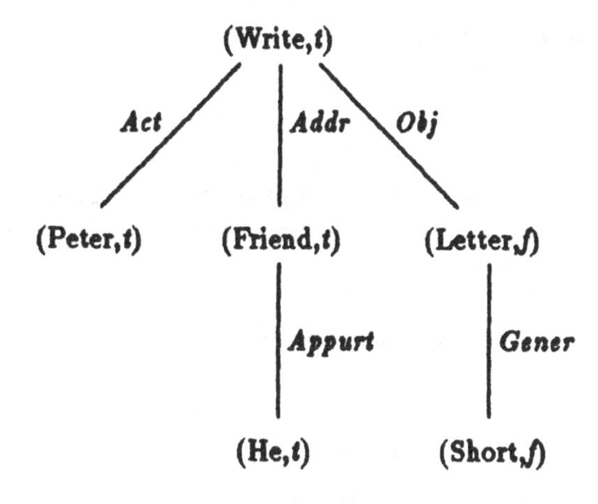

Figure 1: The tectogrammatical tree of the sentence *Peter writes a short letter to his friend*

What is the universe U? It is a set of objects chosen in such a way that whenever an object x has the property φ_S named by the noun S then $x \in U$. This can be constructed e.g. as follows: Let K be a set of generic elements called the kernel space. For example, K can be a union of all the objects described in our dictionary, of those we have regarded during the last week, of those we see in our flat etc. In short, K should contain all the specific objects we have met or imagine. Let $\mathcal{F}(K)$ be a set of all the fuzzy sets on K and put

$$\mathcal{F}^n(K) = \underbrace{\mathcal{F}(\ldots(\mathcal{F}(K)\ldots)}_{n\text{-times}}$$

Let E_K be the smallest set closed with respect to all the Cartesian powers of K, of $\mathcal{F}^n(K), n = 1,\ldots,n$ and all the Cartesian products of these elements. This set is called the *semantic space*. Then the universe of S is a sufficiently big subset $U \subseteq E_K$.

A certain problem is the determination of the membership function. There are several methods proposed in the literature (cf. [13]). The membership function corresponding to the object nouns (e.g. *table, car, donkey* etc.) could be constructed on the basis of the outer characteristics of elements. For example, we may use proportions of some geometric patterns contained in objects etc. A very often used method is statistical analysis of expert (subjective) estimations. Several experiments have been described in the literature. Let us mention that fuzzy methods are rather robust and thus exact determination of the membership function is not as important as it might seem at first glance.

The experience suggests that even individual estimation works well when it is done carefully and seriously.

In practical applications, e.g. in artificial intelligence, it is not very useful to model the meaning of nouns because we would have to find proper representation of its elements in the computer which, in fact, we do not need. The most successful applications are based on modelling of the meaning of adjectives and the syntagms of the form

(quantifier —) adverb — adjective (— noun)

where the syntagm

$$\textit{adverb} - \textit{adjective} \tag{10}$$

plays the crucial role. The most important (and very freequent) adjectives are those inducing an ordering $\leq$ in the universe U. We will assume that $\leq$ is linear. According to lingustitic considerations as well as experiments (cf. [10]), there are certain points $m, s, v, \in U$ where $m < s < v$. The point s is called the *semantic center*. The adjectives inducing an ordering in U usually form antonyms which can be characterized as follows. Let $\mathcal{A}^-, \mathcal{A}^+$ be antonyms (e.g. small – big, cold – hot etc.). Then their meanings are fuzzy sets

$$\begin{aligned} M(\mathcal{A}^-) &= A^- \\ M(\mathcal{A}^+) &= A^+ \end{aligned}$$

such that $\mathrm{Supp}A^- \subseteq \langle m, s)$ and $\mathrm{Supp}A^+ \subseteq (s, v\rangle$ where $\mathrm{Supp}A = \{x \in U; Ax > 0\}$.[5] We will often call $\mathcal{A}^-$ a *negative* and $\mathcal{A}^+$ a *positive* adjective, respectively. There are also couples of antonyms such that a third member $\mathcal{A}^0$ exist. Its meaning is

$$M(\mathcal{A}^0) = A^0$$

where the membership function A^0 has the property

$$s \in \mathrm{Ker}A^0 = \{x \in U; A^0 x = 1\}^6.$$

A typical example of the adjective $\mathcal{A}^0$ is $\mathcal{A}^0 :=$ *average*. In the sequel we will call $\mathcal{A}^0$ a *zero* adjective.

The curves corresponding to the fuzzy sets A^-, A^+, A^0 have characteristic shapes depicted on Fig. 2. Note that they are sometimes called the S^-, S^+ and Π fuzzy sets, respectively.

[5] This set is called the *support* of the fuzzy set A.

[6] This set is called the *kernel* of the fuzzy set A^0.

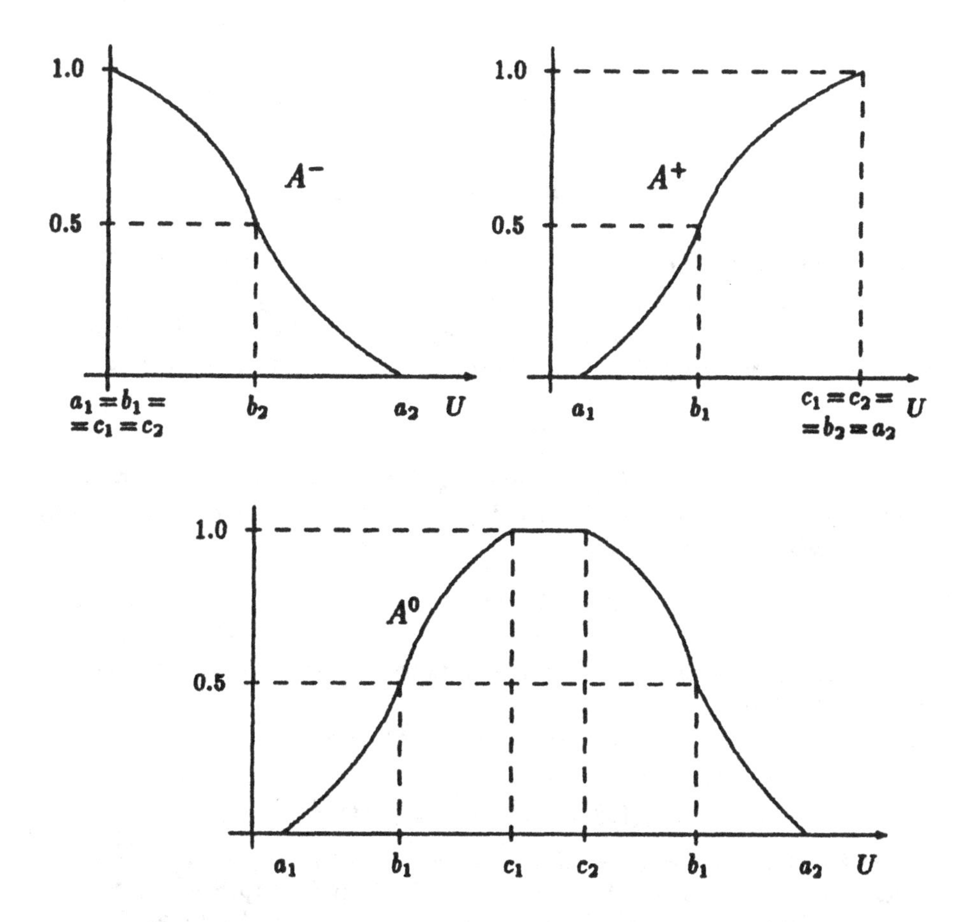

Figure 2: The membership functions corresponding to the meaning of the negative, positive and zero syntagms.

A general formula for all the three fuzzy sets is the following:

$$F(x, a_1, b_1, c_1, c_2, b_2, a_2) = \begin{cases} 0 & \text{if } x < a_1 \text{ or } x > a_2 \\ 1 & \text{if } c_1 \leq x \leq c_2 \\ \frac{1}{2}\left(\frac{x-a_1}{b_1-a_1}\right)^2 & \text{if } a_1 \leq x < b_1 \\ 1 - \frac{1}{2}\left(\frac{x-c_1}{c_1-b_1}\right)^2 & \text{if } b_1 \leq x < c_1 \\ 1 - \frac{1}{2}\left(\frac{x-c_2}{b_2-c_2}\right)^2 & \text{if } c_2 < x \leq b_2 \\ \frac{1}{2}\left(\frac{x-a_2}{a_2-b_2}\right)^2 & \text{if } b_2 < x \leq a_2 \end{cases}$$

æ The meaning of the points $a_1, b_1, c_1, a_2, b_2, c_2 \in U$ is clear from Fig. 2.

The adverb in (10) is an intensifying one (e.g. *very, highly, absolutely, slightly* etc.) and it is usually called the *linguistic modifier* in fuzzy set theory. In general, the meaning of the intensifying adverb m is a pair of functions

$$M(m) = < \zeta_m, \nu_m >$$

where $\zeta_m : U \longrightarrow U$ is a displacement function and $\nu_m : L \longrightarrow L$ is a unary operation fitting L [7]. Hence, the meaning of the syntagm (10) is obtained using the composition of functions

$$M(m\mathcal{A}) = \nu_m \circ A \circ \zeta_m. \tag{11}$$

A typical example, widely used in fuzzy set theory, is the modifier *very* defined as follows:

$$\nu_{very}(a) = \text{CON}(a), \; a \in < 0, 1 >$$

and

$$\zeta_{very}(x) = x + (-1)^k \cdot d \cdot \| \text{Ker}(A) \|$$

where $k = 1$ for A^+ or for A^0 if $x \leq s$, and $k = 2$ for A^- or for A^0 if $x \geq s$. $\| \text{Ker}(A) \|$ is the length of the interval $< \inf(\text{Ker}(A)), \sup(\text{Ker}(A)) >$. The parameter d was experimentally estimated to be a number $d \in < 0.25, 0.40 >$.

Examples of some other linguistic modifiers can be found in the literature.

The meaning of verbs is a very complicated problem and so far, only the copula "to be" in syntagms as

$$p := \mathcal{P} \text{ is } \mathcal{A} \tag{12}$$

[7] Some authors simplify this model by putting $\zeta_m = \text{id}_U$ (an identical function on U).

is modelled where $\mathcal{A}$ is usually the syntagm (10). However, (12) is interpreted rather as a simple assignement than a verb. The $\mathcal{P}$ in (12) is a noun but it is usually not treated as such. Thus, we obtain two ways how the meaning of (12) can be modelled.

a) We put

$$M(p) = M(\mathcal{A}) = A,$$

i.e. the meaning of p is set equal to the meaning of the syntagm $\mathcal{A}$. This is quite reasonable since, as was stated above, we usually need not know the meaning of the noun $\mathcal{P}$ in the applications.

b) Let $M(\mathcal{P}) = P \underset{\sim}{\subseteq} V$. Then we put

$$M(p) \subseteq P \times A \tag{13}$$

where $P \times A$ is the Cartesian product of fuzzy sets defined by

$$(P \times A) < x, y >= Px \wedge Ay.$$

The inclusion in (13) may be proper or improper dependingly on the kind of the noun $\mathcal{P}$. The relation (13) means that each element x from the universe V (a representative of the noun $\mathcal{P}$) is asigned an attribute y from the universe U where $A \underset{\sim}{\subseteq} U$ [8].

As we are in the fuzzy environment it seems reasonable to take the resulting membership degree of the couple $< x, y >$ as minimum of Px and Ay.

L. A. Zadeh [22,21] and some other authors following him suggest to interpret the membership degree $(P \times A) < x, y >$ as a *possibility degree* of the fact that x is A. However, the possibility degree concerns uncertainty which, in our oppinion, does not reflect the vagueness phenomenon contained in the semantics of natural language.

Very important are the conditional sentences of the form

$$C := \text{IF } p \text{ THEN } q \tag{14}$$

where p and q are syntagms of the form (12). In fuzzy set theory we usually put

$$M(C) = M(p) \ominus M(q), \tag{15}$$

i.e. we interpret the implication (14) using the residuum operation between fuzzy sets $M(p)$ and $M(q)$.

The interpretation of the conditional sentences (14) plays the crucial role in the so called approximate reasoning which is one of the most successfully

[8] U is usually the real line. For example, in the syntagms such as *Peter is tall*, the element *Peter* is assigned its height being a real number.

applied areas of fuzzy set theory (see e.g. [8,9]). Let us remark that many authors interpret (14) as the Cartesian product

$$M(C) = M(p) \times M(q). \tag{16}$$

In the applications of the approximate reasoning, this may work since all the fuzzy methods are very robust. However, putting the meaning of the implication (14) equal to the Cartesian product is linguistically as well as logically incorrect since (16) is symmetric and the implication is not. Another reason why (16) often works in the practical applications may be the fact that the implication (14) often describes only some kind of a relation between the input and output and it is not, in fact, understood to be the implication. This discrepancy needs still more analysis.

Let us also mention the problem of linguistic quantifiers (we will denote them by the letter $\mathcal{Q}$). They do not form a uniform group form the linguistic point of view. We place among them numerals including the indefinite ones (e.g. *several*), some adverbs (e.g. *many, few, most*), some pronouns (e.g. *every*), some nouns (e.g. *majority, minority*) and others. From the point of fuzzy set theory, their meaning generally is a fuzzy number, i.e. a fuzzy set $M(\mathcal{Q}) = Q \underset{\sim}{\subseteq} Re$ in the real line. It is proposed in [22] how to interpret the syntagms of the form

$$\mathcal{Q}\mathcal{A} \tag{17}$$

or

$$\mathcal{Q}\mathcal{A}'s \text{ are } \mathcal{B}'s. \tag{18}$$

In (17), the quantifier $\mathcal{Q}$ is interpreted as a fuzzy characterization of the absolute fuzzy cardinality (7) of the fuzzy set

$$A = M(\mathcal{A})$$

while in (18), it characterizes the relative fuzzy cardinality (8) of $M(\mathcal{A})$ with respect to $B = M(\mathcal{B})$. More exactly, we put

$$M(\mathcal{Q}\mathcal{A}) = Q \cap \mathrm{FCard}(A) \tag{19}$$

and

$$M(\mathcal{Q}\mathcal{A}'s \text{ are } \mathcal{B}'s) = Q \cap \mathrm{FCard}_A(B). \tag{20}$$

However, the problem is not finished yet since (19) and (20) have sense only for fuzzy sets with finite support.

In the literature on fuzzy set theory (see e.g. [19,20,22,13] and others), one may find the semantics of the compound syntagms of the form

$$\mathcal{A} \text{ and } \mathcal{B} \tag{21}$$

and

$$\mathcal{A} \text{ or } \mathcal{B} \tag{22}$$

defined using the operations of intersection and union of fuzzy sets, respectively. However, this is only a tentative solution since the syntagms (21) and (22) are special cases of the very complicated phenomenon known in linguistics as the *coordination*. The use of the operations of intersection and union of fuzzy sets may work in some special cases of the close coordination in syntagms e.g. *Peter and Paul ..., Old and dirty car ...* etc.

Even worse situation is encountered with negation. The simple use of the operation of complement of fuzzy sets works only with some kinds of adjectives and nouns. However, negation contained in more complex syntagms is incidental to the phenomenon of the *topic-focus articulation* when only focus is being negated. Fully comprehensive description of this phnomenon in linguistics is not, however, still done.

4 FEW COMMENTS TO THE APPLICATIONS AND LINGUISTIC APPROXIMATION

The theory presented so far has found many interesting applications, especialy in the models connected with the so called *approximate reasoning*. However, we are quite far from grasping of the semantics of natural language more comprehensively and much work has still to be done.

A very important concept which deserves to be mentioned here is that of a *linguistic variable* [20]. This concept made us possible to see a certain part of linguistics from a more technical point of view.

A linguistic variable is, in general, a quintuple

$$< \mathcal{X}, \mathcal{T}(\mathcal{X}), U, G, M >$$

where $\mathcal{X}$ is a name of the variable, $\mathcal{T}(\mathcal{X})$ is its term-set, U is the universe, G the syntactic and M the semantic rules, respectively. For example,

$$\mathcal{X} := size,$$

$$U := < 0, 1000 >,$$

G is a certain, usually context-free grammar generating the set $\mathcal{T}(\mathcal{X})$ of terms such as *big, small, very big, rather average* etc. and M is the syntactic rule assigning to each term $\mathcal{A} \in \mathcal{T}(\mathcal{X})$ its meaning being a fuzzy set

$$M(\mathcal{A}) \subsetapprox U.$$

Lingustic variables play important role in applications. For example, the parameters of a technical system such as *temperature, speed, weight* etc. can be understood to be lingustic variables. As its values can be also crisp, e.g. *exactly 165.2* etc., the concept of a linguistic variable is general enough to capture also the classical concept of the variable.

In applications, one may also meet the problem of a *linguistic aproximation*. We may lay it down as follows. Let $\mathcal{T}$ be a set of syntagms of natural language e.g. the term-set of a linguistic variable and let us be given a fuzzy set $A'_0 \underset{\sim}{\subseteq} U$. Our task is to find a syntagm $\mathcal{A}_0 \in \mathcal{T}$ such that its meaning $M(\mathcal{A}_0) = A_0$ is as close to A'_0 as possible.

There are many ways how to solve this task. However, no sufficiently efficient and general method is known till now. One of the possible procedures is the following.

Let $f :< 0,1 > \longrightarrow < 0,1 >$ be a smooth, increasing, and measurable function. Put

$$R < A, B > = 1 - \frac{\int_{Supp(A) \cup Supp(B)} df(Ax - Bx)}{\int_{Supp(A)} df(Ax) + \int_{Supp(B)} df(Bx)} \tag{23}$$

Then we may find $\mathcal{A}_0 \in \mathcal{T}$ such that

$$R < A_0, A'_0 >$$

is maximal one.

If $\mathcal{T}$ has small number of elements then we may also find $\mathcal{A}_0$ such that

$$\sum_{x \in U} | A_0 x - A'_0 x |^p \tag{24}$$

is minimal for some suitable, previsously set number p. This method is often used in thecnical applications.

A quite effective procedure was proposed by P. Esragh and E. H. Mamdani in [5]. This procedure is suitable for the syntagms of the form (21) and (22) where $\mathcal{A}$ and $\mathcal{B}$ may consist of an adjective, a noun and a linguistic modifier, and the universe U is ordered. According to this method, the membership function A'_0 is divided into parts by the effective turning points (i.e. special points where the membership function changes its course), the parts are approximated by the above partial syntagms $\mathcal{A}, \mathcal{B}$ using (24) and the resulting syntagm is obtained by joining $\mathcal{A}$ and $\mathcal{B}$ using the corresponding connective. In particular, if the two neighbouring parts of the membership function form a "hill' then the corresponding syntagms are joined by the connective *and*, and if they form a "valley" then they are joined by the connective *or*. Note that this works only in the case when the connective *and* is interpreted as the intersection and *or* as the union of fuzzy sets.

5 CONCLUSION

We have briefly presented the main ideas of the modelling of natural language semantics using fuzzy set theory. We attempted to demonstrate how the semantics of some of the basic units can be interpreted, namely the semantics of nouns, adjectives, selected adverbs and the copula "to be". Moreover, the semantics of some cases of the close coordination (the use of connectives) was also touched along with the semantics of conditional sentences. Let us stress that we are still far from grasping of the meaning of more complex syntagms, and even simple clauses when they contain a verb. The reason consists in an extreme complexity of verb semantics since verbs represent the most important units of our language stepping towards the human's recording the surrounding world on the highest level of his intellectual capability.

Some work in this respect is done in [14] where, however, the new world of mathematics called the *alternative set theory* (AST) is used. Fuzzy set theory serves there as a special technical tool which is used at a second stage after the semantics of a sentence (syntagm) in the frame of AST is formed.

Despite the above facts, the use of fuzzy sets in modelling of natural language semantics has already found many successful applications. This is a convincing argument in favour of the usefullness of fuzzy set theory.

References

[1] Bezdek, J. (ed.), **Analysis of Fuzzy Information - Vol. 1: Mathematics and Logic**, CRC Press, Boca Raton, Fl. 1987.

[2] Bezdek, J. (ed.), **Analysis of Fuzzy Information - Vol. 2: Artificial Intelligence and Decision Systems**, CRC Press, Boca Raton, Fl. 1987.

[3] Bezdek, J. (ed.), **Analysis of Fuzzy Information - Vol. 3: Applications in Engineering and Science**, CRC Press, Boca Raton, Fl. 1987.

[4] Dubois, D., Prade, H., **Fuzzy Sets and Systems:Theory and Applications**, Academic Press, New York 1980.

[5] Esragh, F., Mamdani, E.H. *A general approach to linguistic approximation*, Int. J. Man-Mach. Stud., 11(1979), 501-519.

[6] Gaines, B.R., Boose, J.H. (eds.), **Machine Learning and Uncertain Reasoning**, Academic Press, London 1990.

[7] Gärdenfors, P. (ed.), **Generalized quantifiers**, D. Reidel, Dordecht, 1987.

[8] Gupta, M.M., Yamakawa, T. (eds.), **Fuzzy Computing: Theory, Hardware and Applications**, North–Holland, Amsterdam 1988.

[9] Gupta, M.M., Yamakawa, T. (eds.), **Fuzzy Logic in Knowledge–Based Systems, Decision and Control**, North–Holland, Amsterdam 1988.

[10] Kuz'min, V. B., *About semantical structure of linguistic hedges:an experimental hypothesis*, BUSEFAL **24**, 1985, 118 - 125, Université Paul Sabatier, Toulouse.

[11] Lakoff, G., *Hedges: A study in meaning criteria and logic of fuzzy concepts*, J. Philos. Logic **2**(1973), 458 - 508.

[12] Mamdani, E.H., Gaines, B.R. (eds.), **Fuzzy Reasoning and its Applications**, Academic Press, London 1981.

[13] Novák, V., **Fuzzy Sets and Their Applications**, Adam–Hilger, Bristol, 1989.

[14] Novák, V., **The Alternative Mathematical Model of Natural Language Semantics**. Manuscript. Mining Institute, Ostrava 1989. (To be published by Cambridge University Press)

[15] Novák, V., Pedrycz, W., *Fuzzy sets and t-norms in the light of fuzzy logic*, Int. J. Man-Mach. Stud., **29**(1988), 113 - 127.

[16] Pavelka, J., *On fuzzy logic I, II, III*, Zeit. Math. Logic. Grundl. Math. **25**(1979), 45-52, 119-134, 447-464.

[17] Sgall, P. (ed.), **Contributions to functional syntax, semantics, and language comprehension**, Academia, Prague 1984.

[18] Sgall, P., Hajičová, E., Panevová, J., **The meaning of the sentence in its semantic and pragmatic aspects**, D. Reidel, Dordecht 1986.

[19] Zadeh, L.A., *Quantitative Fuzzy Semantics*, Inf.Sci.,**3**(1973), 159-176.

[20] Zadeh, L.A., *The concept of a linguistic variable and its application to approximate reasoning I, II, III*, Inf.Sci.,**8**(1975), 199-257, 301-357;**9**(1975), 43-80.

[21] Zadeh, L.A., *PRUF — a Meaning Representation Language for Natural Languages*, Int.J.Man–Mach.Stud.**10**(1978), 395-460.

[22] Zadeh, L.A., *A computational approach to fuzzy quantifiers in natural languages*, Comp. Math. with Applic. **9**(1983), 149-184.

9

FUZZY-SET-THEORETIC APPLICATIONS IN MODELING OF MAN-MACHINE INTERACTIONS

By Waldemar Karwowski
Center for Industrial Ergonomics
University of Louisville
Louisville, KY 40292, USA
and
Gavriel Salvendy
School of Industrial Engineering
Purdue University
West Lafayette, IN 47907, USA

INTRODUCTION

According to Harre (1972) there are two major purposes of models in science: 1) logical, which enables to make certain inferences which would not otherwise be possible to be made; and 2) epistemiological, to express and extend our knowledge of the world. Models are helpful for explanation and theory formation, as well as simplication and concretization. Zimmermann (1980) classifies models into three groups: 1) formal models (purely axiomatic systems with purely fictitious hypotheses), 2) factual models (conclusions from the models have a bearing on reality and they have to be verified by empirical evidence), and 3) prescriptive models (which postulate rules according to which people should behave). The quality of a model depends on the properties of the model and the functions for which the model is designed (Zimmermann, 1980). In general, good models must have three major properties: 1) formal consistency (all conclusions follow from the hypothesis), 2) usefulness, and 3) efficiency (the model should fulfill the desired function at a minimum effort, time and cost).

Although the usefulness of the mathematical language for modeling purposes is undisputed, there are limits of the possibility of using the classical mathematical language which is based on the dichotomous character of set theory (Zimmermann, 1980). Such restriction applies especially to the man-machine systems. This is due to vagueness of the natural language, and the fact that in empirical research natural language cannot be substituted by formal languages. Formal languages are rather simple and poor, and are useful only for specific purposes. Mathematics and logic as research languages widely applied today in natural sciences and engineering are not very useful for modeling purposes in behavioral sciences and especially in human factors studies. Rather, a new methodology, based on the theory of fuzzy sets and systems is needed to account for the ever present fuzziness of man-machine systems.

As suggested by Smithson (1982), the potential advantages for applications of a fuzzy approach in human sciences are: 1) fuzziness, itself, may be a useful metaphor or model for human language and categorizing processes, and 2) fuzzy mathematics may be able to augment conventional statistical techniques in the analysis of fuzzy data. Fuzzy methods are useful supplements for statistical techniques such as reliability analysis and regressions, and structurally oriented methods such as hierarchical clustering and multidimensional scaling.

HUMAN FACTORS

Human factors discipline is concerned with "the consideration of human characteristics, expectations, and behaviors in the design of the things people use in their work and everyday lives and of the environments in which they work and live" (McCormick, 1970). The "things" that are designed are complex man-machine systems. According to Pew and Baron (1983) the ultimate reasons for building models in general, and man-machine models in particular, are to provide for:

1. A systematic framework that reduces the memory load of the investigator, and prompts him not to overlook the important features of the problem,
2. A basis for extrapolating from the information given to draw new insights and new testable or observable inferences about system or component behavior,
3. A system design tool that permits the generation of design solutions directly,
4. An embodiment of concepts or derived parameters that are useful as measures of performance in the simulated or real environment,
5. A system component to be used in the operational setting to generate behavior, for comparison with the actual operator behavior to anticipate a display of needed data, to introduce alternative strategies or to monitor operator performance, and
6. Consideration of otherwise neglected or obscure aspects of the problem.

According to Topmiller (1981), research in man-machine systems poses an important methodological challenge. This is due to the complexity of such systems, and a need for simultaneous consideration of a variety of interacting factors that affect several dimensions of both individual and group performance. Chapanis (1959) argues that "we do not have adequate methods for finding out all the things we need to know about people. Above all, we need novel and imaginative techniques for the study of man. This is an area in which behavioral scientists can learn much from the engineering and physical sciences."

Research techniques applied in man-machine research typically include the following methods: 1) direct observation (operator opinions, activity sampling techniques, process analysis, etc.), 2) accident study method (risk analysis, critical-incident technique) 3) statistical methods, 4) experimental methods (design of experiments), 5) psychophysical methods (psychophysical scaling and measurement), and 6) articulation testing methods (Chapanis, 1959). Today we are still at the beginning stage of building robust mathematical models for the analysis of complex human-machine systems. This is partially due to lack of appropriate design theory, as well as complexity of human behavior (Topmiller, 1981). The human being is too complex a "system" to be fully understood or describable in all his/her properties, limits, tolerances, and performance capabilities, and no comprehensive mathematical tool has been available up to now to describe and integrate all the above mentioned measures and findings about human behavior (Bernotat, 1984).

FUZZY MODELS

Human work taxonomy can be used to describe five different levels ranging from primarily physical tasks to primarily information processing tasks (Rohmert, 1979). These are:

1) producing force (primarily muscular work),
2) continuously coordinating sensory-monitor functions (like assembling or tracking tasks),
3) converting information into motor actions (e.g. inspection tasks),
4) converting information into output information (e.g. required control tasks), and
5) producing information (primarily creative work).

Regardless of the level of human work, three types of fuzziness are present and should be accounted for in modeling of man-machine systems, i.e.,: 1) fuzziness stemming from our inability to acquire and process adequate amounts of information about the behavior of a particular subsystem (or the whole system), 2) fuzziness due to vagueness of the relationships between people and their working environments, and complexity of the rules and underlying principles related to such systems, and finally, 3) fuzziness inherent in human thought processes and subjective perceptions of the outside world (Karwowski and Mital, 1986). Figure 1 illustrates the above thesis. Traditional man-machine interfaces, which include: 1) information sensing and receiving, 2) information processing, 3) decision-making, 4) control actions, and 5) environmental and situational variables, are represented in two blocks, i.e., human interpretation block and a complex work system block.

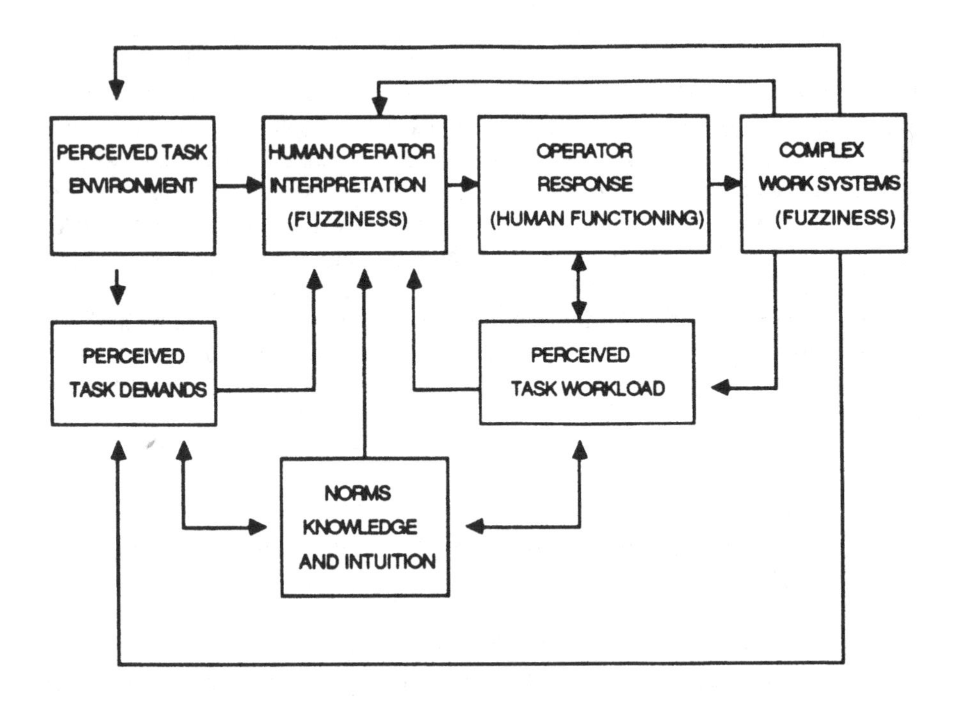

Figure 1. Fuzziness in man-machine interfacing (after Karwowski and Mital, 1986).

Uncertainty, (looked upon in the context of mental workload) which causes unpredictability in one's stimulus and/or response, enters a work situation from several sources (Audley et al., 1979). These are: 1) external disturbance model, 2) varying parameters of the system structure external to the human operator, 3) human produced noise in observing the task stimuli, 4) lack of good internal model of the external system, 5) human-produced distortions in interpreting the externally stipulated criterion of performance, and 6) human-produced motor noise.

In view of the above, the theory of fuzzy sets offers a useful approach when the task demands are vague, with the main advantage being its ability to model imprecise task situations and, therefore, a potential to develop a framework for implementation of workload measures.

FUZZINESS AND HUMAN-MACHINE SYSTEMS

Man-machine studies aim to optimize work systems with respect to physical and psychological characteristics of the users, and investigate complex and ill-defined relationships between people, machines, and physical environments. The main goal of such investigation is to remove the incompatibilities between humans and tasks, and to make the workplace healthy, productive, comfortable and satisfying.

Human-centered systems, which are the objects of man-machine studies, are very complex and difficult to analyze. There are at least three different types of uncertainty inherent to such systems; i.e., inaccuracy, randomness, and vagueness (Bezdek, 1981). Uncertainties due to inaccuracy are related to observations and measurements (representations), while those due to randomness (of events) are independent from observations and constitute an objective property of some real process. Uncertainty due to vagueness (or fuzziness) has to do with the complexity of the system under investigation and the human thought and perception processes (Zadeh, 1973).

A new methodology in the area of man-machine is needed to account for imprecision and vagueness of such relationships. Zadeh (1974) points out that "Although the conventional mathematical techniques have been and will continue to be applied to the analysis of humanistic systems, it is clear that the great complexity of such systems call for approaches that are significantly different in spirit as well as in substance from the traditional methods -- methods which are highly effective when applied to mechanistic systems, but are far too precise in relation to systems in which human behavior plays an important role."

In the past, most of the traditional methodologies disregarded the system complexities, and assumed that the formal properties of mathematics correspond to existing relationships characteristic to the system under investigation (Zadeh, 1974). For example, an uncertainty due to vagueness was often modeled as being of stochastic nature. Such treatment appears to defeat the purpose of any formal man-machine systems' analysis and modeling efforts.

The concept of fuzziness

Fuzziness relates to the specific kind of vagueness having to do with gradations in categories, i.e., degree of vagueness (Smithson, 1982). Uncertainty measured by fuzziness refers to the gradation of membership of an element in some class (category). Although such uncertainty arises at all levels of cognitive processes, people have the abilities to understand and utilize vague and imprecise concepts which are difficult to analyze within the framework of traditional scientific thinking (Hersh et al., 1976; Kramer, 1983; Karwowski and Mital, 1986). Therefore, awareness of vagueness and inexactness, implicit in human behavior, should be the basis of any man-machine studies.

According to Zadeh (1965), the theory of fuzzy sets represents an attempt for constructing a conceptual framework for a systematic treatment of vagueness and uncertainty due to fuzziness in both quantitative and qualitative ways. Such framework is much needed in the human-machine interaction area. As pointed out by Singleton (1982) "most human characteristics have very complex contextual dependencies which are not readily expressible in tabulations of numbers even in multivariate equations." Yet, there is growing evidence that people comprehend vague concepts, such as concepts of a natural language, as if those concepts were represented by fuzzy sets, can manipulate them according to the rules of fuzzy logic (Oden, 1977 and Brownell et al., 1978). Recent research in semantic memory and concept formation indicates that natural categories are fuzzy sets with no clear boundaries separating category members from nonmembers (McCloskey et al., 1978). One can certainly understand the meaning of such concepts as "excessive workload," "low illumination," "heavy weight," "high level of stress," and "tall man," to name a few commonly used descriptors of the human-environment relationship.

As noted by Singleton (1982), "no one has yet developed a comprehensive set of crude and approximate but simple and inexpensive techniques finding solutions to ergonomics problems." Fuzzy set theory, which allows interpretation and manipulation of imprecise (vague) information and recognition and evaluation of uncertainty due to fuzziness (in addition to randomness), may be the closest solution to the above stated need available today.

Conventional versus fuzzy set theory and logic

In a conventional (classical) set theory, an element x either belongs or does not belong to a set X, and the characteristic (membership) function f_x can be represented as follows:

$$\begin{cases} 1 \text{ if } x \in X \text{ (truth value = 1: true)} \\ 0 \text{ if } x \notin X \text{ (truth value = 0: false)} \end{cases}$$

The concept of fuzzy set extends the range of membership values for f_x, and allows graded membership, usually defined on an interval [0, 1]. Consequently, an element may belong to a set with a certain degree of membership, not necessarily 0 or 1. The "excluded middle" concept is then abandoned, and more flexibility is given in specifying the characteristic function. In view of the above, the mathematical logic can also be modified. Interestingly, the classical logic was actually extended as early as 1930 by Lukasiewicz, who proposed the infinite-valued logic. As stated by Giles (1981), "Lukasiewicz logic is exactly appropriate for the formulation of the 'fuzzy set theory' first described by Zadeh; indeed, it is not too much to claim that is related to fuzzy set theory exactly as classical logic is related to ordinary set theory."

The theory of fuzzy sets has been successfully applied in the modeling of ill-defined systems in a variety of disciplines (cognitive psychology, information processing and control, decision-making sciences, biological and medical sciences, sociology and linguistics, image processing and pattern recognition, and artificial intelligence).

Willaeys and Malvache (1979) investigated the perception of visual and vestibular information in a "watch and decision" or industrial inspection (control) tasks. The imprecise nature of the human problem solving procedures was related to the "shaded" strategy of the operator's perception and to the "hard-to-predict" environment of the man-machine environment. The labels of fuzzy sets used by the operator to describe different physical variables of the task were identified, and the fuzzy model of the process-control task was formulated.

Benson (1982) developed an interactive computer graphics program for analytical tasks which are not well defined or utilize imprecise data. Color scales were used to model subjectively defined categories under investigation. Such fuzzy categories were then presented to the analyst. The use of a linguistic approach allowed the identification of membership for different categories of description of visual inspection. The perceptual properties of color proved to be useful in selective focus attention and in distinguishing or disregarding variations between imprecisely defined categories.

Karwowski and others (1988, 1984a and 1984b) developed a fuzzy set based model to assess the acceptability of stresses in manual lifting tasks. Measures of acceptability were expressed in terms of membership functions which described the degrees to which the combined effect of biomechanical and physiological stresses were acceptable to the human operator. The combined acceptabilities of a lifting task were similar to the subjective estimations of the overall task acceptability established by the subjects in a psychophysical experiments.

Terano et al. (1983) introduced a fuzzy set approach into fault-tree analysis, and studied the fuzziness of a human-reliability concept from the man-machine systems safety point of view. Kramer and Rohr (1982) developed a fuzzy model of driver-behavior based on simulated visual pattern processing in lame control. Saaty (1977) distinguished two types of fuzziness in layman perception (for example, perception of illumination intensity) and fuzziness in meaning, advocating that fuzziness is a basic quality of understanding. Hirsh et al. (1981) used a fuzzy dissimilitude relation to describe human vocal patterns.

FUZZY-SET THEORETIC MODELING OF HUMAN-COMPUTER INTERACTION

The interaction between people and computers reflects the cognitive imprecision of the data and uncertainty exhibited in the user's perception of the computing environment, including the limitations of the computer software used. Since human reasoning is not precise, the human-computer interaction (HCI) should be imprecision-tolerant, and should allow for the inexact mode of communication (Karwowski et al., 1990). Recent developments in fuzzy methodologies, fuzzy computing, and fuzzy hardware (computers based on fuzzy logic processing units), created a set of new possibilities for the development of vagueness-tolerant human-computer interfaces.

Human-computer interaction system

The human-computer interaction system (HCIS) can be formally defined (Karwowski et al., 1990) as a quintuple:

$$HCIS = (T, U, C, E, I) \qquad (1)$$

where: T - task requirements (physical and cognitive)
U - user characteristics (physical and cognitive)
C - computer characteristics (hardware and software including computer interfaces)
E - an environment
I - a set of interactions.

The set of interactions **I** embodies all possible interactions between **T**, **U**, **C** in **E** regardless of their nature or strength of association. For example, one of the possible interactions can relate to the data stored in the computer memory and the corresponding knowledge, if any, of the user. The interactions **I** can be elemental, i.e. one to one association, or complex, such as an interaction between the user, the particular software used to achieve the desired task, and available physical interface with the computer. Also, the elemental interactions do not have to directly involve the user. For example, the interaction may involve only **T** and C components. It should be pointed out that the elemental interaction between **U** and **C** reflects the narrow concept of the traditional human-computer interface.

In human-computer interaction, the uncertainty and imprecision due to vagueness (or fuzziness) stems from the high complexity of human-computer systems as well as the nature of computer user's perception and thought processes. As pointed out by Zadeh (1973), the key elements in human thinking are linguistic descriptors, or labels, of classes of objects with gradation of membership of their elements, i.e., fuzzy sets. Furthermore, the human reasoning is approximate rather than exact, and is based upon a logical system with fuzzy truths, connectives, and fuzzy rules of inference (Lakoff, 1973; Kochen, 1975; Hersh and Caramazza, 1976; Mamdani and Gaines, 1981; Schmucker, 1984; Karwowski and Mital, 1986; Smithson, 1987).

Fuzziness in HCI research

Recently, there have been some initial attempts to incorporate fuzziness in the HCI research. Simcox (1984) presented a method to determine compatibility functions that describe the degree of implied attribute of the visual display and the linguistic category that summarizes values of this attribute. Such compatibility functions were postulated to be useful in the construction of the computer graphs as a communication mode. Boy and Kuss (1986) have proposed a fuzzy method for modeling of human-computer interactions in information retrieval tasks, and implemented their method in the computer-based library retrieval system (BIBLIO). Recently, Hesketh et al. (1988) developed a computerized method for fuzzy graphic rating scale using the FUZRATE program which feeds back to the user his/her fuzzy ratings, and then presents the results of combining these ratings.

THE GOMS MODEL

One of the recently proposed models of computer user's information-processing is the GOMS concept (Card et al., 1983). According to the **GOMS** model, the user's cognitive structure consists of four components 1) a set of Goals, 2) a set of Operators, 3) a set of Methods for achieving the goals, 4) a set of Selection Rules for choosing among competing methods for goals. These components can be further defined as follows:

1) *Goals*:A goal is a symbolic structure that defines a state of affairs to be achieved and determines a set of possible methods by which it may be accomplished.
2) *Operators*:Operators are elementary perceptual, motor, or cognitive acts, whose execution is necessary to change any aspect of the user's mental state or to affect the task environment.
3) *Methods*: Methods describe procedures used by the user to accomplish a goal. Methods have a chance of success distinctly less than certain, because of the user's lack of knowledge or appreciation of the task environment. This uncertainty is a prime contributor of the problem-solving character of a task; its absence is a characteristic of a cognitive skill.
4) *Selection Rules*:Rules for predicting from knowledge of the task environment which of several possible methods will be selected by the user in order to accomplish a specific goal.

In 1983, Card et al. devised a text-editing experiment to show the validity of GOMS model. In one experiment subjects were told to perform simple line location tasks and the methods that each subject used to locate a line was recorded. From sample editing sessions, the methods and the associated selection rules for locating a line were inferred. The study concluded that the GOMS knowledge representations were valid for such tasks.

FUZZY GOMS MODELING

Card et al., (1983) noted possible extensions of the GOMS model. Among these were "*the assurance that a GOMS description can be given for a display oriented editor*" and methods for improving the accuracy of the predictions of user's actions. It was also suggested that "*the probabilistic selection rules and conditionalities for predicting which method the user will employ and for expressing probabilistic conditionality within those methods*" be explored.

Another enhancement to the GOMS model would be the ability to account for uncertainty within selection rules (Karwowski et al., 1989). The original GOMS study inferred, from user behavior, rules such as "If the number of lines to the next modification is less than 3 then use the LF-METHOD; else use the QS-METHOD." This type of rule assumes perfect knowledge and absolute certainty of the user's cognitive ability to observe, at a glance, the number of lines to the next change.

In order to account for natural fuzziness of the above human-computer interactions, the GOMS model was recently extended by allowing its components to assume precise, probabilistic or fuzzy values. Such preliminary generalization of the GOMS model proposed by Karwowski et al. (1990) is given in Table 1. The Goals, Operators and Methods components can either be precise or fuzzy, while the Selection Rules are expressed in either probabilistic or fuzzy (as possibilistic or linguistic inexactness) manner.

Table 1. Generalized computer user's cognitive structure based on GOMS model and nature of model components (after Karwowski et al, 1990).

Structure category	Goals (description)	Operators (nature of acts)	Methods (description)	Selection rules (reasoning processes)
1	Precise	Precise	Precise	Probabilistic
2	Precise	Precise	Precise	Fuzzy
3	Precise	Precise	Fuzzy	Probabilistic
4	Precise	Precise	Fuzzy	Fuzzy
5	Precise	Fuzzy	Precise	Probabilistic
6	Precise	Fuzzy	Precise	Fuzzy
7	Precise	Fuzzy	Fuzzy	Probabilistic
8	Precise	Fuzzy	Fuzzy	Fuzzy
9	Fuzzy	Precise	Precise	Probabilistic
10	Fuzzy	Precise	Precise	Fuzzy
11	Fuzzy	Precise	Fuzzy	Probabilistic
12	Fuzzy	Precise	Fuzzy	Fuzzy
13	Fuzzy	Fuzzy	Precise	Probabilistic
14	Fuzzy	Fuzzy	Precise	Fuzzy
15	Fuzzy	Fuzzy	Fuzzy	Probabilistic
16	Fuzzy	Fuzzy	Fuzzy	Fuzzy

FUZZY GOMS MODEL: PILOT STUDY

The example presented below refers to the generalized GOMS structure category #4, where the set of Goals and the set of Operators are precisely defined, while the (predicted) Methods used by the subjects as well as specific Selection Rules applied to accomplish the editing task were based on fuzzy modeling concepts, including application of the linguistic values, fuzzy connectives and fuzzy logic, and possibilistic measures of uncertainty. Such model is referred to as the **Fuzzy GOMS** model.

Karwowski et al. (1989) reported an experiment performed to validate the fuzzy-based GOMS model for text editing task. The experiment was a variation of the manuscript editing experiment by Card et al. (1983). The experiment consisted of the following steps:

1. The subject performed a familiar text editing task using a screen editor (VI)
2. The methods by which the subject achieved his goals (word location) as well as selection rules were elicited
3. It was established that many of the rules had fuzzy components
4. Several compatibility functions for fuzzy terms used by the subject were derived
5. The possibility measure was used to predict the methods that the subject would use
6. The selected methods were compared to **non-fuzzy** predictions and actual experimental data

The subject did not know the file to be edited. The task was performed from the subject's own office and desk. The subject was familiar with and regularly used the VI screen editor.

Knowledge elicitation

The knowledge engineers can use sample runs to infer the rules by which the subjects select their preferred methods of editing text. An additional benefit from a GOMS perspective would be in structuring knowledge elicitation. For example, the expert could be prompted to present the methods and the selection rules and respond in the following manner: "IF the condition X exists and the condition Y exists, THEN use method Z." For example, while performing a task, the subject could be asked to describe why he chose a particular method:

[Subject: The word (to be changed) is more than half of a screen down, so I will use the control-D method and then return-key to the word."]

[Knowledge Engineer: "How strongly do you feel that it is more than half?"]

[Subject: "Very strong, say 0.8."]

The actual distance to the word was measured directly and found to be, 39 lines. So the degree of membership of belonging to the *more than half* class was 0.8 for 39 lines. By having the subject perform many tasks while verbalizing the rules, the methods used, and membership of fuzzy quantifiers can be found.

Experimental methods for pilot study

The results of a pilot study reported by Karwowski et at. (1989) are discussed here in detail. The subject utilized the following five methods to place the cursor on the word(s) to be changed: *1) Control - D: scrolls down one half of a screen; 2) Control-F: jumps to the next page; 3) Return Key: moves the cursor to the left side of the page and down one line; 4) Arrow Up or Down: moves the cursor directly up or down; and 5) Pattern Search: places the cursor on the first occurrence of the pattern.*

The subject verbalized five cursor placement rules and seven fuzzy descriptors. The following rules were used: 1) If the word is *more than half of a screen* from the cursor and on the same screen or if the word is *more than half of a screen* from the cursor and across the printed page then use method #1; 2) If the word is *more than 70 lines* and the pattern is *not distinct* then use method #2; 3) If the word is *less than half of a screen* and *on the left half of the page* use method #3; 4) If the word is *less than half of a screen* and *on the right half of the page* use method #4; and 5) If the word is distinct and *more than 70 lines away* use method #5.

An example of the compatibility functions for the "right hand side of the screen" descriptor elicited in the experiment is given in Figure 2. The knowledge engineer assumed that the subject did not have the perfect cognitive ability to divide a screen directly in half, and rather elicited the knowledge as fuzzy knowledge. For all descriptors, the membership functions were perceived numbers of lines or characters, except the *distinct* and *non-distinct* descriptors. The *distinct* and *non-distinct* descriptors were given as counts of failed pattern recognitions and served, basically, to predict the patience of the user.

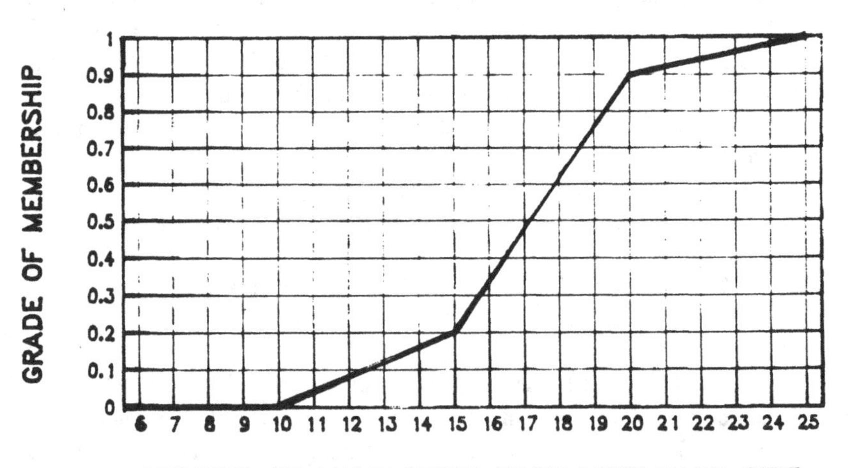

Figure 2. Fuzzy descriptor for the "right hand side of the screen" (after Karwowski et al., 1990).

Example of rule selection procedure

Once all the rules, methods, and corresponding membership functions have been elicited, the theory of possibility (Zadeh, 1978) was used to model the expert's rule selection process. For this purpose, each of the potential rules was assigned a possibility measure equal to the membership value(s) associated with it during the elicitation phase of experiment. The possibility measure $\pi(A)$ was defined after Zadeh (1978) as follows:

$$\pi(A) = \text{Poss}\,\{X \text{ is } A\} \Rightarrow \sup \min\,\{f_A(u), \pi_x(u)\}, \quad (2)$$

where π_x (u) is the possibility distribution induced by the proposition (X is Z), and A is a fuzzy set in the universe U.

The following sub-task is used to illustrate the process of predicting the rule selection based on the linguistic inexactness of expert's actions. Sub-Task: **Move down 27 lines to a position in column 20.** The following rules (**R**) apply:

Rule #1:Membership value of *more than a half of the screen* = 0.4
[The possibility that the rule applies is 0.4.]

Rule #2:Membership value of *more than 70 lines* = 0
[The possibility that the rule applies is 0].

Rule #3:Membership value of *less than half of the screen* = 0.3, and
Membership value of *left hand side of the line* = 0.4
[The possibility that the rule applies is 0.3 and 0.4.]

Rule #4:Membership value of *right half of line* = 0.9, and
Membership value of *less than half of the screen* = 0.3
[The possibility that the rule applies is 0.3 and 0.9.]

Rule #5:Membership value of *more than 70 lines* = 0
[The possibility that the rule applies is 0].

The possibility measure of the possibility distribution of **X** that the subject would select a given rule form the universe of available rules **R** was defined after Zadeh (1978). In case of the example cited above, the most applicable rule was derived based on the possibility measure of **{X is Rule #}**as follows:
Poss **{X** is Rule **#}**=MAX [{(Rule#1, 0.4)}, {(Rule#2, 0)}, MIN {(Rule#3, 0.3), (Rule#3, 0.4)}, MIN { (Rule#4, 0.3), (Rule#4, 0.9)}, {(Rule#5, 0)}]=MAX [{(Rule#1, 0.4)}, {(Rule#2, 0)},{(Rule#3, 0.3)}, {(Rule#4, 0.3)}, {(Rule#5, 0)}] ={(Rule#1, 0.4)}.

Given the set of five applicable rules (**R**), the possibility of selecting Rule#1 as the most applicable one is 0.4. It was predicted based on the possibilistic measure of uncertainty that the subject would use Rule #1, i.e. the CONTROL-D method. All fuzzy model predictions in the experiment were checked against the selection rule decisions made by the subjects.

Results of the pilot study

In the pilot study reported by Karwowski et al. (1989), one model was run using the fuzzy GOMS approach to the cursor placement task. Out of seventeen decisions, the fuzzy GOMS model predicted 13, or 76% correctly. Another run was made by replacing fuzzy quantifiers with the **non-fuzzy** rules. The **non-fuzzy** GOMS model predicted only 8, or 47% of the cursor placement decisions correctly.

Table 2. Sample #1: cursor placement rules for the pilot study (after Karwowski et al., 1989).

WORD LOCATION		METHODS		
Number of lines down	Word's column number	Method used	Fuzzy prediction	Non-fuzzy prediction
8	15	3	3,4	3
27	20	3	1	3
12	14	4	4	3
21	20	4	4	3
44	21	5	1	1
11	24	1	3	3
10	29	3	3	3
31	29	1	1	3
26	18	1	1	3
7	24	4	4	3
29	22	1	1	3
101	25	2	2	2
100	22	5	5	5
7	5	4	3	3
4	42	4	4	4
70	21	1	1	1
12	20	4	4	3

It was also observed that the use of fuzzy concepts seemed very natural within the knowledge elicitation process. It seemed much easier to ask for fuzzy memberships in the linguistic terms, than it would be to try and ascertain exact cut-offs for selection rules. This observation supports the results of the study by

Kochen (1975) who concluded that a higher degree of consistency in subjects response was found if they were allowed to give imprecise (verbal) descriptors of fuzzy concepts.

FUZZY INTERACTIONS: MORE RESULTS

Five subjects, graduate engineering students, participated in the main laboratory experiment reported by Karwowski et al. (1990). Subjects were asked to perform word placement while explaining what and why they were choosing their particular methods and the associated selection rules. If these selection rules appeared to have fuzzy components, these components were quantified by asking the subject to verbalize a membership value for the applicability of the rule. It was noted that fuzziness was based upon the participants cognitive ability to measure terms such as "about one half of a page".

For example, if the rule displayed a fuzzy component, either the paper or the screen (depending upon how the subjects referenced the fuzzy term) was pointed to, and the subjects were asked questions such as: "From 0 to 100 how much does this case belong to the class of FAR?" The resulting value was used to define the corresponding membership functions.

One antecedent, universally identified, was: "If the word to be located is **distinct**, then." The subjects were asked to determine whether the word to be located was distinct or not (binary decision). Later, each participant was asked to rate the distinctness as a fuzzy number from 0 to 1. A somewhat surprising result was that, on the whole, participants were more correct in choosing the fuzzy "distinctness" as opposed to the binary, **non-fuzzy** "distinct" category. Once the methods and the selection rules were elicited, the subjects were asked to perform similar word placement tasks on a different file. The methods which they used were recorded, and this case served to test the validity of both the **fuzzy** and **non-fuzzy** models.

For each of the fuzzy components of the text editing task, the uncertainty was quantified by presenting the subject with different scenarios. The curser would be placed on the screen and pointed to a word asking the degree such a scenario belonged to one of the fuzzy sets. An exhaustive collection of points was not conducted, but rather only a few points taken and interpolated (graphically) between the points.

The results of the main study are first illustrated using one subject only. Subject #2 utilized two methods to place the cursor at the given word. This reflected the subject's perception of the task as not an editing task, but rather a word location task. The methods used were: 1) *Search for the pattern (/xxx would search for the next occurrence of pattern xxx), and 2) Search for a near pattern*. The two rules utilized by the subject were simple: *1) If the word is "distinct" then use method #1, otherwise, 2) Use method #2.*

The subject was asked to rate whether each word was distinct (for **non-fuzzy** analysis), and then later asked for the "distinctness" or a fuzzy number for each word (in essence giving a fuzzy rating). For simplicity, the subject was only asked to rate the 20 words to be searched for and not the words located nearby.

This was not ideal because another rule was noted (but not verbalized): "If there is a 'very distinct' word 'near' the word to be located, search for that pattern instead."

Table 3 shows the word number, distinctness ratings, methods actually used, and the **non-fuzzy** model predictions (differentiated by using the concept of distinctness to predict the subject's keystrokes). It is obvious that in the case of subject #2, the results were not conclusive, and did not imply that fuzziness helps in the GOMS modeling. The **non-fuzzy** model correctly predicted 55% of the keystrokes, while the **fuzzy model** predicted 60% of the keystrokes. This low rating may be due to the fact that the rules elicited were not those used, and that the relationship between concept of distinctness and the methods used could depend on the distance to the searched word.

Table 3. Example of results for subject #2 (after Karwowski et al., 1990).

Word number	Number of lines	Distinct (yes/no)	Grade of distinctness	Method used	Non-fuzzy prediction	Fuzzy prediction
1	34	Y	0.75	S	S*	S*
2	12	Y	0.8	S	S*	S*
3	52	Y	0.75	SN	S	S
4	116	N	0.6	SN	SN*	S
5	8	N	0.3	S	SN	SN
6	30	N	0.35	S	SN	SN
7	44	N	0.65	S	SN	S*
8	118	Y	0.4	S	S*	SN
9	12	N	0.55	S	SN	S*
10	54	N	0	SN	SN*	SN*
11	13	N	0	SN	SN*	SN*
12	16	N	0.35	S	SN	SN
13	171	N	0	SN	SN*	SN*
14	25	N	0.35	S	SN	SN
15	4	N	0	SN	SN*	SN*
16	4	Y	0.4	S	S*	SN
17	38	N	0.6	S	SN	S*
18	198	Y	0.45	SN	S	SN*
19	16	N	0	SN	SN*	SN*
20	14	Y	0.8	S	S*	S*
		Rate of correct model predictions			11/20 (55%)	12/20 (60%)

S = Direct pattern search (method #1)
SN = Search pattern near word (indirect pattern search: method #2)
* = Correct prediction

Model prediction comparison

Table 4 shows a summary of prediction performance for both models and all subjects. Overall, across all subjects and trials, the **non-fuzzy** GOMS model successfully predicted 58.7% of the responses, while the **fuzzy** GOMS model predicted 82.3% of the subjects decisions. The Wilcoxon test showed that this difference was highly significant ([chi-square statistics] $X^2 = 9.95$, $p < 0.01$).

Table 4. Summary of experimental results for the main study (after Karwowski et al., 1990).

		Success rate (correct prediction)	
Subject number	Number of trials	Non-fuzzy GOMS prediction rate	Fuzzy GOMS prediction rate
1	20	11 (55.0%)	12 (60.0%)
2	74	35 (47.0%)	63 (85.1%)
3	26	19 (73.0%)	22 (84.6%)
4	27	19 (70.4%)	21 (76.9%)
5	153	92 (60.1%)	129 (84.3%)
Total	300	176 (58.7%)	247 (82.3%)

Several interesting observations were made through this expansion of the experimental data. The most important one was that in many cases adding the fuzzy functions helped tremendously in clarifying the meaning of rules. Specifically, a fuzzy definition of "distinctness" proved to be superior (in many cases) to its binary definition. Although the addition of fuzziness to the model structure could be seen as a "fine tuning" taking place in the elicitation process, this was not always the case (for example see results for subject #2).

CONCLUSIONS

Fuzzy methodologies can be very useful in the analysis and design of man-machines systems in general, and human-computer interaction systems in particular, by allowing to model vague and imprecise relationship between the user and computer. In order for this premise to succeed, one must identify the sources of fuzziness in the data and communication schemes relevant to the human-computer interaction. By incorporating the concept of fuzziness and linguistic inexactness based on possibility theory into the model of system performance, better performance prediction for human-computer system may be achieved.

The imprecision-tolerant communication scheme for human-computer interaction tasks should be based on fuzzy-theoretic extension of the GOMS model. In order to realize the potential benefits of fuzzy communication scheme, the natural fuzziness of the *Operators, Methods* and *Selection Rules* of the GOMS model should be modeled in order to allow the user to communicate with the computer system in a vague but intuitively comfortable way.

Since fuzziness plays an essential role in human cognition and performance, more research is needed to fully explore the potential of this concept in the area of human factors. It is believed that the theory of fuzzy sets and systems will allow one to account for natural vagueness, nondistributional subjectivity, and imprecision of man-machine systems which are too complex or too ill-defined to admit the use of conventional methods of analysis.

A formal treatment of vagueness is an important and necessary step toward more realistic handling of imprecision and uncertainty due to human and behavior through process at work. It is our view that the theory of fuzzy sets will prove successful in narrowing the gap between the world of the precise or "hard" sciences and the world of the cognitive or "soft" sciences. This can be achieved by providing a mathematical framework in which vague conceptual phenomena where fuzzy descriptors, relations, and criteria are dominant (Zimmermann, 1985) can be adequately studied and modeled.

ACKNOWLEDGEMENTS

We are indebted to Mrs. Laura Abell, Secretary at the Center for Industrial Ergonomics, University of Louisville, for her work on preparation of the manuscript.

REFERENCES

AUDLEY, R. J., ROUSE, W., SENDERS, T., and SHERIDAN, T. 1979, Final report of mathematical modelling group, in N. Moray (ed.), *Mental Workload, Its Theory and Measurement,* (Plenum Press, New York), 269-285.

BENSON, W. H. 1982, in *Fuzzy Sets and Possibility Theory,* R. R. Yager (ed.), (Pergamon Press, New York).

BERNOTAT, R. 1984, Generation of ergonomic data and their application to equipment design, in H. Schmidtke (ed.), *Ergonomic Data for Equipment Design,* (Plenum Press, New York), 57-75.

BEZDEK, J. 1981, *Pattern Recognition with Fuzzy Objective Function Algorithms* (Plenum Press: New York).

BOY, G. A., and KUSS, P. M. 1986, A fuzzy method for modeling of human-computer interactions in information retrieval tasks, in W. Karwowski and A. Mital (eds.), *Applications of Fuzzy Set Theory in Human Factors,* (Elsevier: Amsterdam), 117-133.

BROWNELL, H. H. and CARAMAZZA, A. 1978, Categorizing with overlapping categories, *Memory and Cognition,* **6**, 481-490.

CARD, S. K., MORAN, T. P., and NEWELL, A. 1983, *The Psychology of Human-Computer Interaction* (London: Lawrence Erlbaum Associates.

CHAPANIS, A. 1959, *Research Techniques in Human Engineering*, (The John Hopkins Press, Baltimore).

GILES, R. 1981, *Fuzzy Reasoning and Its Applications*, E. H. Mamdani and B. R. Gaines (eds.), (Academic Press, London).

HARRE, R. 1972, *The Philosophies of Science* (Oxford University Press, London).

HERSH, H. M., and CARAMAZZA, A. 1976, A fuzzy set approach to modifiers and vagueness in natural language, *Journal of Experimental Psychology: General*, 3, 254-276.

HESKETH, B., PRYOR, R., GLEITZMAN, M., and HESKETH, T. 1988, Practical applications of psychometric evaluation of a computerized fuzzy graphic rating scale, in T. Zeteni (ed.), *Fuzzy Sets in Psychology*, (North-Holland: Amsterdam), 425-454.

HIRSH, G., LAMOTTE, M., MASS, M. T., and VIGNERON, M. T. 1981, Phonemic classification using a fuzzy dissimilitude relation, *Fuzzy Sets and Systems*, 5, 267-276.

KARWOWSKI, W. and AYOUB, M. M. 1984a, Fuzzy modelling of stresses in manual lifting tasks, *Ergonomics*, 27, 641-649.

KARWOWSKI, W., AYOUB, M. M., ALLEY, L. R., and SMITH, T. L. 1984b, Fuzzy approach in psychophysical modeling of human operator-manual lifting system, *Fuzzy Sets and Systems*, 14, 65-76.

KARWOWSKI, W., and MITAL, A., (Editors), 1986, *Applications of Fuzzy Set Theory in Human Factors* (Elsevier: Amsterdam).

KARWOWSKI, W., MAREK, T. and NOWOROL, C. 1988, Theoretical basis of the science of ergonomics, in *Proceedings of the 10th Congress of the International Ergonomics Association*, Sydney, Australia, (Taylor & Francis, London) 756-758.

KARWOWSKI, W., KOSIBA, E., BENABDALLAH, S., and SALVENDY, G. 1989, Fuzzy data and communication in human-computer interaction: for bad or for good, in G. Salvendy and M.J. Smith (eds.), *Designing and Using Human-Computer Interfaces and Knowledge Based Systems*, (Elsevier: Amsterdam), 402-409.

KARWOWSKI, W., KOSIBA, E., BENABDALLAH, S. and SALVENDY, G. 1990, A Framework for development of fuzzy GOMS model for human-computer interaction, *International Journal of Human-Computer Interaction*, 2, 287-305.

KOCHEN, M. 1975, Applications of fuzzy sets in psychology, in L. A. Zadeh, K. S. Fu, K. Tanaka and M. Shimuro (eds.), *Fuzzy Sets and Their Applications to Cognitive and Decision Processes*, (Academic Press: New York), 395-408.

KRAMER, U. 1983, in *Proceedings of the Third European Annual Conference on Human Decision Making and Manual Control*, (Roskilde, Denmark), 313.

KRAMER, U. and ROHR, R. 1982, in *Analysis, Design and Evaluations of Man-Machine Systems*, G. Johannsen and J. E. Rijnsdorp (eds.), (Pergamon Press, Oxford), 31-35.

LAKOFF, H. 1973, A study in meaning criteria and the logic of fuzzy concepts, *Journal of Philosophical Logic*, **2**, 458-508.

MAMDANI, E. H., and GAINES, B. R., (Editors), 1981, *Fuzzy Reasoning and Its Applications*, (Academie Press: London).

MCCLOSKEY, M. E. and GLUCKSBERG, S. 1978, *Memory and Cognition*, **6**, 462-472.

MCCORMICK, E. J. 1970, *Human Factors Engineering*, (McGraw Hill, New York).

ODEN, G. C. 1977, Human perception and performance, *Journal of Experimental Psychology*, **3**, 565-575.

PEW, R. W. and BARON, S. 1983, *Automatica*, **19**, 663-676.

ROHMERT, W. 1979, in N. Moray (Ed.), *Mental Workload, Its Theory and Measurement*, (Plenum Press, New York), 481.

SAATY, S. L., 1977, Exploring the interface between hierarchies, multiple objectives and fuzzy sets, *Fuzzy Sets and Systems*, **1**, 57-68.

SCHMUCKER, K. J. 1984, *Fuzzy Sets, Natural Language Computations, and Risk Analysis*, (Computer Science Press, Maryland).

SIMCOX, W. A. 1984, A method for pragmatic communication in graphic displays, *Human Factors*, **26**, 483-487.

SINGLETON, W. T. 1982, *The Body at Work. Biological Ergonomics*, (University Press, Cambridge).

SMITHSON, M. 1982, Applications of fuzzy set concepts to behavioral sciences, *Mathematical Social Sciences*, **2**, 257-274.

SMITHSON, M. 1987, *Fuzzy Set Analysis for Behavioral and Social Sciences* (Springer-Verlag: New York).

TERANO, T., MURAYAMA, Y., AKIJAMA, N. 1983, Human reliability and safety evaluation of man-machine systems, *Automatica*, **19**, 719-722.

TOPMILLER, D.A. 1981, in *Manned Systems Design: Methods, Equipment and Applications*, J. Moraal and K. F. Kraiss (eds.), (Plenum Press, New York), 3-21.

WILLAEYS, D. and MALVACHE, N. 1979, in *Advances on Fuzzy Set Theory and Applications*, in M. M. Gupta, R. K. Ragade and R. R. Yager (eds.), (North Holland, Amsterdam).

ZADEH, L. A. 1965, Fuzzy sets, *Information and Control*, **8**, 338-353.

ZADEH, L. A. 1973, Outline of a new approach to the analysis of complex systems and decision processes, *IEEE Trans. Systems, Man, and Cybernetics*, SMC-3, 28-44.

ZADEH, L. A. 1974, Numerical versus linguistic variables, *Newspaper of the Circuits and Systems Society*, **7**, 3-4.

ZADEH, L. A. 1978, Fuzzy sets as a basis for a theory of possibility, *Fuzzy Sets and Systems: 1*, **3**.

ZIMMERMANN, H. J. 1980, Testability and meaning of mathematical models in social sciences, *Mathematical Modeling*, **1**, 123-139.

ZIMMERMANN, H. J. 1985, *Fuzzy Set Theory and Its Applications*, (Kluwer-Nijhoff Publishing, Boston).

10

QUESTIONNAIRES AND FUZZINESS

Bernadette Bouchon-Meunier
CNRS, LAFORIA, Université Paris VI, Tour 46
4 place Jussieu, 75252 Paris Cédex 05, France

INTRODUCTION

Questionnaires represent hierarchical processes disjoining the elements of a given set by using successive tests or operators [12]. They involve the probabilities of the results of the tests, or the probabilities of the modalities of the operators. In the case where the tests or operators depend on imprecise factors, such as the accuracy of physical measurements or the linguistic description of variables, the questionnaires take into account coefficients evaluating the fuzziness of the data. The construction of such questionnaires is submitted to several kinds of constraints and requires appropriate algorithms.

When the questionnaire is only characterized by probabilistic elements, its average length is generally interesting to minimize in order to improve the efficiency of the process it represents, with respect to some basic constraints. The tests or operators can be chosen with regard to the quantity of information they process. The construction of the most efficient questionnaire is either holistic [11, 12], taking into account all the tests or operators which must be used, or it is selective, based on the choice of the most significative tests or operators with regard to the purpose of the process [2, 13, 14].

If fuzzy criteria are involved, the efficiency of the questionnaire deals with the specificity of the results it provides. A trade-off must be obtained between the preservation of some fuzziness in the tests or operators allowing a flexibility in the management of the available data, and the reliability of the results obtained through the questionnaire [3, 4].

The support of a questionnaire is a finite, directed and valuated graph without circuit, where every vertex is connected with one of them, called the root, by at least one path or series of edges (exactly one in the case of arborescent questionnaires). No edge ends in the root. There exist terminal vertices from which no edge is descending. Several systems of valuations can be defined for the edges and the vertices, for instance probabilistic valuations, utility values, coefficients of fuzziness. The tests or operators are connected to the non-terminal nodes and their possible results or modalities are associated with the edges descending from this node.

The simplest case of a questionnaire is arborescent. Such a model is extensively used in various fields and it corresponds to weighted trees. In the classical probabilistic framework, where the data are associated with a given uncertainty, applications of arborescent questionnaires exist in the study of search trees, decision trees, fault trees, species identification, hierarchical classification, diagnosis assistance, decision-making, preference elicitation, knowledge acquisition for instance.

When the involved tests or operators are not precisely described, arborescent questionnaires must take into account both uncertainty and imprecision and they must lead to conclusions which are acceptable in spite of the imprecision. We present here several utilizations of questionnaires in a fuzzy framework.

QUESTIONNAIRES WITH LINGUISTIC VARIABLES

Let us consider a given set $D = \{d_1, ..., d_n\}$ of elements to identify, for instance a set of decisions to make, of diagnosis to identify, of classes to recognize, which are supposed to be defined without any ambiguity. We suppose that the probability distribution $P = \{p_1, ..., p_n\}$ is available, with p_j the probability of d_j to be present in the considered world, for $1 \leq j \leq n$.

We also consider a set $Q = \{q_1, ..., q_m\}$ of so-called "questions", which represent tests or operators. A question q_i is a link between a linguistic variable X_i defined on a universe U_i and a family of a(i) labels, denoted by $q_i^1, ..., q_i^{a(i)}$, and associated with possibility distributions $f_i^1, ..., f_i^{a(i)}$, defined on U_i and lying in [0, 1] (see Figure1).

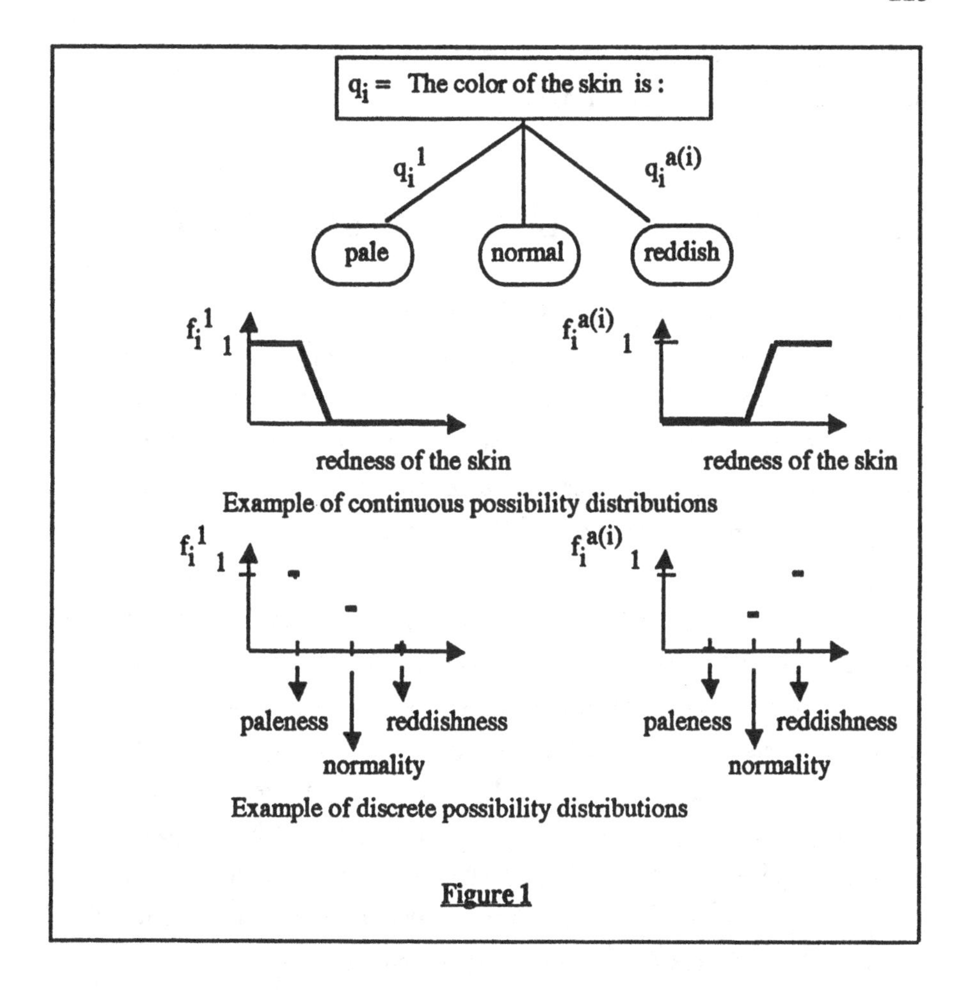

Figure 1

Two different types of problems can be regarded, depending on the fact that the questions of Q are deterministic or not with regard to the elements of D.

- either there is a possibilisitic relationship between lists of answers to questions of Q and the elements of D, yielding the possibility of $d \in D$ to be concerned in a studied situation, according to the obtained answers, and the certainty we can have in this assertion [7]. We construct a questionnaire by successively choosing questions of Q bringing as much information as possible on the elements of D and we stop asking new questions when an element of D is sufficiently well identified (selective construction).

- or there is a precise relationship between lists of answers to questions of Q and elements of D, and we construct a questionnaire by ordering the questions of Q in such a way that every element of D can be associated with a terminal vertex of the questionnaire (holistic construction) [1, 3].

SELECTIVE CONSTRUCTION OF QUESTIONNAIRES

(See annex 1 for technical details about this section).

In a probabilistic study, the probabilities prob (q_i^k / d_j), $1 \leq i \leq m$, $1 \leq k \leq a(i)$, $1 \leq j \leq n$, of obtaining every label associated with a question would be given for every element of D. In many cases, there is no means of knowing these probabilities and the only knowledge we have regarding the simultaneous presence of a given label and an element of D is possibilistic.

Let us suppose given the possibility $\pi(d_j / q_i^k)$ that we are in front of the case d_j of D, for $1 \leq j \leq n$, when we obtain the label q_i^k for question q_i, for $1 \leq i \leq m$, $1 \leq k \leq a(i)$. As there is no absolute certainty that this answer implies that d_j must be identified, we also suppose given the necessity $N(d_j / q_i^k)$ quantifying this certainty.

We can also suppose given some knowledge about the fact that the element d_j can be thought of, when an answer different from q_i^k is obtained to question q_i : let $\pi(d_j / \neg q_i^k)$ and $N(d_j / \neg q_i^k)$ denote the possibility and the certainty that d_j is acceptable when q_i^k is not obtained. If these values are not precisely known, they will be replaced [10] by the interval [0, 1] to which they belong.

We fix thresholds s and t in [0, 1], defining the acceptable values [s, 1] and [t, 1] for the lowest acceptable possibility and the lowest acceptable certainty of an element of D to be satisfying when given labels are obtained for a question.

The problem we consider is the following :

- first of all, how to determine the sequence of questions necessary and sufficient to identify every element of D as reliably and efficiently as possible, (step 1)
- secondly, how to use this sequence of questions every time we have to recognize a particular case under study. (step 2)

Applications of this model can be found in knowledge acquisition, in diagnosis assistance, in species identification for instance. The first step corresponds to the construction of the sequence of questions providing the best recognition of classes on a training set of examples, the second step is associated with the identification of the convenient class for an example not belonging to the training set.

Step 1

It is obvious that the element d_j of D will be immediatly recognized if there is a question q_i yielding an answer q_i^k such that $N(d_j / q_i^k) = \pi(d_j / q_i^k) = 1$. No further question will be necessary in this case, but at least one other question must be asked in the general case.

The first question to be asked will be q_i, for $1 \leq i \leq m$, processing the most efficient information about the elements of D and we propose to evaluate this efficiency by means of the average certainty $Cer(q_i)$ provided by q_i on the recognition of any element of D

Then, the first question to be asked will be q_i such that $Cer(q_i)$ is maximum.

Now, let us suppose that a sequence $S_r = (x_1, ..., x_r)$ of questions is not sufficient to determine an element d_{jo} of D such that its possibility to be present, given the answers it provides to questions $x_1, ..., x_r$, is sufficiently high and the certainty available on its identification is acceptable (see Figure 2).

If labels $x_1^{k(1)}, ..., x_r^{k(r)}$ are respectively obtained for this sequence S_r of questions, we evaluate the possibility $Pos\ (d_{jo} / x_1^{k(1)}, ..., x_r^{k(r)})$ that d_{jo} should be identified, and the certainty $Nec\ (d_{jo} / x_1^{k(1)}, ..., x_r^{k(r)})$ that this identification is satisfying.

As the sequence S_r of questions is not sufficient to identify an element of D with the list of obtained labels $x_1^{k(1)}, ..., x_r^{k(r)}$, a new question q_i must be asked.

For an obtained label q_i^k, we evaluate the average certainty $Cer(x_1^{k(1)}, ..., x_r^{k(r)}, q_i^k)$ provided by these (r+1) questions on any element of D.

We choose the question q_i which processes the most efficient information

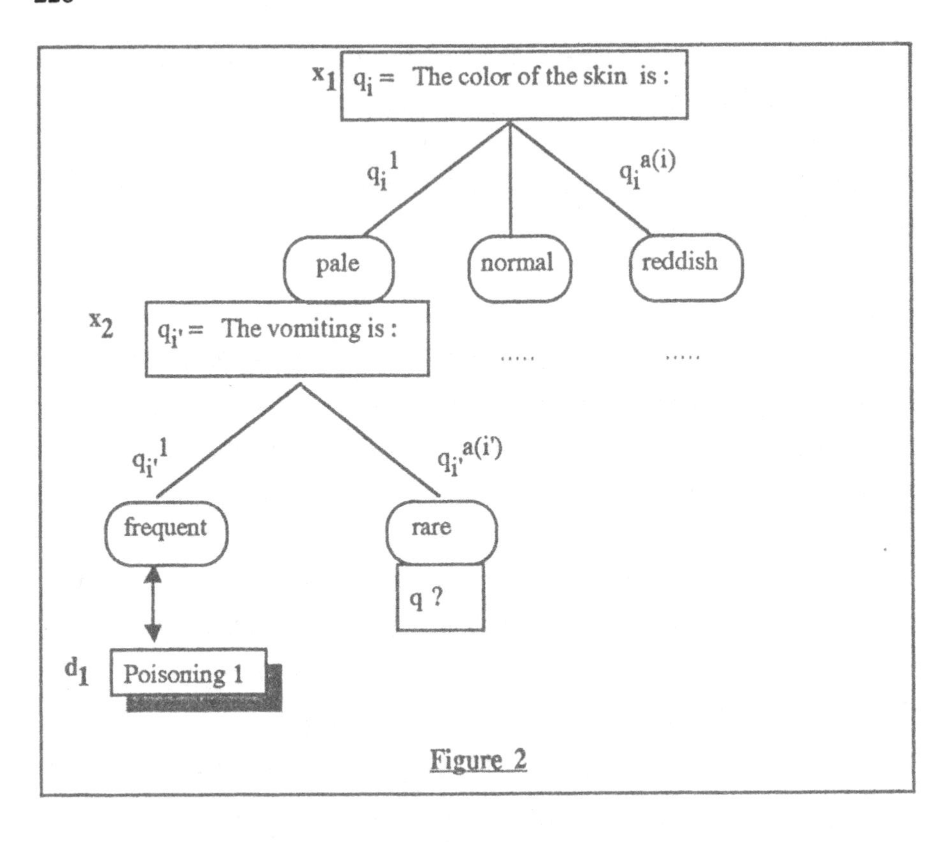

Figure 2

about D, with regard to all its possible labels, or, equivalently, which gives the highest <u>absolute certainty</u> $C(q_i)$.

No further question will be asked when a sequence of labels $x_1^{k(1)}, \ldots, x_r^{k(r)}$ is obtained for questions $x_1, \ldots, x_r$, and there exists an element d_{jo} of D such that $Pos(d_{jo} / x_1^{k(1)}, \ldots, x_r^{k(r)}) \geq s$, and $Nec(d_{jo} / x_1^{k(1)}, \ldots, x_r^{k(r)}) \geq t$. Then d_{jo} will be associated with S_r, which is called <u>terminal</u>.

Step 2 :

For a new given particular situation c_o, an element of D must be identified from the answers to the various questions of the questionnaire we have constructed.

Let S_r be a terminal sequence of questions of Q in this questionnaire, to which c_o provides answers $x_1^{k(1)}$, ..., $x_r^{k(r)}$ chosen, for every question, in the list of available labels. Then, we clearly identify the element of D associated with S_r.

As the labels associated with every question are not precise, we must accept that an answer is provided in a way somewhat different from the expression we expect in the list of authorized labels. Let us denote by q'_i the label obtained as an answer to question q_i, $1 \leq i \leq m$, more or less different from all the q_i^k, $1 \leq k \leq a(i)$, and by g_i the possibility distribution describing q'_i, defined on U_i and lying in [0, 1]. (See Figure 3)

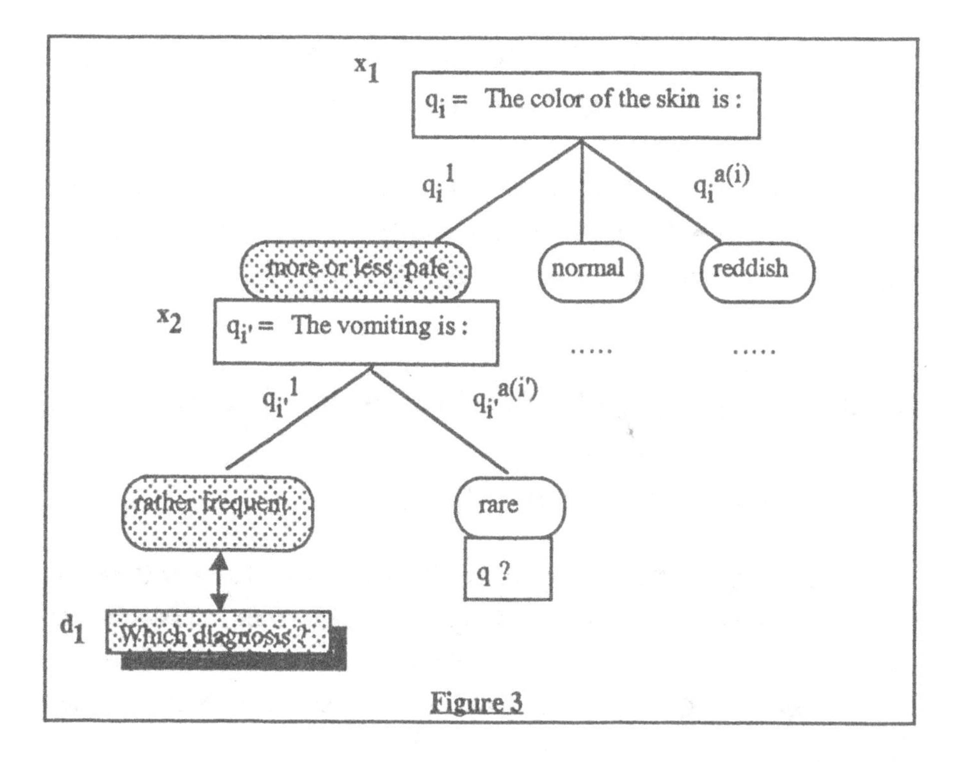

Figure 3

The compatibility of this answer q'_i with one of the labels q_i^k proposed for q_i, , with $1 \leq k \leq a(i)$, is measured by the classical possibility and necessity measures of adequation [9] respectively denoted by $\pi(q_i^k; q'_i)$ and $N(q_i^k; q'_i)$.

We deduce the possibility $\pi^k(d_j)$ that d_j is concerned by the particular situation c_o, according to the proximity of its answer with q_i^k, and the certainty of this assertion $N^k(d_j)$. This evaluation will be performed for the labels q_i^k such that $\pi(q_i^k; q'_i) \geq s$ and $N(q_i^k; q'_i) \geq t$. It is then possible to have several sequences of questions to use, i.e. several pathes of the questionnaire to follow before the recognition of a particular element of D.

More generally, let us consider again a terminal sequence S_r of questions leading to the identification of d_{jo} in the questionnaire. Because of the differences which may exist between the expected answers to these questions, and the labels obtained from the particular case c_o, the possibility and certainty of d_{jo} will be the following :

$$Pos\,(d_{jo} / x_1^{k(1)}, \ldots, x_r^{k(r)}) = \min_{1 \leq i \leq r} \pi^{k(i)}(d_{jo}),$$
$$Nec\,(d_{jo} / x_1^{k(1)}, \ldots, x_r^{k(r)}) = \max_{1 \leq i \leq r} N^{k(i)}(d_{jo}).$$

The element d_{jo} will be definitely identified for the situation c_o, by means of the sequence of questions S_r, if there exists labels $x_1^{k(1)}, \ldots, x_r^{k(r)}$ yielding Pos $(d_{jo} / x_1^{k(1)}, \ldots, x_r^{k(r)}) \geq s$ and Nec $(d_{jo} / x_1^{k(1)}, \ldots, x_r^{k(r)}) \geq t$.

HOLISTIC CONSTRUCTION OF QUESTIONNAIRES

(See annex 2 for technical details about this section).

Let us suppose that we want to use all the tests or operators of Q, and we have to order them in such a way that the questionnaire we construct associates an element of D with each terminal node. The questionnaire could be arborescent or not. We suppose that Q and D are compatible, which means that such a construction is possible.

The problem we consider is the choice of the questions providing the most efficient questionnaire with regard to the recognition to make. Its quality can be evaluated [3] with respect to the fuzziness which is involved in the characterizations deduced from the fuzzy tests or operators, and improved, when several constructions of questionnaires are possible, by an appropriate choice of a the order of some questions when possible. Applications can be found in search trees, in species identification, for instance.

Several aspects of such a choice can be proposed [3, 4, 6] and we propose one method hereunder.

Let us suppose that the labels q_i^k associated with the questions q_i of Q are conveniently defined in such a way that they determine a fuzzy partition $\prod_i$ of the universe U_i on which the concerned linguistic variable X_i is defined. The classes of this fuzzy partition are fuzzy subsets of U_i defined by membership functions equal to f_i^k, $1 \leq k \leq a(i)$, in every point of U_i. We suppose given the probability distribution P_i of the variable X_i, for the studied population.

The problem we consider is the identification of a crisp (non-fuzzy) partition of U_i, able to represent the information contained in $\prod_i$. We may think of several applications of this problem : in knowledge acquisition, if the training set deals with crisp data and then non-fuzzy tests or operators, and the new examples are described by means of fuzzy questions ; in decision-making, when a crisp decision must be taken from fuzzy test s or operators or from the answers provided by the inquired personto a crisp question q_i by indicating preference grades for the elements q_i^k which are proposed to her; in preference elicitation, when the inquirer makes a choice between two fuzzy questions about the same variable.

For a given threshold r in [0, 1], we associate with $\prod_i$ a crisp partition $\prod_i^*$ of level r, by defining crisp classes as $q_i^{k*} = \{ u \ / \ f_i^k(u) \geq r \}$, $1 \leq k \leq a(i)$. Obviously, such a crisp partition does not exist for any value of r and some thresholds correspond to several possible crisp partitions. We suppose that the tests or operators are defined in such a way that there always exist a value r providing a crisp partition. We can consider the average weight of each fuzzy label by introducing its r-probability $P_i^r(q_i^k)$ as the average value of its associated possibility distribution, for the values at least equal to r.

This generalization of the concept of probability to a fuzzy subset of the universe allows to measure the fuzzy information $I_i^{r*}(\prod_i)$ processed by $\prod_i$ for the threshold r with respect to the crisp partition $\prod_i^*$. We use this tool as a measure of the proximity between $\prod_i$ and $\prod_i^*$.

Let us consider the case where we are given a set of fuzzy operators Q and we look for the crisp partition associated with each of them, loosing as little information as possible when passing from fuzzy descriptions to crisp descriptions.

For every q_i^k associated with the fuzzy partition $\prod_i$ of U_i, we choose the crisp partition $\prod_i^*$ such that the fuzzy information $I_i^{r*}(\prod_i)$, processed by $\prod_i$ for the threshold r with respect to $\prod_i^*$, is maximum.

If several tests or operators are available for the same linguistic variable X_i on U_i, the most interesting is the one processing the greatest absolute fuzzy information with regard to all the possible crisp partitions which could be associated with it.

REFERENCES

[1] AKDAG, H., BOUCHON, B. (1988) - Using fuzzy set theory in the analysis of structures of information, Fuzzy Sets and Systems, 3, 28.

[2] AURAY J.P., DURU G., TERRENOIRE M., TOUNISSOUX D. ZIGHED A. (1985) - Un logiciel pour une méthode de segmentation non arborescente, Informatique et Sciences Humaines, vol. 64.

[3] BOUCHON B. (1981) - Fuzzy questionnaires, Fuzzy Sets and Systems 6, pp. 1-9.

[4] BOUCHON B. (1985) - Questionnaires in a fuzzy setting, in Management decision support systems using fuzzy sets and possibility theory, eds. J. Kacprzyk and R.R. Yager, Verlag TUV Rheinland, 189-197.

[5] BOUCHON, B. (1987) - Preferences deduced from fuzzy questions, in Soft optimization models using fuzzy sets and possibility theory (J. Kacprzyk and S.A. Orlovski, eds), D. Reidel Publishing Company pp. 110-120.

[6] BOUCHON, B. (1988) - Questionnaires with fuzzy and probabilistic elements, in Combining fuzzy imprecision with probbilistic uncertainty in decision making (J. Kacprzyk, M. Fedrizzi, eds.), Springer Verlag, pp. 115-125.

[7] BOUCHON, B. (1990) - Sequences of questions involving linguistic variables; in Approximate reasoning tools for artificial intelligence, (M. Delgado, J.L. Verdegay, eds.), Verlag TUV Rheinland.

[8] BOUCHON B., COHEN, G. (1986) - Partitions and fuzziness, J. of Mathematical Analysis and Applications, vol. 113, 1986.

[9] DUBOIS, D. PRADE, H. (1987) - Théorie des possibilités, applications à la représentation des connaissances en informatique, Masson.

[10] FARRENY H., PRADE H., WYSS E. (1986) - Approximate reasoning in a rule-based expert system using possibility theory : a case study, in Information Processing (H.J. Kugler, ed.), Elsevier Science Publishers B.V..

[11] PAYNE R. (1985) - Genkey : a general program for constructing aids to identification, Informatique et Sciences Humaines, vol. 64.

[12] PICARD C.F. (1980) - Graphs and questionnaires, North Holland,

Amsterdam.

[13] TERRENOIRE M. (1970) - Pseudoquestionnaires et information, C.R. Acad. Sc. 271 A, pp. 884-887.

[14] M. TERRENOIRE (1970) - Pseudoquestionnaires, Thèse de Doctorat d'Etat, Lyon.

[15] E. WYSS (1988) - TAIGER, un générateur de systèmes experts adapté au traitement de données incertaines et imprécises, Thèse, Institut National Polytechnique de Toulouse.

Annex 1 :

Possibility and necessity coefficients associated with every element d_j of D, when the label q_i^k is obtained as an answer to the test or the operator q_i of Q, are respectively denoted by $\pi(d_j / q_i^k)$ and $N(d_j / q_i^k)$. They belong to [0, 1] and they are such that $N(d_j / q_i^k) \leq \pi(d_j / q_i^k)$, with $N(d_j / q_i^k) = 0$ if $\pi(d_j / q_i^k) < 1$ and $\pi(d_j / q_i^k) = 1$ if $N(d_j / q_i^k) \neq 0$.

The average certainty $Cer(q_i)$ provided by a single test or operator q_i on the recognition of any element of D is defined as follows :

$$Cer(q_i) = \Sigma_{1 \leq k \leq a(i)} \Sigma_{1 \leq j \leq n} N(d_j / q_i^k)\, p_j . \quad (1)$$

The possibility $Pos\,(d_{jo} / x_1^{k(1)}, \ldots, x_r^{k(r)})$ that the element d_{jo} of D must be identified, and the certainty $Nec\ (d_{jo} / x_1^{k(1)}, \ldots, x_r^{k(r)})$ that this identification is satisfying, when labels $x_1^{k(1)}, \ldots, x_r^{k(r)}$ are obtained as answers to tests or operators $x_1, \ldots, x_r$, are evaluated by means of the following coefficients :

$$Pos\,(d_{jo} / x_1^{k(1)}, \ldots, x_r^{k(r)}) = \min_{1 \leq i \leq r} \pi(d_{jo} / x_i^{k(i)}), \quad (2)$$

$$Nec\ (d_{jo} / x_1^{k(1)}, \ldots, x_r^{k(r)}) = \max_{1 \leq i \leq r} \pi(d_{jo} / x_i^{k(i)}). \quad (3)$$

We define as follows the average certainty $Cer(x_1^{k(1)}, \ldots, x_s^{k(s)})$, provided about the recognition of any element of D, by a sequence of labels $x_1^{k(1)}, \ldots, x_s^{k(s)}$ obtained as answers to tests or operators $x_1, \ldots, x_s$ of Q :

$$Cer(x_1^{k(1)}, \ldots, x_s^{k(s)}) = \Sigma_{1 \leq j \leq n} \Delta_{i,j}^{k}\, Nec(d_j / x_1^{k(1)}, \ldots, x_s^{k(s)})\, p_j, \quad (4)$$

with $\Delta_{i,j}^k = 1$ if $Pos\,(d_j / x_1^{k(1)}, \ldots, x_s^{k(s)}) \geq s$, and 0 otherwise.

The absolute certainty of a test or an operator q_i of Q, after the sequence of labels $x_1^{k(1)}, \ldots, x_r^{k(r)}$ is obtained as answers to tests or operators $x_1, \ldots, x_s$ is defined as follows :

$$C(q_i) = (1/a(i))\, \Sigma_{1 \leq k \leq a(i)}\, Cer(x_1^{k(1)}, \ldots, x_r^{k(r)}, q_i^k) \quad (5)$$

Possibility measure of the adequation of any answer q'_i with a given label q_i^k, for a question (test or operator) q_i of Q:

$$\pi(q_i^k; q'_i) = \sup\,\{u \text{ in } U_i\}\, \min(f_i^k(u), g_i(u)),\ 1 \leq k \leq a(i), \quad (6)$$

Necessity measure of this adequation :

$$N(q_i^k; q'_i) = \inf\{u \text{ in } U_i\} \max(1 - f_i^k(u), g_i(u)), 1 \leq k \leq a(i). \qquad (7)$$

For the particular situation c_o, the possibility $\pi^k(d_j)$ that d_j is concerned according to the proximity of the obtained answer q'_i with q_i^k, and the certainty of this assertion $N^k(d_j)$, will be evaluated by the following coefficients [10, 15] :

$$\pi^k(d_j) = \max[\min\{\pi(d_j / q_i^k), \pi(q_i^k; q'_i)\}, \min\{\pi(d_j j / \neg q_i^k), 1 - N(q_i^k; q'_i)\}], \qquad (8)$$

$$N^k(d_j) = \min[\max\{N d_j / q_i^k), 1 - \pi(q_i^k; q'_i)\}, \max\{N(d_j / \neg q_i^k), N(q_i^k; q'_i)\}]. \qquad (9)$$

As indicated in [10], the values can be replaced by the interval to which they belong in the case where they are not precisely known.

Annex 2

A fuzzy partition of the universe U_i on which the concerned linguistic variable X_i is defined satisfies :

$\Sigma_{1 \leq k \leq a(i)} f_i^k(u) = 1$, for every point u in U_i,

and $\Sigma\{u \text{ in } U_i\}\ f_i^k(u) > 0$, for every k, $1 \leq k \leq a(i)$.

The r-probability $P_i^r(q_i^k)$ of a fuzzy label q_i^k with regard to the crisp class q_i^{k*} is defined by :

$$P_i^r(q_i^k) = \Sigma\{u \text{ in } q_i^{k*}\} f_i^k(u) P_i(u).$$

The fuzzy information $I_i^{r*}(\Pi_i)$ processed by a fuzzy partition Π_i of U_i for the threshold r with respect to the crisp partition Π_i^* is defined as follows :

$$I_i^{r*}(\Pi_i) = \Sigma_{1 \leq k \leq a(i)} L(P_i^r(q_i^k)) / [\Sigma_{1 \leq k \leq a(i)} P_i^r(q_i^k)],$$

with the function $L(x) = -x \log(x)$.

Properties of this fuzzy information lead to its mazimization in order to have the best compatibility between a fuzzy partition and any possible associated crisp partition for a given threshold r.

The absolute fuzzy information processed by a fuzzy partition Π_i^* with regard to all the possible crisp partitions which could be associated with it equals :

$\max\{I_i^{r*}(\Pi_i) / \Pi_i^* \text{ associated with } \Pi_i\}$.

11

FUZZY LOGIC KNOWLEDGE SYSTEMS AND ARTIFICIAL NEURAL NETWORKS IN MEDICINE AND BIOLOGY

Elie Sanchez
Faculty of Medicine, University of Marseille, and
***Neurinfo Research Department**
Institut Méditerranéen de Technologie
13451 Marseille Cedex13, France

ABSTRACT

This tutorial paper has been written for biologists, physicians or beginners in fuzzy sets theory and applications. This field is introduced in the framework of medical diagnosis problems. The paper describes and illustrates with practical examples, a general methodology of special interest in the processing of borderline cases, that allows a graded assignment of diagnoses to patients. A pattern of medical knowledge consists of a tableau with linguistic entries or of fuzzy propositions. Relationships between symptoms and diagnoses are interpreted as labels of fuzzy sets. It is shown how possibility measures (soft matching) can be used and combined to derive diagnoses after measurements on collected data.

The concepts and methods are illustrated in a biomedical application on inflammatory protein variations. In the case of poor diagnostic classifications, it is introduced appropriate ponderations, acting on the characterizations of proteins, in order to decrease their relative influence. As a consequence, when pattern matching is achieved, the final ranking of inflammatory syndromes assigned to a given patient might change to better fit the actual classification. Defuzzification of results (i.e. diagnostic groups assigned to patients) is performed as a non fuzzy sets partition issued from a "separating power", and not as the center of gravity method commonly employed in fuzzy control.

It is then introduced a model of fuzzy connectionist expert system, in which an artificial neural network is designed to build the knowledge base of an expert system, from training examples (this model can also be used for specifications of rules in fuzzy logic control). Two types of weights are associated with the connections : primary linguistic weights, interpreted as labels of fuzzy sets, and secondary numerical weights. Cell activation is computed through MIN-MAX fuzzy equations of the weights. Learning consists in finding the (numerical) weights and the network topology. This feedforward network is described and illustrated in the same biomedical domain as in the first part.

***Address for correspondence**

Keywords : Fuzzy Logic, Linguistic Model, Fuzzy Propositions, Medical Knowledge Representation, Medical Diagnosis, Soft Matching, Relative Importance, Defuzzification, Separating Power, Artificial Neural Networks, Fuzzy Connectionist Expert Systems, Linguistic Weights.

INTRODUCTION

In many situations, physicians use subjective or intuitive judgments. They cannot always logically, or in simple terms, explain how they derive conclusions, because of the complex mental processes inherent to the nature of the cases to be diagnosed, or to the difficulty of recalling their years of training and experience.

Interpretation of biological analyses suffers from some arbitrariness, particularly at the boundaries of the quantities that are measured, or evaluated. It is customary to use symbols like +++, ++, +, N, -, - -, - - -, or, ↑↑↑, ↑↑, ↑, N, ↓, ↓↓, ↓↓↓ to denote variations ('N' stands for 'Normal'). In general, limits of values that characterize *abnormalities*, or *normality*, define numerical intervals that are used to describe standards in variations. First of all, normal or non pathological states, have to be determined. They constitute the reference to which abnormalities are specified.

Biologists are familiar with normal variation ranges that are a prerequisite to a proper interpretation of all laboratory tests. Notions of *statistical normality* are usually derived from frequency distributions, not always confined to Gaussian distributions. But, depending on the measurement procedures of a given laboratory, on the epidemiologist, the biologist or the clinician who manipulates and interprets measurements, but also on the nature of the populations under study, and on conditions of *physiological (biological) normality*, one has to commonly rely on fiducial limitts (see for example [1,2] for discussions on *normality*).

The main drawback in working with intervals to represent normality, or ranges of variations for abnormalities, is the weak reliability on thresholds. Moreover, such boundaries are more or less physician dependent in practice. For example [3], "the normal base-line value for a given individual's lactic acid dehydrogenase may be at the extreme low point of the normal range for the general populations. Thus he (the physician) could develop an elevation due to a disease process that is significant and still within the normal range of the population." Still in [3], under a table defining the range of normal values for blood chemistry, one may read : "these ranges are a guide to the normal concentrations of blood constituents. For accurate interpretations, always refer to normal values established by individual laboratories, since individual differences in procedures may affect the actual ranges."

A problem that is often posed lies in the ill-definition and in the treatment of the boundaries of the intervals. To cope with borderline cases, fuzzy set theory provides very natural and appropriate tools. So it is here assumed that imprecision in the description of variations is of a fuzzy type and terms like "Normal, Slightly Decreased, Very Increased, etc.," will be treated as labels of fuzzy sets in (possibly different) universes of discourse. These fuzzy sets represent linguistic intervals, and around cutoff boundaries, very close points will not be totally accepted or rejected like in yes-or-no procedures, according to their position with respect to the frontier.

A coding with ↑'s or ↓'s is sometimes too restrictive : it is not always possible to choose between ↑ and ↑↑ for example, and in some patterns one may find "from ↑ to ↑↑." A scale with degrees ranging from 0 to 1 is very convenient. Note that it is not needed to set up precise values in [0,1] : in interpreting patterns, it is sufficient to have a rough idea of the curve expressing the compatibility between measurements and concepts.

It will now be described a general methodology, illustrated with an application, of special interest in the processing of borderline cases, and which offers the physician, practical assistance in obtaining the same results in the same abnormal profiles.

PATTERN (MEDICAL KNOWLEDGE)

In this paper, a pattern of *Medical Knowledge* consists of a tableau with linguistic entries. These linguistic associations are supposed to be given by experts, having in mind that different experts may provide somehow different characterisations for a same pattern.

This *Medical Knowledge* can be interpreted in terms of fuzzy propositions, like "Temperature is Slightly_Increased," i.e. of the form "S is F," where S is a variable (referred to as the name of a Sign, of a Symptom, or generally of an Attribute) taking values in a universe of discourse U, and F is a fuzzy subset of U. The tableau expresses relationships between attributes (S) such as temperature, plasma lipids, arterial pressure, serum proteins, etc. , and diagnoses (Δ) or groups, types, syndromes, diseases, etc. The linguistic entries are assumed to be labels of fuzzy sets (F), or more specifically, fuzzy intervals. Note that the term "diagnosis" is more or less arbitrary, it is a convenient way to summarize or synthetize information. In decision processes, symptoms can be viewed as diagnoses and vice-versa. Characterizations of diagnoses appear in the rows of a tableau as shown in fig. 1.

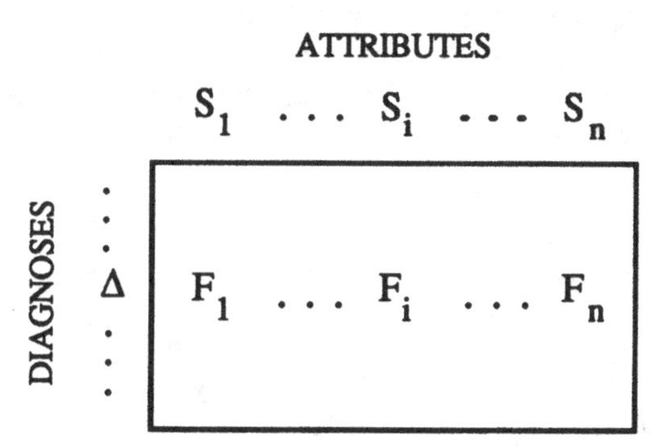

Fig. 1 - Tableau with linguistic entries represented by fuzzy sets.

In this tableau, S_i (i=1,n) is the name of a variable (Sign, Symptom, or Attribute) taking values in a universe of discourse U_i, and F_i is a fuzzy subset of U_i. For example, in a typical serum protein pattern [3], one may find the tableau of fig. 2.

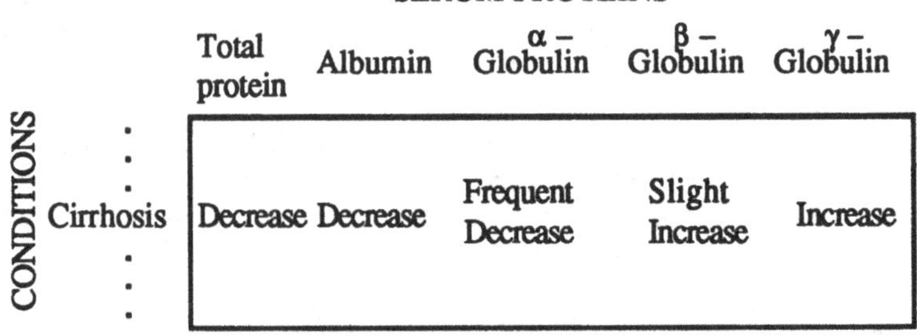

SERUM PROTEINS
Total protein
Albumin
α – Globulin
β – Globulin
γ – Globulin
CONDITIONS
Cirrhosis
Decrease
Decrease
Frequent Decrease
Slight Increase
Increase

Fig. 2 - Part of a serum protein pattern.

The *Medical Knowledge* represented by a generic Diagnosis Δ in the tableau of fig. 1 is interpreted as conjunctions (ANDs) of elementary propositions :

$$\Delta \; \underline{IF} \; P_1 \; \underline{AND} \; ... \; \underline{AND} \; P_i \; ... \; \underline{AND} \; P_n,$$

where for i=1,n, P_i takes the form "S_i is F_i." For example (see fig. 2) :

Cirrhosis IF Total proteins (S_1) are Decreased (F_1)
AND ... AND α-Globulins (S_i) are Frequently Decreased (F_i)
AND ... AND γ-Globulins (S_n) are Increased (F_n).

Here is another example [4], in the framework of inflammatory protein variations

Vasculitis IF C3-Complement Fraction is Decreased or Normal
AND Alpha-1-Antitrypsine is Decreased or Normal
AND Orosomucoid is Increased
AND Haptoglobin is Very Increased
AND C-Reactive Protein is Very Increased.

In the characterisation of Vasculitis, one has for example "Haptoglobin is Very Increased," where "Very Increased" is the label of a fuzzy set "VERY INCREASED", depicted in fig.3.

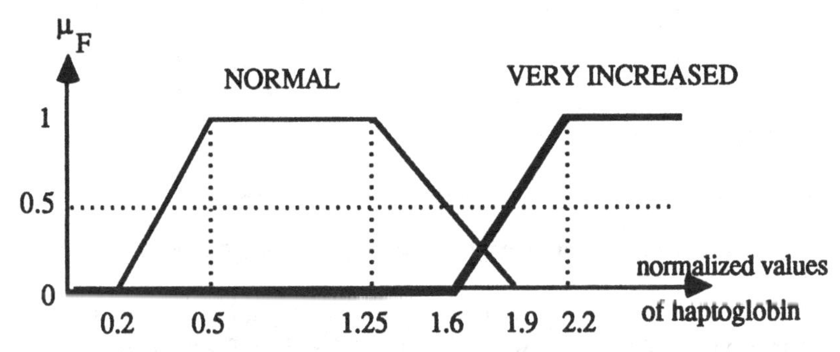

Fig. 3 - Illustration of "Haptoglobin is Very Increased."

The information contained in "Haptoglobin is Very Increased" does not provide a precise characterisation of the numerical values to be assigned to a variable named "Haptoglobin," but it indicates a soft constraint on its possible values. In the pattern of *Medical Knowledge* , the fuzzy sets are fuzzy intervals that extend the definition of usual (crisp) intervals. Fuzzy intervals are here of three types, they fuzzify crisp intervals and they mean "fuzzily greater (or smaller)" than a given value a, or "fuzzily between" two values b and c (for example fuzzy intervals representing NORMAL ranges are usually of this last type). Values like a, b, c, have a grade of membership equal to 0.5. In particular, a fuzzy [b,c]-type interval can reduce to a fuzzy number D, meaning "around a value d" (see fig.4). In this case, the ***bandwidth*** is the separation between the two values having a 0.5 grade of membership, it is a convenient fuzziness indicator of the fuzzy number D.

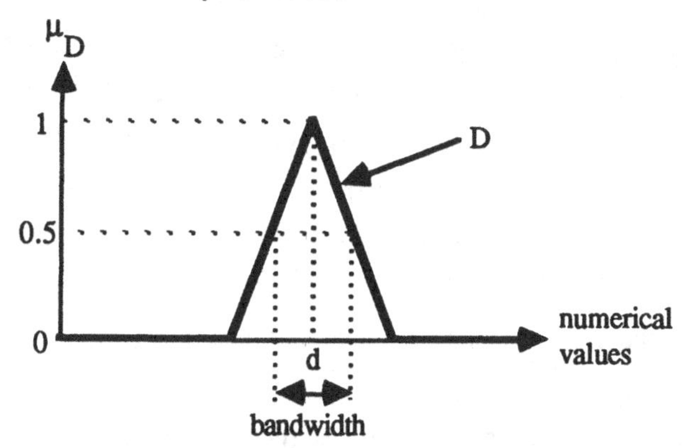

Fig. 4 - Fuzzy number D, meaning "around d."

It is very important for these membership functions to be easily modifiable during the training phase, for their evaluation. For simplicity, their shapes have been chosen here, as trapezoidal or triangular ones. Practically, it is not very important to set, for example, 0.7 or 0.75 as grades of membership when the curves are empirically designed. What mostly matters, is the monotonicity of the function and the position of strategic values, i.e. values with grades of membership equal to 0, 0.5 or to 1. Usually, for each type of laboratory analysis, the biologist determines for a specific purpose or, more generally refers to a variation range in which should fall the *normal* quantitative measurements. He/she has a rough idea of the limits for abnormalities, having in mind more or less well-defined intervals. To determine a patient's condition, it is then sufficient to check in which interval the measured value falls. If we consider a non fuzzy proposition of the form : "S_i is a number in the interval [2,5]," we mean that any number in the interval [2,5] is a possible value to be assigned to the variable S_i and it is not possible for a number outside this interval to be assigned to S_i. In other words, for u_i in the universe of discourse U_i :

$$\begin{aligned} \text{Possibility}\{S_i = u_i\} &= 1 \quad \text{for } 2 \leq u_i \leq 5 \\ &= 0 \quad \text{for } u_i < 2 \text{ or } u_i > 5. \end{aligned}$$

Returning now to the fuzzy case, the proposition "Haptoglobin is Very Increased" (i.e. of the form "S_i is F_i") means that :

$$\text{Possibility}\{\text{Haptoglobin} = u_i\} = \mu_{\text{VERY\_INCREASED}}(u_i)$$

or $$\text{Possibility}\{S_i = u_i\} = \mu_{F_i}(u_i).$$

ASSIGNMENT OF DIAGNOSES TO PATIENTS

A given patient will be assigned each diagnosis, a grade between 0 and 1. In typical cases, one diagnosis will have a grade equal (or close) to 0 and all the other diagnoses will have a grade equal (or close) to 1. The interesting cases will be the intrinsically fuzzy ones, i.e. several diagnoses assigned to a patient, with grades between 0 and 1.

Let us consider a diagnosis Δ, characterised by "(S_1 is F_1) AND ... AND (S_i is F_i) AND ... AND (S_n is F_n)". The attributes S_1, ..., S_i, ..., S_n have to be measured on the patient, yielding the values :

S_1(patient) = d_1 in U_1, ..., S_i(patient) = d_i in U_i, ..., S_n(patient) = d_n in U_n.

Then, Possibility{S_1(patient) = d_1, ..., S_n(patient) = d_n, GIVEN "(S_1 is F_1) AND ... AND (S_n is F_n)"} = MIN($\mu_{F_1}(d_1)$, ..., $\mu_{F_n}(d_n)$), where the MIN operator usually translates the conjunction AND. Finally, such minimum of the above numbers provides a grade of compatibility of the patient's condition, for diagnosis Δ. The same operations are performed for all diagnoses, yielding a ranking in diagnoses for the patient.

In fact, the measured data are often fuzzy in at least two aspects :

i) imprecision in measurements,

ii) interpretation of the values,

so that it is natural to transform each measured (numerical) value into a fuzzy number (like in fig. 4), e.g. "S_i(patient) = d_i" is transformed into "S_i(patient) is D_i." The patient's condition is now expressed as a conjunction of fuzzy propositions involving fuzzy numbers, so that now, one has the following.

Possibility{S_1(patient) is D_1 AND ... AND S_n(patient) is D_n, GIVEN "(S_1 is F_1) AND ... AND (S_n is F_n)"} = MIN($\pi(F_1,D_1)$, ..., $\pi(F_n,D_n)$), where for i = 1,n,

$$\pi(F_i,D_i) = \mathrm{SUP}(F_i \cap D_i),$$

$$\text{i.e. } \forall u_i \in U_i,\ \pi(F_i,D_i) = \mathrm{SUP}_{u_i}\ \mathrm{MIN}[\mu_{F_i}(u_i),\mu_{D_i}(u_i)].$$

$\pi(F_i,D_i)$ = Possibility{D_i GIVEN F_i} is called a possibility measure [5]. It is illustrated in fig.5, where its numerical value indicates a weak compatibility of "around d_i" with the fuzzy interval representing "Very Increased."

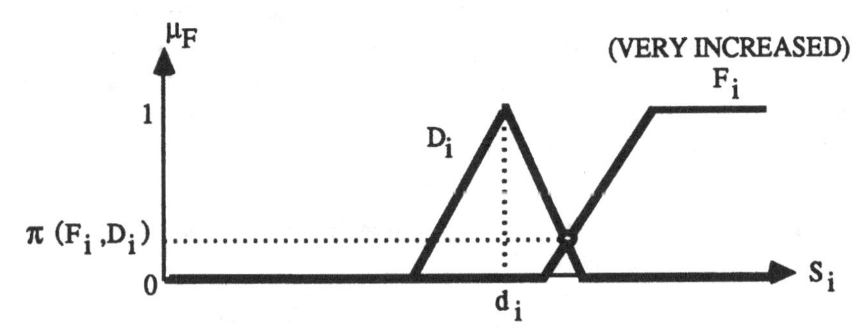

Fig. 5 - Possibility measure of D_i with respect to F_i.

Finally, the patients are assigned a ranking in all diagnostic profiles, by means of grades lying between 0 and 1. For each patient, the set of all diagnoses, associated with their grades of assignment derived from the possibility measures (they are numbers in the interval [0,1]), can be considered as a discrete fuzzy set, $\mathfrak{D}$. For each diagnosis (Δ), one has for example $\mu_{\mathfrak{D}}(\Delta) = MIN[\pi(F_1,D_1), ..., \pi(F_n,D_n)]$. $\mathfrak{D}$ wil be defuzzified, as shown in the sequel.

RELATIVE WEIGHTING

Practically, some attributes might be *less important* than others in the characterization of a diagnosis. For a given diagnosis, relative importance among attributes can be translated by means of weights (α, β, γ ...) ranging in [0,1]. A value "0" weight assigned to an attribute means that this attribute is not important at all in the evaluation of the diagnosis and hence it can be deleted, whereas a value "1" weight does not modify the importance of the protein. Intermediate grades of importance can be tuned by adjusting values of weights within the unit interval.

In the pattern, fuzzy propositions ("S is F," in the generic form) characterizing a given group, appear as conjunctions (ANDs). Assignement of a weight α to take into account the relative importance of protein variations, can assume the following form [6,7], for F fuzzy set in a universe of discourse U :

$$F^{\alpha} = MAX\,(1-\alpha, F),$$
$$\text{i.e. } \forall x \in U, \mu_{F^{\alpha}}(x) = MAX[(1-\alpha), \mu_F(x)].$$

Generally, a t-conorm could replace the MAX operator in the above formula [8]. Limit cases have the following meanings.

$\alpha = 0$: $\forall x \in U, \mu_{F^0}(x) = 1$, i.e. F^0 is neutral for conjunctions and therefore, it can be deleted,

$\alpha = 1$: $\forall x \in U, \mu_{F^1}(x) = \mu_F(x)$, i.e. the weight has no effect.

In the case of Vasculitis, the following weights have been assigned, yielding the modified rule :

Vasculitis IF C3-Complement Fraction is (Decreased or Normal)$^{0.1}$
AND Alpha-1-Antitrypsine is (Decreased or Normal)$^{1.0}$
AND Orosomucoid is (Increased)$^{0.8}$
AND Haptoglobin is (Very Increased)$^{0.3}$
AND C-Reactive Protein is (Very Increased)$^{0.8}$.

Note that C3-Complement Fraction could have been neglected (weight close to 0) and that no weight might be assigned to "Decreased or Normal" in Alpha-1-Antitrypsine (weight equal to 1, i.e. no effect of the weight). For example, the modified fuzzy variations of Haptoglobin (with weight 0.3 to "Very Increased"), and the corresponding modified fuzzy measure are presented in figure 6.

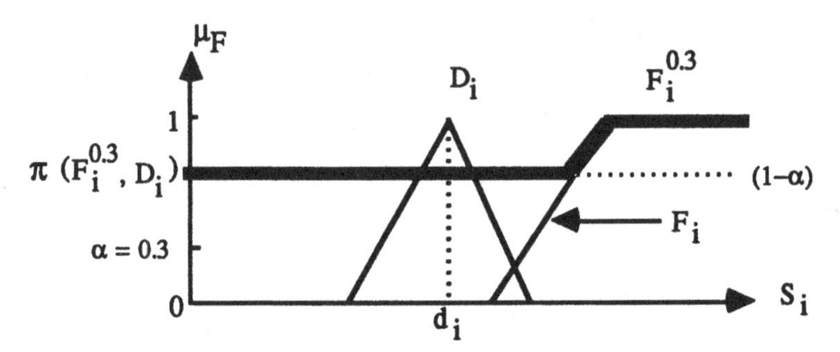

Fig. 6 - Possibility measure of D_i with respect to a weighted F_i.

DEFUZZIFICATION

If the patients are to be assigned non fuzzy diagnoses, we must defuzzify the fuzzy set $\mathfrak{D}$ of diagnoses, that has been evaluated following a MIN aggregation. For this purpose, one may use the concept of separating power [9], which is different from the center of gravity method commonly employed in fuzzy control. The separating power s($\mathfrak{D}$) allows to evaluate to which extent a fuzzy set, like $\mathfrak{D}$, of a universe of discourse U (U is here the set of the given diagnoses under study), *separates* optimally U into a non fuzzy partition (A,A'), where A' is the complement set of A. The set A is defined as follows :

$$s(\mathfrak{D}) = \mathfrak{D} * A = \sup \{\mathfrak{D}*B \text{ such that } U \supseteq B, B \neq \emptyset\}, \text{ in which}$$

$$\mathfrak{D} * B = |\text{card}(\mathfrak{D}_B) / \text{card}(B) - \text{card}(\mathfrak{D}_{B'}) / \text{card}(B')|,$$

where $\mathfrak{D}_B$ denotes the restriction of $\mathfrak{D}$ to B, Card(B) is the cardinality of B, and card($\mathfrak{D}_B$) is the fuzzy cardinality of $\mathfrak{D}_B$; for example, $\text{card}(\mathfrak{D}_B) = \Sigma_{\Delta \in B}\, \mu_{\mathfrak{D}}(\Delta)$.

Applying the separating power to the fuzzy set $\mathfrak{D}$, it is derived the optimal partition ((A,A') above) to $\mathfrak{D}$. A is finally the (non fuzzy) set of diagnoses assigned to patients.

APPLICATION TO INFLAMMATORY PROTEIN VARIATIONS

This application is reported from [4]. The following five proteins, involved in biological inflammatory reactions, have been chosen.

- C3 (C3-Complement Fraction)
- A1AT (Alpha-1-Antitrypsine)
- Om (Orosomucoid)
- Hpt (Haptoglobin)
- CRP (C-Reactive Protein).

The Protein-Biological_Inflammatory_Syndrome (P.B.I.S.) pattern contains eleven groups :

- Normal condition

- Eight Biological Inflammatory Syndromes :
  . Bacterial Infections
  . Viral Infections
  . Vasculitis
  . Nephrotic syndromes
  . Acute Glomerular Nephritis
  . Intravascular Hemolysis with inflammation
  . Collagen Diseases non Lupus and without infection
  . Lupus
- Intravascular Hemolysis without inflammation
- Glomerular Renal Insufficiency without inflammation

The protein variations can be easily interpreted in linguistic terms by physicians, so that the P.B.I.S. pattern is well adapted to a fuzzy sets representation. The fuzzy propositions in this pattern have been interpreted in a linguistic tableau form (one of its rows is reproduced in Table 7).

BIOL. INFL. SYND.	PROTEINS				
	C3	A1AT	Om	Hpt	CRP
. . .	. . .				
Vasculitis	Decreased or Normal	Decreased or Normal	Increased	Very Increased	Very Increased
. . .	. . .				

Table 7 - Linguistic characterisation of Vasculitis in the P.B.I.S. pattern.

The fuzzy sets corresponding to this linguistic pattern have been established for each entry of the tableau (one of its rows is in Table 8).

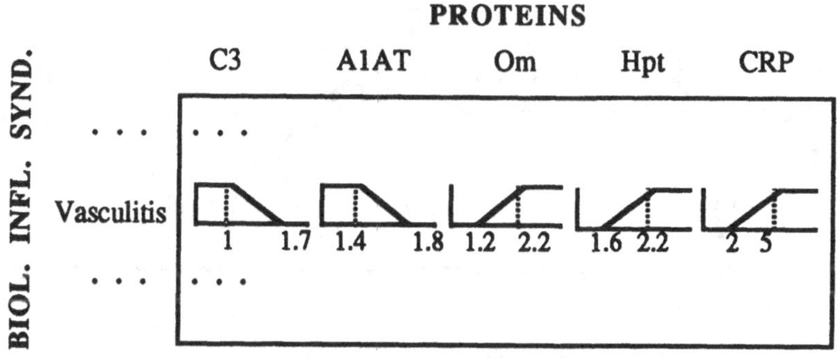

Table 8 - Fuzzy sets characterisation of Vasculitis in the P.B.I.S. pattern.

In this study, fuzzy numbers issued from measurements over patients have been compared with the corresponding fuzzy sets in the P.B.I.S. pattern, by means of

three measures or indexes hereafter defined : possibility measure (π), necessity measure (ν), truth-possibility index (ρ).
For each protein (S), let F be a fuzzy set characterizing S in a diagnostic group, and let D be the fuzzy number issued from the seric level of S, measured over a patient,

i) Possibility measure [5]. By definition, $\pi(F,D) = \mathrm{Sup}\,(F \cap D)$.

ii) Necessity measure [10]. By definition, $\nu(F,D) = 1 - \pi(F',D)$, where F' denotes the fuzzy complement of F, i.e. $F' = 1 - F$. Note that $\nu(F,D) = 1 - \mathrm{Sup}\,(F' \cap D) = \mathrm{Inf}\,(F \cup D')$.

iii) Truth-possibility index [11,12]. By definition, $\rho(F,D) = \pi(\tau_0,\tau_1)$, where τ_0 and τ_1 are related to truth-qualification [5], according to the semantic entailment :

$$[(S \text{ is } F) \text{ is } \tau_1] \rightarrow X \text{ is } D \rightarrow [(S \text{ is } F) \text{ is } \tau_0].$$

With the special case of fuzzy sets in this study, one simply shows that $\rho(F,D) = \mu_D(d)$, where D means "around d". Moreover, one can show that the following ranking holds [13] (see figure 9 for an illustration) :

$$\nu \leq \rho \leq \pi,$$

so that these indexes can be chosen according to optimistic or pessimistic considerations.

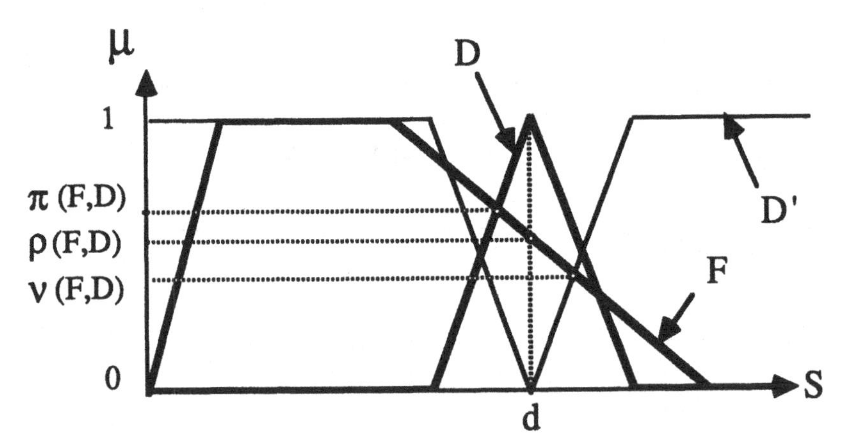

Fig. 9 - Compatibility measures or indexes.

For each of the eleven groups, comparison of a patient's condition with the pattern yields five (one for each protein) triples of numbers (ν_i, ρ_i, π_i), $i = 1,5$, which are aggregated by means of the MIN operator, expressing conjunctions :

$(\nu, \rho, \pi) = (\mathrm{MIN}_i\ \nu_i, \mathrm{MIN}_i\ \rho_i, \mathrm{MIN}_i\ \pi_i)$. Finally for each patient, one has three different rankings of diagnoses derived from ν, ρ, π.

ILLUSTRATIVE EXAMPLE

This patient case we report from [4], has been medically diagnosed as Vasculitis. The protein profile of this patient is given as follows.

	C3	A1AT	Om	Hpt	CRP
raw data (g/l)	1.50	2.26	1.75	10.0	0.060
normalized data	1.85	1.00	1.99	5.59	10.0

For simplicity, we only present the matching results from the possibility measure (the π_i's) and the corresponding crisp partition (A,A') that has been found to be associated with the fuzzy set of diagnostic groups $\mathfrak{D}$.

A = {Collagen Diseases} $(\min_i \pi_i = 0.43)$ $s(\mathfrak{D}) = 0.40$.

A' consists of all the remaining diagnostic groups. Vasculitis does not appear here for one of the possibility measures, $\pi_1 = \pi(\text{Vasculitis,C3})$, is nearly equal to zero. Hence, the MIN operator acting on the π_i's produces a value practically equal to zero, whatever values are computed from the other possibility measures associated with Vasculitis. In fact, for Vasculitis, the possibility measure results are as follows.

	C3	A1AT	Om	Hpt	CRP	$\min_i \pi_i$
π_i's	0.04	1	0.82	1	1	0.04

The four proteins (A1AT, Om, Hpt and CRP) have a high grade of matching and Vasculitis is rejected because of the only mismatch due to C3. But as already pointed out, in the case of Vasculitis, C3 can be nearly neglected (weight equal to 0.1). Hence, in a weighted process, one will derive :

A = {Vasculitis} $(\min_i \pi_i = 0.85)$ $s(\mathfrak{D}) = 0.78$.

The right diagnostic group of Vasculitis appears now, and it is computed with a better separating power (0.78) than in the case of a non-weighted process (0.40).

We recall now the weights of importance associated with the five proteins in the characterisation of Vasculitis. For this patient's case, we also give the matching results in the non-weighted process, followed by the ones of the weighted process, using only the π_i's.

	C3	A1AT	Om	Hpt	CRP	$\min_i \pi_i$
Weights (Vasculitis)	0.1	1.0	0.8	0.3	0.8	
π_i's (non weighting)	0.04	1	0.82	1	1	0.04
π_i's (weighting)	0.9	1	0.85	1	1	0.85

In an automatic classification process, the aggregation we have presented for Vasculitis (Δ), has to be performed for all of the eleven diagnostic groups, yielding for each patient a fuzzy set ($\mathfrak{D}$) of diagnostic groups that can be defuzzified by means of the separating power.

FUZZY LOGIC AND ARTIFICIAL NEURAL NETWORKS

It is now introduced a model of fuzzy connectionist expert system, in which an artificial neural network is designed to build the knowledge base of an expert system, from training examples (this model can also be used for specifications of rules in fuzzy logic control).

Expert systems have shown some weaknesses, for example in the process of eliciting knowledge from experts, in learning capabilities or in producing poor results at the limits of the system's domain of expertise. Neural networks are offering noticeable contributions to expert systems such as : training by example, dynamical adjustment of changes in the environment, ability to generalize, tolerance to noise, graceful degradation at the border of the domain of expertise, ability to discover new relations between variables. Fuzzy logic, supporting interpolative reasoning [14], is playing a key role in human cognitive systems, it lies at the base of pattern classification, qualitative reasoning, analogical reasoning, case-based reasoning, neural modeling, system identification and related fields. The standards of accuracy and precision prevailing in traditional computers are presently questioned or discarded, especially while narrowing the gap between human reasoning and machine reasoning. In the context of approximate reasoning, expert systems and fuzzy logic control on one side, and artificial neural networks, on the other side, share common features and techniques [15]. Connectionist networks (or artificial neural networks) tools are now used in learning control problems like the cart-pole balancing system [16-18]. Combination of fuzzy logic with neural networks theory is enhancing the capability of intelligent systems to learn from experience and adapt to changes in an environment with qualitative, imprecise, uncertain or incomplete information.

FUZZY CONNECTIONIST EXPERT SYSTEMS

Fuzzy logic has been used in conjunction with artificial neural networks in a variety of recent papers [17-31]. In the spirit of S.I. Gallant's model [19] of connectionist expert system (CES), we proposed in [31] an expert classification system in which a connectionist model is used to extract or to tune the knowledge from a training set of examples. An important feature of this model is its fuzzy nature with an intrinsic treatment of fuzziness. Nevertheless, unlike in the CES model, fuzzy sets are not considered from their crisp representations.

Inputs to the neural system are weighted, but we assume that weights are of two types : *primary* weights, in general followed by *secondary* weights. Primary weights express the main information on knowledge. They have a linguistic form and they are interpreted as labels of fuzzy sets, meaning for example : *Increased, Decreased, Very-increased, Normal*, etc., like in the application we just described. Depending on applications, these fuzzy sets are defined over universes of discourses related to the nature of the input cells or, like in fuzzy control, they can be members of a given partition of the interval [-1,+1], with triangular shaped membership functions (fig.10) typically meaning *"Negative Large (NL), Negative Medium (NM), Negative Small (NS), Approximately Zero (ZR), Positive Small (PS), Positive Medium (PM), Positive Large (PL)"*, or more simply having the only three linguistic values *"Decreased, Normal, Increased"*. Secondary weights are numbers in [0,1], they reflect the grade of weakness of the corresponding connection (the weaker the connection, the closer to 1 the weight) and they do not necessarily act on connections but when they do so, they follow a primary weight they are combined with.

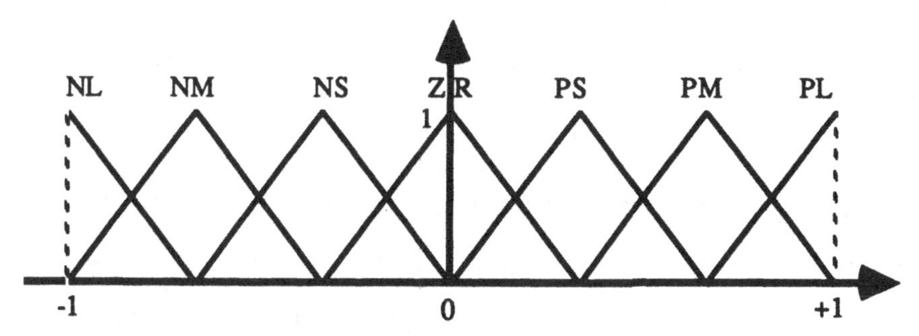

Fig. 10 - A fuzzy partition of [-1,+1]

The neuro fuzzy system is a feedforward network with no thresholds : fuzzy sets avoid the use of thresholds, by considering graded transitions from one state to the other. There are no directed cycles, no feedback, one iteration is sufficient for inferencing. The training phase is not performed from methods involving weighted sums of inputs, but from intrinsically fuzzy equations, using MIN and MAX operators. This phase consists in finding the numerical weights from examples. It is not asked to find the membership functions of the primary weights in the general case, for any universe of discourse. It is assumed that a human expert has a rough idea of the shapes, the task is to tune the curves according to the information provided from input-output examples : this is a general remark to keep in mind when designing models of fuzzy connectionist expert systems (FCES).

Learning now mainly consists in finding the numerical secondary weights and the network topology : numerical weights close to "1" will indicate an absence of the corresponding primary weight, whereas numerical weights close to "0" will not influence at all the corresponding primary weight. Then the primary linguistic weights might be adjusted, when appropriate, by moving the slopes of the curves in the intrinsically fuzzy zone (grades of membership different from 0 and from 1).

The neuro fuzzy network consists of connections between input cells (S_j), output cells (Δ_i), and possible hidden cells (H_{ij}). Primary weights (w_{ij}) are linguistic labels of fuzzy sets, characterizing the variations of the input cells ("S_j is w_{ij}") in relation with the output cells (see fig. 11).

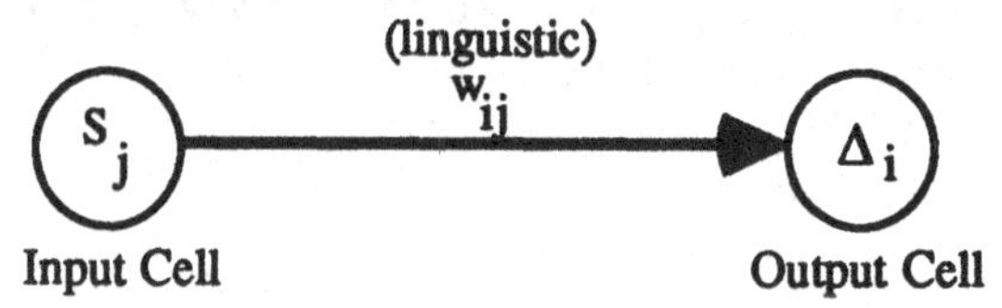

Fig. 11 - Connection with an only primary (linguistic) weight

We assume, depending on the context, that w_{ij} denotes indifferently (as no confusion arises) a linguistic weight or the associated fuzzy set. Secondary weights (b_{ij}) are numbers in the unit interval. In the network, input cells have connections pointing either to hidden cells and followed by connections towards output cells (fig. 12) or, directly, to output cells (this case corresponds to a numerical weight equal to 0), but not necessarily to all output cells (no connection at all corresponds to a numerical weight equal to 1). As soon as a connection is issued from an input cell, a linguistic weight exists, but not always a numerical weight does, in the case of no hidden cell. Hidden cells have only numerical weights associated with connections towards output cells.

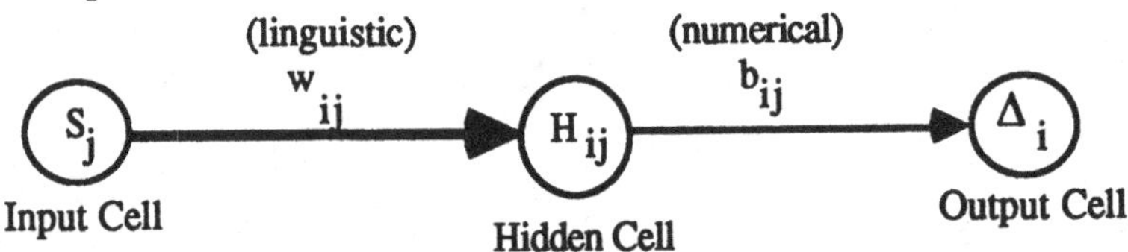

Fig. 12 - General connection with a primary (linguistic) weight and a secondary (numerical) weight

Input cells can take on numerical values or fuzzy numbers, in their underlying universe of discourse. When the input cells S_j's are given, output cells Δ_i's are computed according to the following formula (combination of weights for inferencing) :

$$\Delta_i = \mathrm{MIN}_j \ \mathrm{MAX}\ [b_{ij}, \mu_{w_{ij}}(d_j)] \qquad \text{for numerical } d_j\text{'s}$$

or else, $\Delta_i = \mathrm{MIN}_j \ \mathrm{MAX}\ [b_{ij}, \pi(w_{ij}, D_j)]$ for fuzzy numbers D_j's,

where :
- d_j's are numerical value assigned to S_j's,
- D_j's are fuzzy numbers meaning "around d_j's,"
- $\mu_{w_{ij}}(d_j)$ is the grade of membership of d_j in w_{ij},
- $\pi(w_{ij}, D_j)$ is the possibility measure of D_j GIVEN w_{ij}.

Of course, a mixed formula for a Δ_i can involve both numerical d_j's and fuzzy numbers D_j's, and in the above formula, t-norms and t-conorms could replace MIN and MAX operators, respectively.

Let us consider now training examples, i.e. for a Δ_i, it is given the corresponding S_j's connected to it.

1st case.- The w_{ij}'s are assumed to be known, at least as a rough approximation, so that the unknown are the b_{ij}'s. How to solve this type of equation was early presented in [32] (see also [33] for extensions and more developments) in the general case of complete dually Brouwerian lattices, in which the set of x's such that $\mathrm{MAX}(a,x) \geq b$ contains a least element, denoted $a\varepsilon b$ (note that $a\varepsilon b$ is also defined in [0,1] as being equal to b if $a < b$ and to 0 if $a \geq b$). In case of poor solutions, membership functions of the w_{ij}'s are adjusted by shifting or changing the slopes (tuning).

2nd case.- Neither the w_{ij}'s, nor the b_{ij}'s are known, but the w_{ij}'s are supposed to be members of a known finite fuzzy partition of [-1,+1], like in fuzzy logic control (see fig.10). Again, for each w_{ij} of the fuzzy partition, the above equation has to be solved.

BIOMEDICAL APPLICATION

We now illustrate the fuzzy connectionist network, using the same previous biomedical domain : inflammatory protein variations. We consider the same five proteins : C3-Complement Fraction (C3), Alpha-1-Antitrypsine (A1AT), Orosomucoid (Om), Haptoglobin (Hpt), C-Reactive Protein (CRP) and, for simplicity, only four diagnostic groups composed of the Normal condition and of three biological inflammatory syndromes : Bacterial Infection, Vasculitis, Nephrotic Syndromes.

The P.B.I.S. network we present here, is depicted in fig. 13, in which the five proteins correspond to the input cells S_1,..., S_5 and the four groups to the output cells Δ_1,..., Δ_4. There are seven hidden cells associated with numerical weights. The linguistic weights have the following meaning.

w_{11} : normal,

w_{12} : normal, w_{22}: increased, w_{32} : decreased or normal, w_{42} : decreased or normal,

w_{13} : normal, w_{23} : increased, w_{33} : increased, w_{43} : decreased or normal,

w_{14} : normal, w_{24}: increased, w_{34} : very increased, w_{44} : slightly increased or increased,

w_{15} : normal, w_{25} : very increased, w_{35}: very increased.

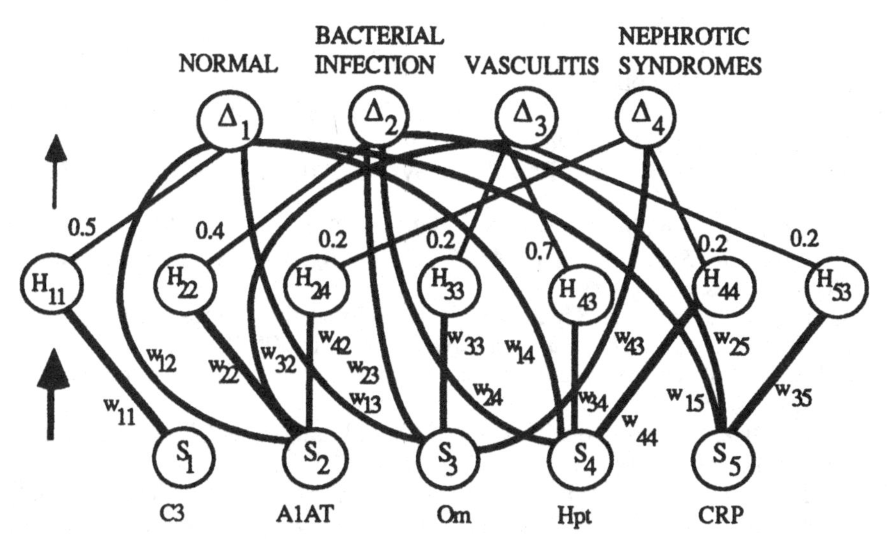

Fig. 13 - P.B.I.S. neuro fuzzy network.

For example, in this network, Vasculitis (Δ_3) is connected with :

- A1AT (S_2) : decreased-or-normal (w_{32}),
- Om (S_3) : increased (w_{33}), with weight 0.2 (b_{33}),
- Hpt (S_4) : very-increased (w_{34}), with weight 0.7 (b_{34}),
- CRP (S_5) : very-increased (w_{35}), with weight 0.2 (b_{35}).

There is no connection with C3 (S_1), corresponding to a numerical weight 1 (b_{31}).

We are now practically exploring this biomedical application (computed results will be presented in a forthcoming extended version of the last section of this paper) and we are studying an application of this method to handwritten character recognition.

ACKNOWLEDGEMENTS. The author wishes to thank Dr. R. Bartolin, for his participation and earlier collaboration on medical aspects of fuzzy sets applications.

REFERENCES

[1] J.L. Beaumont, L.A. Carlson, G.R. Cooper, Z. Fejfar, D.S. Fredrikson and T. Strasser, "Classification of Hyperlipidemias and Hyperlipoproteinemias," *Bull. W.H.O.*, 43 (1970), pp. 891-915.

[2] D.S. Fredrickson, R.I. Levy and R.S. Lee, "Fat transport in lipoproteins," *N. Engl. J. Med.*, 276 (1967), pp. 32-44, 94-103, 148-156, 215-226, 273-281.

[3] R.M. French, "Guide to Diagnostic Procedures," *Mc Graw-Hill*, New-York (1975).

[4] E. Sanchez and R. Bartolin, "Fuzzy Inference and Medical Diagnosis, a Case Study," *Proc. of the First Annual Meeting of the Biomedical Fuzzy Systems Association*, Kurashiki, Japan (1989), in *J. of the Biom. Fuzzy Syst. Ass.*, Vol.1, N.1 (1990), pp. 4-21.

[5] L.A. Zadeh, "Fuzzy Sets as a Basis for a Theory of Possibility," *Fuzzy Sets and Systems*, 1 (1978), pp. 3-28.

[6] E. Sanchez, "Soft Queries in Knowledge Systems," *Proc. of the Second IFSA World Congress*, Tokyo (1987), pp. 597-599.

[7] E. Sanchez, "Importance in Knowledge Systems," *Information Systems*, Vol. 14, N°6 (1989), pp. 455-464.

[8] E. Sanchez, "Handling Requests in Intelligent Retrieval," in *Contributions on Approximate Reasoning and Artificial Intelligence*, M. Delgado and J.L. Verdegay, eds. (to appear).

[9] C. Dujet,"Valuation et Séparation dans les Ensembles Flous," *Structures de l'Information, 18 - Publications du CNRS* (1980), pp. 95-105.

[10] M. Cayrol, H. Farreny and H. Prade, "Fuzzy Pattern Matching," *Kybernetes*, 11 (1982), pp. 103-116.

[11] E. Sanchez, "On Truth-qualification in Natural Languages," *Proc. Int. Conf. on Cybernetics and Society*, Tokyo (1978), pp. 1233-1236.

[12] E. Sanchez, "Mesures de Possibilité, qualifications de Vérité et Classification de Formes Linguistiques en Médecine," in *Actes Table Ronde C.N.R.S.*, Lyon (1980).

[13] R. Bartolin, "Aide au Diagnostic Médical par Mesures de Comparaisons Floues et Pouvoir Séparateur. Approche Linguistique des Profils Protéiques Inflammatoires Biologiques," *Thèse d'Etat en Bilogie Humaine*, Marseille (1987).

[14] L.A. Zadeh , "Interpolative Reasoning Based on Fuzzy Logic and its Application to Control and Systems Analysis," invited lecture, abstract in the *Proc. of the Int. Conf. on Fuzzy Logic & Neural Networks*, Iizuka, Japan (1990).

[15] E. Sanchez, "Connectionism, Artificial Intelligence and Fuzzy Control," invited lecture, abstract in the *Proc. of the Second Annual Meeting of the Biomedical Fuzzy Systems Association*, Kawasaki Medical School, Kurashiki, Japan (1990).

[16] A.G. Barto, R.S. Sutton and C.W. Anderson, "Neuronlike Adaptive Elements that Can Solve Difficult Learning Control Problems," *IEEE Trans. S.M.C.*, vol. 13, N°5 (1983) pp.834-846.

[17] C.C. Lee(1989), "A Self-learning Rule-based Controller Employing Approximate Reasoning and Neural Net Concepts," *Memo U.C. Berkeley*, N°UCB/ERL, M89/84 (1989), to appear in the *Int. J. of Intelligent Systems.*

[18] C.C. Lee, "Intelligent Control Based on Fuzzy Logic and Neural Network Theory," *Proc. of the Int. Conf. on Fuzzy Logic & Neural Networks*, Iizuka, Japan (1990) pp.759-764.

[19] S.I. Gallant S.I. (1988), "Connectionist Expert Systems," *Com. of the ACM*, vol. 31, N°2 (1988) pp.152-169.

[20] M. Frydenberg and S.I. Gallant S.I. (1987), "Fuzziness and Expert System Generation," *Lect. Notes in Computer Science* (B. Bouchon and R.R. Yager, Eds), Springer-Verlag, vol. 286 (1987) pp.137-143.

[21] B. Kosko, "Fuzzy Associative Memories," in *Fuzzy Expert Systems* (A. Kandel, Ed.), Addison-Wesley, Reading, Mass. (1986).

[23] M. Togai, "Fuzzy Neural Net Processor and its Programming Environment," *Preprints of the 1988 first joint technology workshop on neural networks and fuzzy logic*, NASA, Johnson Space Center, Houston, TX (1988).

[24] H. Takagi and I. Hayashi, "Artificial_Neural_Network-Driven Fuzzy Reasoning," *Proc. of the Int. workshop on fuzzy systems applications*, Kyushu Institute of Technology, Iizuka, Japan (1988) pp.217-218.

[25] R.R. Yager, "On the Interface of Fuzzy Sets and Neural Networks," *Proc. of the Int. workshop on fuzzy systems applications*, Kyushu Institute of Technology, Iizuka, Japan (1988) pp. 215-216.

[26] D.L. Hudson, M.E. Cohen and M.F. Anderson, "Determination of Testing Efficacy in Carcinoma of the Lung Using a Neural Network Model," in *Computer applications in medical care* (R.A. Greenes, Ed.), vol. 12 (1988) pp.251-255.

[27] S.S. Chen, "Knowledge Acquisition on Neural Networks," *Lect. Notes in Computer Sciences* (B. Bouchon, L. Saitta and R.R. Yager, Eds.), Springer-Verlag, vol. 313 (1988) pp.281-289.

[28] T. Yamakawa and S. Tomoda, "A Fuzzy Neuron and its Application to Pattern Recognition," *Proc. of the Third IFSA Congress*, Seattle, WA (1989) pp.30-38.

[29] K. Yoshida, Y. Hayashi and A. Imura A. (1989), "A Connectionist Expert System for Diagnosing Hepatobiliary Disorders," *Proc. of MEDINFO 89*, Beijing and Singapore (1989) pp. 116-120.

[30] J. Yen, "Using Fuzzy Logic to Integrate Neural Networks and Knowledge-based Systems," *Proc. of the Neural networks and fuzzy logic workshop* , NASA, Johnson Space Center, Houston, TX (1990).

[31] E. Sanchez, "Fuzzy Connectionist Expert Systems," *Proc. of the Int. Conf. on Fuzzy Logic & Neural Networks*, Iizuka, Japan (1990) pp.31-35.

[32] E. Sanchez, "Resolution of Composite Fuzzy Relation Equations," *Information and Control*, vol. 30, N°1 (1976) pp.38-48.

[33] A. Di Nola, W. Pedrycz, E. Sanchez and S. Sessa, "Fuzzy Relation Equations and their Applications to Knowledge Engineering," *Kluwer Acad. Pub.*, Dordrecht (1989).

12

THE REPRESENTATION AND USE OF UNCERTAINTY AND METAKNOWLEDGE IN MILORD

R. López de Mántaras, C. Sierra, J. Agustí

Centre d'Estudis Avançats de Blanes
CSIC
17300 Blanes, Spain
e-mail: mantaras@ceab.es

INTRODUCTION

One of the most interesting aspects of Expert Systems research is to gain some insights about human problem solving strategies by trying to emulate them in programs. Experts in a domain are better than novices in performing problem solving tasks. This is due to their greater experience in solving problems that provides them with better strategies. Such strategies are knowledge about how to use the knowledge they have in their domain of expertise. This kind of knowledge is called metaknowledge and is represented by means of meta-rules in the MILORD system for diagnostic reasoning. Diagnostic reasoning heavily involves metaknowledge to focuss attention on the most plausible hypotheses or goals in a given situation and to control the inference process. Furthermore, uncertainty also plays an important role at the control level, for example, decisions are taken depending on the uncertainty of the facts supporting them.

On the other hand, psycological experiments (Kuipers et al., 1989) show that human problem solvers do not use numbers to deal with uncertainty but symbolic descriptions expressing categorical and ordinal relations and that in complex situations, the propagation and combination of uncertainty is a local context dependend process. MILORD has a modular structure that allows to represent and manage uncertainty by means of local operators defined over a set of ordered linguistic terms defined by the expert.

In this paper we describe the MILORD system foccusing in the metaknowledge and in role that uncertainty plays in such modular system, that is, its role in the local deductive mechanisms within each module and as a control feature in the task of selecting and combining modules to achieve a solution.

Before describing MILORD, the paper starts by presenting fundamental concepts on control structures for rule-based systems.

INFERENCE CONTROL FOR RULE-BASED SYSTEMS

Inference control in problem solving is the aspect where the use of metaknowledge has more strongly been used (Aiello & Levi, 1988). Problem solving consists in the activation of rules starting from a set of known facts. The application of one rule may cause the activation of another one resulting it what is known as Rule Chaining. This can happen when the conclusions of one rule match the conditions for another. In general there is more than one rule that may be applicable at the same time but only one must be selected. This situation is known as the comflict resolution problem. Most expert systems make arbitrary choices such as the first rule in the list of applicable rules is the one selected, or the one containing more conditions, etc. On the other hand, having control knowledge represented by meta-rules allows to reason about which rule should be applied, that is, the system can dynamically decide which is the best object-level inference to perform.

Another important aspect in problem solving is the control flow, that is, in which order the modules and submodules will be executed. In traditional software the control structure is fixed: one module calls other modules to execute its subtasks and the calling sequence imposes the order of execution of the tasks. Control structures in expert systems can not be as rigid because often the expert has to adapt the order of execution of modules based on opportunities or obstacles that may arise. Such opportunistic problem solving behaviour is driven by what we call strategic knowledge and it is also part of the control meta-level. This control knowledge is extremely important because allows to closely emulate human's problem solving behaviour and therefore increases the credibility of the expert system.

An example of a meta-rule representing strategic problem solving knowledge is (Godo et al., 1989):

> IF pneumonia is suspected and patient has AIDS
> THEN consider first the modules: P-CARINII, TBC,
> CITOMEGALOVIRUS,
> CRIPTOCOCCUS

It is important to make clear that we have two levels of reasoning: object-level and control-level.

The object-level is where the inferences about the problem domain are performed. At this level we have the rules that represent knowledge about the domain as well as descriptions of objects, properties and relations in the domain.

The control-level is concerned with the problem solving strategies, that is, it controls in which order the tasks and subtasks will be executed. More sophisticated expert systems may have several control levels like in MILORD where there is a level whose goal is to combine different sequences of goals resulting from the application of more than one control meta-rule as we will see later.

The overall problem solving control flow jumps back and forth between those levels. Part of the reasoning takes place at the control-level to deduce the next task (module) to be executed. Then, reasoning will proceed at the object-level (inside the module) to deduce new domain facts. As a result of that, new control meta-rules might be applied that could suggest a new sequence of goals to be considered and combined with the previous one. This combined strategy will then be executed and so on. Later in the paper we will describe in more detail this process.

UNCERTAINTY MANAGEMENT

Most AI research on reasoning under uncertainty is concerned with normative methods to propagate and combine certainty values and there is some disagreement between the proponents of the different methods (Bayesians, Dempster-Shaferians, Fuzzy logicians, etc.). Hoewever, these methods do not really claim to closely mimic human problem solving under uncertainty. Although human problem solvers are almost always uncertain about the possible solution in complex domains, they often achieve their goals despite uncertainty by using methods that are particularized to the type of problem solving that they are performing at a given time. In fact, like (Cohen et al., 1987) puts it, managing uncertainty consists in selecting actions that simultaneously achieve solutions and reduce their uncertainty. This view leads to consider uncertainty as playing an important role at the control level because it is useful to constrain the focuss of attention (which part of the problem to work next) and action selection (how to work on it) as will be shown in the framework of MILORD.

Furthermore, we belive that large complex expert systems draw their problem solving capabilities more from the power of the structure and control of their knowledge bases than from the particular uncertainty management formalism they use. On the other hand, the structure in the knowledge bases makes the propagation and combination of uncertainty a local, context dependent process.

MODULARITY AND LOCALITY

A knowledge base (KB) is a large set of knowledge units that covers a domain of expertise and provides solutions to problems in that domain of expertise.

When faced with a particular case, human experts use only a subset of their knowledge for two reasons: adequacy of the general knowledge - the theory - to the particular problem and availability and cost of data. For example, the suspicion of a bacterian disease will rule out all knowledge referring to virical diseases; and also a patient in coma will make useless all the knowledge units that need patient's answers.

The adequation of general knowledge to a particular problem is done at a certain level of granularity, for instance, the expert uses all the knowledge related to the diagnisis of a colon neoplasy or the knowledge related to the radiological analysis of a chest x-ray.

In particular the structuration of KB's is made in MILORD taking into account this granularity in the use of knowledge.

Each structural unit or theory (module from now on) will define an indivisible set of knowledge units (for example rules and predicates). The control will be responsible for the combination of the modules. The combination will represent the particularization of general knoledge to the problem that is being solved. The control will determine which combinations are acceptable.

For example, a module that determines the dosis of penicillin that has to be given to a patient must not be presented in any acceptable combination for a patient allergic to penicillin.

The modularization of KB's leads to the concept of locality in the modules of a KB. It is possible to define the contents of a module independently of the definition of the rest of the modules. This possibility, methodologicaly desirable, allows the use of different local logics and reasoning mechanisms adapted to the subtasks that the system is performing.

MODULARITY OVER MILORD: THE COLAPSES LANGUAGE

The basic units of KB's written in our language, COLAPSES, are the modules. These may be hierarchically organized, and consist of an encapsulated set of import, export, rule, meta-rule and submodule declarations. The declarations of submodules in a module is what structures the hierarchy. The declarations of submodules do not differ from the declaration of modules. We shall briefly outline which is the meaning of the primitive components of a module. A complete definition of the language and its semantics can be found in (Sierra, Agustí, 1990).

Import: determines the non-deducible facts needed in the module to apply the rules. These facts are to be obtained from the user at run time.

Export: defines which facts deduced or imported inside a module are visible from the rest of the modules that include the module as a submodule.

Rule: define the deductive units that relate the import and the export components within a module.

Metarule: defines the meta-logical componentof the module. Thus, the meta-rules of a module will control the execution of the rules in the module and the execution of the submodules in the hierarchy underneath the module.

The syntax of a module definition is as follows:

Module *modid*= modexpr

where *modid* stands for an identifier of the module and *modexpr* for the body of the definition made out of the components specified above. Let us look at an example of module definition.

```
Module gram_esputum=
  begin
        import Class, Morphology
        export morpho, esputum_ok
        deductive knowledge:
                Rules:
                R001 If class>4 then esputum_ok is sure
                ...
        end deductive
end
```

There is also the possibility of defining generic modules that represent functional abstractions of several non genric modules.

LOCAL LOGICS

It is clear that experts use different approaches to the management of uncertainty depending on the task they are performing. Usually expert systems building tools provide a fixed way of dealing with uncertainty proposing a unique and global method for representing and combining evidence. In the COALPSES language it is possible to define different deduction procedures for each one of the modules. If from a methodological point of view a task is associated with a module then, a different logic can be used depending on the task.

The definition of local logics is made by the next primitive in the COLAPSES language:

Inference system:
    Truth values = *list of linguistic terms*
    Renaiming = *morphisms between linguistic terms*
    Connectives:
        Conjunction = *function definition*
        Disjunction = *function definition*
    Inference patterns:
    Modus ponens = *function definition*

This primitive is included as a component of the deductive knowledge of a module.

Next, we shall explain each one of the components of the local logic definition.

Truth values. This component defines the set of linguistic terms that will be used in the logical valuation of facts, rules and meta-rules of the module where this logic is to be used. Different modules can have different sets of linguistic terms.

Renaiming. Modules in a KB define a hierarchy of tasks. Each of the modules can have a different logic, so it is necessary to define a way of interconnecting these different logics. In MILORD this is done in a declarative way. Each module that contains several submodules has a set of morphism definitions that translate the valuations of predicates in the submodules to valuations in the logic of the module.

Module B =
  begin
    Module A =
      begin
        Import C
        Export P
        Deductive knowledge:
        Rules:
            R1 **if** C **then conclude** P **is** possible

```
                Inference system:
                    Truth values= (false, possible, true)
                End deductive
            end
        Import D
        Export Q
        Deductive knowlegdge:
            Rules:
                R1 if A/P and D then conclude Q is
                                quite_possible
            Inference system:
                Truth values= (impossible, moderately_possible,
                                quite_possible, sure)
                Renaming=  A/false = = > impossible
                           A/possible = = > quite_possible
                           A/true = = > sure
        end deductive
    end
```

Notice in the above example that the predicate P exported by the submodule A of B which is used in the rule defined in B will be evaluated with one of the three values: false, possible or true. To use this fact in the module B we need to change that value for a different one which can be used by the logic defined in B. This is done by the via of the renaming definition.

Connectives. This component defines the function that will be used in the deduction process associated with the module. Different multiple-valued functions can be defined or elicited depending on the task defined by the module. Next we explain the connectives elicitation process.

OPERATOR ELICITATION WITH LINGUISTIC TERMS

The elicitation of connective operators has been widely studied when truth values are expressed in the unit interval [0, 1]. On the contrary, little effort has been devoted to study what such operators would be like in the case of a finite number of truth-values. This problem has been encountered in the field of Expert Systems when trying to model expert reasoning by means of linguistically expressed uncertainty about the truth of rules and facts (Godo et al., 1989). Most previous works (López de Mántaras, 1990) in generating operators for lingustic terms used some kind of discretization on the continuous truth-space [0, 1]. In this approach the expert was requiered to give a numerical representation for the linguistic terms (intervals, fuzzy intervals, fuzzy labels). then, a combination function in [0, 1] was selected to model a logical connective. The selection was made according to some properties the function should fulfill. Next, the selected function was applied to the representations of terms, and, whenever the result of a combination lied outside the term set, it was approximated to the "closest" term, in order to keep the term set closed under combinations. This approach has some drawbacks, however;

- often the experts supplying the knowledge are not able to define the meaning of the linguistic values using a numerical scale, although they have no difficulty in ordering them.

- different experts might not agree on the representation of some or all the linguistic values.

- the necessary approximation process does not always ensure that resulting operators satisfy the properties which originally were required to the functions used to generate them.

These disadvantages lead us to propose an alternative approach (López de Mántaras et al., 1990). The central idea consists in treating linguistic terms as mere labels without assuming any underlying numerical repesentation, and then eliciting the connective operators directly on the set of labels. The only *a priori* requirement is that these labels should represent a totally ordered set of linguistic expressions about uncertainty. For each logical connective, a set of desirable properties of the corresponding operator is listed. These properties act as constraints on the set of possible solutions. In this way, all operators fulfilling the set of properties are generated. Afterwards, the domain expert may select the one he thinks fits better his own way of uncertainty management. This approach can be easily implemented by formulating it as a constraint satisfaction problem, and most of the disadvantages of the former approach are avoided.

META-REASONING BY INTROESPECTION USING UNCERTAINTY

Having considered uncertainty as a logical component of the COLAPSES language, i.e. the semantics of formulae, the control of reasoning under the uncertainty must be considered as a component of the meta-logic. Thus the meta-inference over the uncertainty will determine which will the inference control be at the logic level. This meta-inference acts upon the logic component using mechanisms of introspection, that is, the same language represents the uncertainty of the propositions and provides mechanisms both to look at this uncertainty and to determine the control to be followed.

This meta-control is defined as a component of the modules, allowing a local meta-logic definition. This control component acts over the deductive knowledge and over the submodules hierarchy. It determines which rules and submodules are useful for the current case. The mechanism of interaction between both components is a reflexion mechanism: the deductive component reflects on the control component to know which will be the next strategic step, which submodule to execute next, or which rule to use next.

It is not a full reflection mechanism because we allow the meta-logic to see only the valuation of atomic formulae (facts) and the valuation of strategies (sets of modules that combined can lead the system to the solution of the problem), rules and meta-rules can not be consulted by the meta-logic.

This general mechanism is used to guide the inference process in different directions; we are going to discuss some of them.

EVIDENCE INCREASING

The current uncertainty of facts can be used to control the deduction steps in order to increase the evidence of a given hypothesis. So, for example, if we have an alcoholic patient with a cavitation in the chest x-ray and there is low evidence for tuberculosis, then the Ziehl-Nielssen test to determine more clearly whether he has a tuberculosis should not be done. But if he presents a risk factor for AIDS then we shall increase our evidence for tuberculosis and the test will be suggested. This is expressed as follows:

> **If** tuberculosis > moderately_possible
> **then conclude** Test Ziehl-Nielssen
>
> **If** risk_factor_for_AIDS **then conclude** tuberculosis is possible
>
> **If** Alcoholic **and** Cavitation
> **then** tuberculosis **is** almost_impossible

Remark: The first rule is a rule of the meta-logic component of the language whilst the others are rules at the logic level.

STRATEGY FOCUSING

The uncertainty of facts can determine the set of hypothesis to be followed in the sequel.

Example:

> If the pneumonia is bacterial with certainty < quite_possible and
> the pneumonia is atipical with certainty > possible
> **Then consider**
> Mycoplasm, Virus, Clamidia, Tuberculosis, Nocardia,
> Criptococcus, Pneumocistis-Carinii
> **with certainty** quite possible

This example means that the modules to be used in order to find a solution to the current case are those indicated in the conclusion of the meta-rule and should be considered in the order specified there.

Strategies have a certainty degree attached to them. This is useful to differentiate the strategies generated by every especific data, from those generated by general data. As an example consider the case of a patient with AIDS (which is a kind of immunodepression). If we know that the patient suffers from AIDS, a more specific strategy (and also more certain) can be generated. But if just know that the patient has a immunodepression a less certain general strategy would be generated. Since we may have several candidate strategies simultaneously, combining different strategies is a matter of great importance in the control of the system. This is also achived by looking at the uncertainty of the strategies, as the next example shows:

> **If** Strategy (X) **and** Strategy (Y) **and** Certainty (X) > Certainty (Y)
> **and** Goals (X) $\cap$ Goals (Y) $\neq \emptyset$
> **Then** Ockham (X, Y)

where Ockham (X, Y) is a combination of the strategies that favour those moduls found in the intersection of both strategies

KNOWLEDGE ADEQUATION

As indicated at the begining of the paper a KB is a set of knowledge units that have to be adapted to the current case. For example alcoholism is a useful concept when determining a bacterial pneumonia, but it is useless for non-bacterial diseases. Then, for example a possible use of the uncertainty of the fact bacterianicity is to decide about the use of a given concept in the whole KB, i.e. to adequate the general knowledge to the particular problem. Example:

If no bacterian disease
then do not consider alcoholism in the search of the solution

SOLUTION ACCEPTANCE

The degree of uncertainty of a fact can also be used to stop the execution of the system. For example

If Pneumocitis-carinii **and** tuberculosis < possible
and Criptococcus < possible
Then stop

The control tasks we have discussed use uncertainty as a control parameter and are tasks of the meta-logic level. They are represented as a local meta-logic component of each module in what is called the control knowledge component of a module. In the next paragraph we shall describe in some detail this locality.

METACONTROL AND LOCALITY

The structured definition of KB's helps not only in the definition of safe and maintainable KB's but also gives some new features that where impossible to achieve in the previous generation of systems. Among them the most important is the possibility of defining a local meta-logical components for each one of the modules.

The definition of strategies (ordered set of elementary steps to solve a problem) in a previous version of the MILORD system (Godo et al., 1989) was made globally. Only one strategy could be active at any moment. Presently, as many strategies as nodes in the module graph structure can be active. This flexibility is linked with the fact that each module can have a different treatment of uncertainty. So, the uncertainty plays a different role as a control feature depending on the association between module and logic.

Furthermore, given the fact that the system consists of a hierarchy of submodules the meta-logical components act ones upon the others in a pyramidal fashion. This allows us to have as many meta-logic levels as necessary in an application. Further research will be purused along this line. A richer representation of the logic components in the meta-logic will also be investigated and sound semantics from the logic point of view will be defined.

CONCLUSION

One interesting aspect of building expert systems is to learn something about human problem solving strategies by trying to reproduce them in programs. Human problem solver's are uncertain in many situations and do not use a simple normative method to handle uncertainty. Instead they take advantage of a good organization in the problem solving task to obtain good solutions using qualitative approximations. This suggests to consider uncertainty as playing an important role at the control level by guiding the problem solving strategies. In order to illustrate these points, we have described a modular architecture and language that extensively exploits uncertaintyas a control feature and uses local context dependent combination and propagation uncertainty operators.

BIBLIOGRAPHY

1. Agustí J., Sierra C., Sannella D. (1989): "Adding generic modules to flat-rules based languages: a low cost approach", in *Methodologies for Intelligent Systems 4*, (Z. Ras ed), Elsevier Science Pub, pp43-51.
2. Aiello L., Levi G. (1988): "The uses of Metaknowledge in AI Systems", in *Meta-Level Architectures and Reflection* (P. Maes, D. Nardi, ed.), North-Holland, 243-254.
3. Cohen P.R., Day D., De lisio J., Greenberg M., Kjeldsen R., Suthers D., Berman P. (1987): "Management of Uncertainty in Medicine", *International Journal of Approximate Reasoning* 1:103-116.
4. Godo L., López de Mántaras R., Sierra C., Verdaguer A. (1989): "MILORD, the architecture and management of linguistically expressed uncertainty", *Int. Journal of Intelligent Systems*, vol. 4, n.4, pp 471-501.
5. Kuipers B., Moskowitz A. J., Kassirer J. P. (1988): "Critiacl Decisions under Uncertainty: Representation and Structure", *Cognitive Science* 12, 177-210.
6. López de Mántaras R., Godo L., Sangüesa R. (1990): "Connective operators Elicitation for Linguistic Term Sets", *Proc. Intl. Conference on Fuzzy Logic and Neural Networks*, Iizuka, Japan, 729-733.
7. López de Mántaras R. (1990): "Aproximate Reasoning Models", *Ellis Horwood Series in Artificial Intelligence*, London.
8. Sierra C., Agustí J. (1990): "COLAPSES: Syntax and Semantics", CEAB Reserach Report 90/8.

13

FUZZY LOGIC WITH LINGUISTIC QUANTIFIERS IN GROUP DECISION MAKING

Janusz Kacprzyk*, Mario Fedrizzi and Hannu Nurmi*****

*** Systems Research Institute, Polish Academy of Sciences, ul. Newelska 6, 01-447 Warsaw, Poland**
**** Institute of Informatics, University of Trento, Via Rosmini 42, 38100 Trento, Italy**
***** Department of Political Science, University of Turku, SF - 20500 Turku, Finland**

Abstract

We present how fuzzy logic with linguistic quantifiers, mainly its calculi of linguistically quantified propositions, can be used in group decision making. Basically, the fuzzy linguistic quantifiers (exemplified by *most, almost all*, ...) are employed to represent a fuzzy majority which is in many cases closer to a real human perception of the very essence of majority. Fuzzy logic provides here means for a formal handling of such a fuzzy majority which was not possible by using traditional formal apparata. Using a fuzzy majority, and assuming fuzzy individual and social preference relations, we redefine solution concepts in group decision making, and present new «soft» degrees of consensus.

Keywords

Fuzzy logic, linguistic quantifier, fuzzy preference relation, fuzzy majority, group decision making, social choice.

1. INTRODUCTION

Decision making, whose essence is basically to find a best option from among some feasible (relevant, available, ...) ones, is what human beings constantly face in all their activities. In virtually all nontrivial situations decision making does require intelligence.

Due to an increasing complexity of environments in which decisions are to be made today, the human decision maker is often under pressure and stress, and overloaded. Some (computerized) decision support may be therefore of much help. Since, as we mentioned before, intelligence is required, a decision support system should be what might be termed *intelligent*. However, in spite of a considerable progress in broadly perceived artificial intelligence, we are still far from knowing definitely how to devise *intelligent systems*, i.e. in our context how to introduce intelligence into decision support systems.

One of crucial difficulties in this respect is that decision support should rely on some formal decision making models. Unfortunately, though there is an abundance of them, for virtually all imaginable situations, they have been developed within a traditionally perceived mathematical direction where, roughly speaking, «nice» formal properties have had priority over «human consistency». This has led to some crucial problems among which what may be termed an implementation barrier is certainly of primal concern. Basically, its essence is that the human decision makers are often not willing to accept results obtained by formally (mathematically) valid models.

Attempts to incorporate some sort of human consistency (which may be viewed as a first step to the incorporation of intelligence) in decision making models have been undertaken for a long time (see e.g. Braybrook and Lindblom, 1963). For instance, in this perspective we can view various aspiration - level - based approaches in which, say, a strict optimization (which is often contradictory to a real and human perception of the problem's specifics) is replaced by a much milder requirement to attain some levels of satisfaction (see Simon, 1972).

There has also been attempts to attain the above mentioned human consistency by means of fuzzy – logic – based tools. This has mainly involved the use of calculi of linguistically quantified statements. These attempts have concerned multicriteria decision making (cf. Kacprzyk and Yager, 1984a, b, 1990; Yager, 1983a, b, 1984, 1985a, b), multistage decision making (Kacprzyk, 1983; Kacprzyk and Iwanski, 1987), and group decision making and consensus formation which will be discussed in more detail in this paper. For more general papers on issues related to that fuzzy – logic – based perspective on human consistency, see also Kacprzyk (1987b).

In this paper we will consider the problem of how fuzzy logic may be used to attain a higher human consistency of group decision making and consensus models. Such models, adopting our perspective, may help provide a basis of *intelligent* decision support systems for group decision making and consensus formation.

The essence of group decision making may be summarized as follows. There is a set of options and a set of individuals who provide their preferences over the set of options. The problem is basically to find a solution meant to be an option (or a set of options) which is best acceptable by the group of individuals as a whole.

Though the above basic problem formulation seems to be extremely simple, maybe even trivial, it is certainly not. Since its very beginning group decision making has been plagued by negative results exemplified by Arrow's general impossibility theorem, Gibbard's and Satterthwaite's results on the manipulability of social choice functions, McKelvey's and Schofield's findings on the instability of solutions in spatial contexts, etc. (Arrow, 1963; Gibbard, 1973; Safferthwaite, 1975; McKelvey, 1979; Schofield, 1984; see also Nurmi, 1987; Nurmi, Fedrizzi and Kacprzyk, 1990). Basically, all these findings can be summarized as follows: no matter which group choice procedure we will employ, it will satisfy some set of plausible conditions but not another set of equally plausible ones. This general property pertains to all possible choice procedures, so that attempts to develop new, more sophisticated choice procedures do not seem very promising in this respect. Much more promising seems to be to modify some basic assumptions underlying the group decision making process. This line of reasoning is pursued here.

Since the process of decision making, notably of group type, is centered on the human beings, with their inherent subjectivity, imprecision and vagueness in the articulation of opinions, etc., fuzzy sets have been used in this field for a long time. A predominant research direction is here based on the introduction of an *individual* or *social fuzzy preference relation* which is then used to find some choice sets. There is a rich literature on this topic (cf. Tanino, 1984, 1988, or many articles in Kacprzyk and Fedrizzi, 1990), and since this is not explicitly related to the use of fuzzy logic, we will not discuss these issues in more detail here (though we will assume that the preference relations are fuzzy). We will concentrate on other elements of group decision making models where a contribution of fuzzy logic can be explicitly demonstrated.

One of basic elements underlying group decision making is the concept of a *majority* (notice that the *solution* is to be some option(s) best acceptable by the group as a whole, that is by *most* of its members since in no real situation it would be accepted by *all*). Some of the above mentioned problems with group decision making are closely related to a (too) strict perception of *majority* (e.g., at least a half). A natural line of reasoning is to try to somehow make that strict concept of majority closer to its human perception. And here, we find many examples in all kinds of human judgments that what the human beings consider as a required majority to, say, justify the choice of a course of action is often much more vague. A good example in a biological context may be found in Loewer and Laddaga (1985): «... It can correctly be said that there is a *consensus* among biologists that Darwinian natural selection is an important cause of evolution though there is currently *no consensus* concerning Gould's hypothesis of speciation. This means that there is a *widespread agreement* among biologists concerning the first matter but *disagreement* concerning the second...». A rigid majority as, e.g., more than 75% would evidently not reflect the essence of the above statement. It should be noted that there are naturally situations when a strict majority is necessary, for obvious reasons, as in all political elections.

To briefly summarize the above considerations, we can say that a possibility to accommodate a less rigid («soft») majority (as, say, an equivalent of a *widespread agreement* in the above citation) would certainly help make group decision models more human consistent.

It is easy to see that most natural manifestations of such a «soft» majority are the so-called linguistic quantifiers as, e.g., *most, almost all, much more than a half*, etc. One can readily notice that no conventional formal (e.g., logical) apparatus provides means for handling such quantifiers since, e.g., in virtually all conventional logics only two quantifiers, *at least one* and *all* are accounted for.

Fortunately enough, there have been proposed in recent years some fuzzy - logic - based calculi of linguistically quantified propositions (Yager, 1983a, b; Zadeh, 1983) which can make it possible to handle fuzzy linguistic quantifiers. These calculi have been applied by the authors to introduce a fuzzy majority (represented by a fuzzy linguistic quantifier) into group decision making and consensus formation models (Fedrizzi and Kacprzyk, 1988; Kacprzyk, 1984, 1985b, 1986, 1987a; Kacprzyk and Fedrizzi, 1986, 1988, 1989; Kacprzyk, Fedrizzi and Nurmi, 1990; Kacprzyk and Nurmi, 1988; Nurmi and Kacprzyk, 1990; Nurmi, Fedrizzi and Kacprzyk, 1990), and also in an implemented decision support system for consensus reaching (Fedrizzi, Kacprzyk and Zadrozny, 1988; Kacprzyk, Fedrizzi and Zadrozny, 1988).

All that is clearly an example of a contribution fuzzy logic (with linguistic quantifiers) can make to qualitatively improve group decision making models.

We will briefly present below the essence of this approach trying to maintain readability, and referring the reader who might be interested in more detail to a proper literature.

Our notation related to fuzzy sets is standard. A *fuzzy set* A in X, $A \subseteq \mathbf{X}$, is characterized by, and often equated with its *membership function* $\mu_A: \mathbf{X} \to [0, 1]$; $\mu_A(x) \in [0, 1]$ is the *grade of membership* of x in A, from full membership to full nonmembership through all intermediate values. For a finite $\mathbf{X} = \{x_1, ..., x_n\}$ we write $A = \mu_A(x_1)/x_1 + ... + \mu_A(x_n)/x_n$ where '$\mu_A(x_i)/x_i$' is the pair 'grade of membership-element' and «+» is meant in the set – theoretic sense. Moreover, we denote $a \wedge b = \min(a, b)$, $a \vee b = \max(a, b)$, and «$\rightarrow$» stands for an implication operator in multivalued logic. Other, more specific notation will be introduced when needed.

2. FUZZY - LOGIC - BASED CALCULI OF LINGUISTICALLY QUANTIFIED PROPOSITIONS

Linguistically quantified propositions (statements) are commonly used in everyday life and may be exemplified by, say, «*most* experts are convinced» or «*almost all* good cars are expensive».

In general, we can write a linguistically quantified proposition as

$$Qy\text{'s are } F \tag{1}$$

where Q is a *linguistic quantifier* (e.g., most), $\mathbf{Y} = \{y\}$ is a *set of objects* (e.g., experts), and F is a *property* (e.g., convinced).

It is quite natural that we may wish to assign to the particular y's (objects) a different importance (or relevance from the point of view of the fact mentioned in the statement). *Importance*, B, may therefore be added to (1) yielding

$$QBy\text{'s are } F \tag{2}$$

that is, say, «*most* (Q) of the *important* (B) experts (y's) are convinced (F)».

For our purposes, the main problem is now to find the truth of such linguistically quantified statements, i.e. eiTher truth (Qy's are F) or truth (QBy's are F) knowing truth (y_i is F), $\forall\, y_i \in \mathbf{Y}$. This may be done using two basic calculi, one due to Zadeh (1983) and one due to Yager (1983a, b). In the following we will present the essence of Zadeh's calculus since it is simpler and more transparent, hence better suited for the purposes of this volume, though we should bear in mind that in many instances Yager's calculus may be more «adequate» (cf. Kacprzyk, 1986, 1987b; Kacprzyk and Fedrizzi, 1989).

In Zadeh's (1983) method, a *fuzzy linguistic quantifier* Q is assumed to be a fuzzy set defined in [0,1].

For instance, Q = «most» may be given as

$$\mu_{\text{«most»}}(x) = \begin{cases} 1 & \text{for } x \geq 0.8 \\ 2x - 0{,}6 & \text{for } 0.3 < x < 0.8 \\ 0 & \text{for } x \leq 0.3 \end{cases} \tag{3}$$

which may be meant as that if at least 80% of some elements satisfy a property, then most of them certainly (to degree 1) satisfy it, when less than 30% of them satisfy a property, then most of them certainly do not satisfy it (satisfy to degree 0), and between 30% and 80% – the more of them satisfy that property, the higher the degree of satisfaction by most of the elements.

Notice that we will consider here the *proportional quantifiers* exemplified by «most», «almost all», etc. as they are more important for the modelling a fuzzy majority than the *absolute quantifiers* exemplified by «about 5», «much more than 10», etc. The reasoning for the absolute quantifiers is however analogous.

Property F is defined as a fuzzy set in Y. For instance, if Y = {X, Y, Z} is the set of experts and F is a property «convinced», then F may be exemplified by F = «convinced» = 0.1/X + 0.6/Y + 0.8/Z which means that expert X is convinced to degree 0.1, Y to degree 0.6 and Z to degree 0.8. If now $Y = \{y_1, \ldots, y_p\}$, then it is assumed that truth $(y_i \text{ is } F) = \mu_F(y_i)$, i = 1, ..., p.

The value of truth (Qy's are F) is determined in the following two steps (Zadeh, 1983):

$$r = \Sigma\text{Count}(F) / \Sigma\text{Count}(y) = \frac{1}{p}\sum_{i=1}^{p} \mu_F(y_i) \tag{4}$$

$$\text{truth (Qy's are F)} = \mu_Q(r) \tag{5}$$

Basically, (4) determines some mean proportion of elements satisfying the property under consideration, and (5) determines the degree to which this percentage satisfies the meaning of Q.

In the case of *importance* added, B is defined as a fuzzy set in Y, and $\mu_B(y_i) \in [0,1]$ is a degree of importance of y_i: from 1 for definitely important to 0 for definitely unimportant, through all intermediate values. For instance, B = «important» = 0.2/X + 0.5/Y + 0.6/Z means that expert X is important (competent) to degree 0.2, Y to degree 0.5, and Z to degree 0.6.

We rewrite first «QBy's are F» as «Q(B and F)y's are B» which leads to the followings counterparts of (4) and (5) r' = ΣCount(B and F) / ΣCount (B) =

$$= \sum_{i=1}^{p} \mu_B(y_i) \wedge \mu_F(y_i)) / \sum_{i=1}^{p} \mu_B(y_i) \qquad (6)$$

$$\text{truth (QBy's are F)} = \mu_Q (r') \qquad (7)$$

The essence of these two steps is similar as that of (4) and (5).

Example 1. Let Y = «experts" = {X, Y, Z}, F = «convinced» = 0.1/X + 0.6/Y + 0.8/Z, Q = «most» be given by (3), B = «important = 0.2/X + 0.5/Y + 0.6/Z. Then: r = 0.5, r' = 0.8, and truth («most experts are convinced») = 0.4 and truth («most of the important experts are convinced») = 1.

The method presented is simple and efficient and has proven to be useful in a multitude of cases. Sometimes, however, it may lead to somewhat counterintuitive results (cf. Yager, 1983b). An alternate calculus by Yager (1983a, b; 1985a, b) may often be more useful though it is far more complicated, conceptually and numerically, and will not be dealt with here.

3. GROUP DECISION MAKING UNDER FUZZY PREFERENCES WITH A FUZZY MAJORITY REPRESENTED BY A LINGUISTIC QUANTIFIER

The purpose of this section is to redefine some solution concepts of group decision making under fuzzy preference relations by employing Zadeh's fuzzy logic - based - calculus of linguistically quantified propositions to deal with a fuzzy majority.

To set the stage for our next discussion, we will sketch the essence of *group decision making*. We have therefore a set of n options, $S = \{s_1, ..., s_n\}$, and a set of m individuals, $I = \{1, ..., m\}$. Each individual $k \in I$ provides his or her preferences over S. Since these preferences may be not clear - cut, their representation by individual fuzzy preference relations is strongly advocated (see, e.g., the articles in Kacprzyk and Fedrizzi, 1990).

A fuzzy preference relation of individual k, R_k, is given by its membership function $\mu_{R_k} : S \times S \rightarrow [0, 1]$ such that

$$\mu_{R_k}(s_i, s_j) = \begin{cases} 1 & \text{if } s_i \text{ is definitely preferred over } s_j \\ c \in (0.5, 1) & \text{if } s_i \text{ is slightly preferred over } s_j \\ 0.5 & \text{if there is no preference (i.e. indifference)} \\ d \in (0, 0.5) & \text{if } s_j \text{ is slightly preferred over } s_i \\ 0 & \text{if } s_j \text{ is definitely preferred over } s_i \end{cases} \qquad (8)$$

If card S is small enough, as we assume here, R_k,may be represented by a matrix $R = [r^k_{ij}]$, $r^k_{ij} = \mu_{R_k}(s_i, s_j)$; i, j = 1, ... n; k = 1, ... m. R_k is commoly assumed (also here) reciprocal, i.e. $r^k_{ij} + r^k_{ji} = 1$; moreover, $r^k_{ii} = 0$, for all i, j, k.

The fuzzy preference relations, similarly as their nonfuzzy countrparts, are evidently a point of departure for devising a multitude of solution concepts. Basically, two lines of reasoning may be followed here (cf. Kacprzyk, 1986):

- a direct approach
  $\{R_1, ..., R_m\} \rightarrow$ solution
- an undirect approach
  $\{R_1, ..., R_m\} \rightarrow R \rightarrow$ solution

that is, in the first case we determine a solution just on the basis of the individual fuzzy preference relations, and in the second case we form first a social fuzzy preference relation (defined similarly as its individual counterpart but concerning the whole group of individuals) which is then used to find a solution. A solution is here not clearly understood – see, e.g., Nurmi (1983, 1988 a) for diverse solution concepts.

More details related to the use of fuzzy preference relations as a point of departure in group decision making can be found in, e.g., Nurmi (1988) and in other articles in Kacprzyk and Roubens (1988) or Kacprzyk and Fedrizzi (1990).

Here we will show how to redefine some better known solution concepts, the *core* for the direct approach and the *consensus winner* for the indirect approach, using a *fuzzy* majority represented by a linguistic quantifier.

3.1. *Direct derivation of a solution – the core*

Among many solution cncepts proposed in the literature for the direct approach (i.e. for $\{R_1, ..., R_m\} \rightarrow$ solution) the *core* is intuitively appealing and often used. Conventionally, the core is defined as a set of undominated options, i.e. those not defeated in pairwise comparisons by a required majority (strict!) $r \leq m$, i.e.

$$C = \{s_j \in S: \neg \exists\, s_i \in S \text{ such that } r^k_{ij} > 0.5 \text{ for at least r individuals}\} \tag{9}$$

Nurmi (1981) extends the core to the fuzzy α – core defined as

$$C_\alpha = \{s_j \in S: \neg \exists\, s_i \in S \text{ such that } r^k_{ij} > \alpha \geq 0.5 \text{ for at least r individuals}\} \tag{10}$$

i.e. as a set of options not sufficiently (at least to degree α) defeated by the required majority.

Suppose now that the required majority is imprecisely specified as, e.g., given by a fuzzy linguistic quantifier as, say, *most* defined by (3).

While trying to redefine the above concepts of cores under a fuzzy majority, we start by denoting

$$h_{ij}^{k} = 1 \quad \text{if } r_{ij}^{k} < 0.5$$
$$= 0 \quad \text{otherwise} \tag{11}$$

where here and later on in this section, if not otherwise specified, $i, j = 1, \ldots, n$ and $k = 1, \ldots, m$. Thus, h_{ij}^{k} reflects if option s_j defeats s_i or not.

Then

$$h_j^k = \frac{1}{n-1} \sum_{i=1, i \neq j}^{n} h_{ij}^{k} \tag{12}$$

is the extent to which individual k is not against option s_j.

Next

$$h_j = \frac{1}{m} \sum_{k=1}^{m} h_j^k \tag{13}$$

is to what extent all the individuals are not against s_j.

And

$$v_Q^i = \mu_Q(h_j) \tag{14}$$

is to what extent Q (say, *most*) individuals are not against s_j.

The *fuzzy Q - core* is now defined as a fuzzy set

$$C_Q = v_Q^1 / s_1 + \ldots + v_Q^n / s_n \tag{15}$$

i.e. a fuzzy set of options that are not defeated by Q (say, *most*) individuals.

Analogously, by introducing a threshold of the degree of defeat in (11), we can define the fuzzy α/Q - core. First, we denote

$$h_{ij}^{k}(\alpha) = 1 \quad \text{if } h_{ij}^{k} < \alpha \leq 0.5$$
$$= 0 \quad \text{otherwise} \tag{16}$$

and then, following the line of reasoning (12) - (15), and using $h_j^k(\alpha)$, $h_j(\alpha)$ and v_Q^j, respectively, we define the *fuzzy α/Q – core as*

$$C_{\alpha/Q} = v_Q^1(\alpha) / s_1 + \ldots + v_Q^n / s_n. \tag{17}$$

i.e. a fuzzy set of options that are not sufficiently (at least to degree 1 - α) defeated by Q individuals.

We can also explicitly introduce the strength of defeat into (11) and define the fuzzy s/Q – core. Namely, we can introduce a function like

$$\underline{h}_{ij}^{k} = 2\,(0.5 - r_{ij}^{k}) \qquad \text{if } r_{ij}^{k} < 0.5$$
$$= 0 \qquad \text{otherwise} \tag{18}$$

and then, following the line of reasoning (12) - (15), but using $\underline{h}_j^k$, $\underline{h}_j$ and $\underline{v}_Q^j$ instead of h_j^k , h_j and v_Q^j, respectively, we define the *fuzzy s/Q – core*, as

$$C_{s/Q} = \underline{v}_Q^1 \;/ s_1 + \ldots + \underline{v}_Q^n \;/ s_n \tag{19}$$

i.e. as a fuzzy set of options that are not strongly defeated by Q individuals.

Example 2. Suppose that we have four individuals, k = 1, 2, 3, 4, whose fuzzy preference relations are

$$R_1 = i \overset{j}{\begin{bmatrix} 0 & 0.3 & 0.7 & 0.1 \\ 0.7 & 0 & 0.6 & 0.6 \\ 0.3 & 0.4 & 0 & 0.2 \\ 0.9 & 0.4 & 0.8 & 0 \end{bmatrix}} \qquad R_2 = i \overset{j}{\begin{bmatrix} 0 & 0.4 & 0.6 & 0.2 \\ 0.6 & 0 & 0.7 & 0.4 \\ 0.4 & 0.3 & 0 & 0.1 \\ 0.8 & 0.6 & 0.9 & 0 \end{bmatrix}}$$

$$R_3 = i \overset{j}{\begin{bmatrix} 0 & 0.5 & 0.7 & 0 \\ 0.5 & 0 & 0.8 & 0.4 \\ 0.3 & 0.2 & 0 & 0.2 \\ 1 & 0.6 & 0.8 & 0 \end{bmatrix}} \qquad R_4 = i \overset{j}{\begin{bmatrix} 0 & 0.4 & 0.7 & 0.8 \\ 0.6 & 0 & 0.4 & 0.3 \\ 0.3 & 0.6 & 0 & 0.1 \\ 0.7 & 0.7 & 0.9 & 0 \end{bmatrix}}$$

Suppose now that the fuzzy linguistic quantifier is Q = «most» defined by (3). Then, say,

$$C_{\text{«most»}} = \frac{17}{30}\, s_2 + 1 / s_4$$
$$C_{0.3/\text{«most»}} = 0.9/s_4$$
$$C_{s/\text{«most»}} = 0.4/s_4$$

that is, for instance, in case of $C_{\text{«most»}}$ option s_2 belongs to the fuzzy Q – core to the extent 17/30 and option s_4 to the extent 1, and analogously for $C_{0.3/\text{«most»}}$ and $C_{s/\text{«most»}}$. Notice that though the results are different, for obvious reasons, s_4 is clearly the best choice which is evident if we examine the given individual fuzzy preference relations.

3.2. *Indirect derivation of a solution – the consensus winner*

We follow now the scheme $\{R_1, \ldots, R_m\} \to R \to$ solution, i.e. from the individual fuzzy preference relations we determine first a social fuzzy preference relation, which is similar to its individual counterpart but concerns the whole group of individuals, and then find a solution from the social fuzzy preference relation.

We will not deal here with the first step, i.e. $\{R_1, \ldots, R_m\} \to R$, and assume that $R = [r_{ij}]$ is given by

$$r_{ij} = \begin{cases} \frac{1}{m}\sum_{k=1}^{m} a_{ij}^{k} & \text{if } i \neq j \\ 0 & \text{otherwise} \end{cases} \tag{20}$$

where

$$a_{ij}^{k} = \begin{cases} 1 & \text{if } r_{ij}^{k} > 0.5 \\ 0 & \text{otherwise} \end{cases} \tag{21}$$

Notice that R need not be reciprocal (for reciprocal $R_1, \ldots, R_m$). For other approaches to the determination of R, see, e.g., Blin and Whinston (1973).

We will discuss now the second step, i.e. $R \to$ solution, that is how to determine a solution from a social fuzzy preference relation. A solution concept of much intuitive appeal is here the *consensus winner* (Nurmi, 1981) which will be extended here under a fuzzy majority expressed by a fuzzy linguistic quantifier.

We start with

$$g_{ij} = \begin{cases} 1 & \text{if } r_{ij} > 0.5 \\ 0 & \text{otherwise} \end{cases} \tag{22}$$

which expresses whether s_i defeats s_j or not, and then

$$g_j = \frac{1}{n-1} \sum_{j=1, j \neq i}^{n} g_{ij} \tag{23}$$

which is a mean degree to which option s_i is preferred over all the other options options. Next

$$z_Q^i = \mu_Q(g_i) \tag{24}$$

is the extent to which s_i is preferred over Q other options.

Finally, we define the *fuzzy Q – consensus winner* as

$$W_Q = z_Q^1 / s_1 + \ldots + z_Q^n / s_n \tag{25}$$

i.e. as a fuzzy set of options that are preferred over Q other options.

And analogously as in the case of the core, we can introduce a threshold to (22), i.e.

$$g_{ij}(\alpha) \begin{array}{ll} = 1 & \text{if } r_{ij} > \alpha \geq 0.5 \\ = 0 & \text{otherwise} \end{array} \tag{26}$$

and then, following the reasoning (23) and (24), and replacing g_i and z_Q^i by $g_i(\alpha)$, and $z_Q^i(\alpha)$, respectively, we can define the *fuzzy α/Q – consensus winner* as

$$W_{\alpha/Q} = z_Q^1(\alpha) / s_1 + \ldots + z_Q^n(\alpha) / s_n. \tag{27}$$

i.e. as a fuzzy set of options that are preferred over Q (say, *most*) other options.

Furthermore, we can also explicitly introduce the strength of preference into (22) by, e.g., defining

$$g_{ij} \begin{array}{ll} = 2(r_{ij} - 0.5) & \text{if } r_{ij} > 0.5 \\ = 0 & \text{otherwise} \end{array} \tag{28}$$

and then, following the reasoning (23) and (24), and replacing g_i and z_Q^i by $\underline{g}_i$ and $\underline{z}_Q^i$, respectively, we can define the *fuzzy s/Q – consensus winner* as

$$W_{s/Q} = \underline{z}_Q^1 / s_1 + \ldots + \underline{z}_Q^n / s_n \tag{29}$$

i.e. as a fuzzy set of options that are strongly preferred over Q other options.

For more details on the above solution concepts, as well as on some other ones, see, e.g., Kacprzyk (1985b, c; 1986a) and Kacprzyk and Nurmi (1988).

Example 3. For the same individual fuzzy preference relations as in Example 2, and using (20) and (21), we obtain the following social fuzzy preference relation

$$R = i \begin{array}{c} j \\ \begin{bmatrix} 0 & 0 & 1 & 0 \\ 3/4 & 0 & 3/4 & 1/4 \\ 0 & 1/4 & 0 & 0 \\ 1 & 3/4 & 1 & 0 \end{bmatrix} \end{array}$$

If now Q = «most» is given by (3), then we obtain

$$W_{\text{«most»}} = \frac{1}{15} / s_1 + \frac{11}{15} / s_2 + 1 / s_4$$

$$W_{0.8/\text{«most»}} = \frac{1}{15} / s_1 + \frac{11}{15} / s_4$$

$$W_{s/\text{«most»}} = \frac{2}{15} / s_1 + \frac{11}{15} / s_2 + 1 / s_4$$

which is not to be read similarly as for the fuzzy cores in Example 2. Notice that here once again option s_4 is clearly the best choice which is obvious by examining the social fuzzy preference relation.

This concludes our brief exposition of how to employ fuzzy linguistic quantifiers to model the fuzzy majority in group decision making. For readability and simplicity we have only shown the application of Zadeh's calculus of linguistically quantified propositions. The use of Yager's calculus is presented in the source papers by Kacprzyk (1984; 1985b, c; 1986a; 1987a) or in the surveys by Kacprzyk and Nurmi (1989) or Fedrizzi, Kacprzyk and Nurmi (1989). On the other hand, information on some newer solution concepts based on individual and social fuzzy preference relations which are the so-called fuzzy tournaments may be found in Nurmi and Kacprzyk (1990).

4. «SOFT» DEGREES OF CONSENSUS UNDER FUZZY PREFERENCES AND A FUZZY MAJORITY REPRESENTED AS A FUZZY LINGUISTIC QUANTIFIER

In this section we will show how to use fuzzy linguistic quantifiers as representations of a fuzzy majority to define a new «soft» *degree of consensus* as proposed in Kacprzyk (1987), and then advanced in Kacprzyk and Fedrizzi (1986, 1988, 1990), and Fedrizzi and Kacprzyk (1988). This degree is meant to overcome some «rigidness» of conventional degrees of consensus in which full consensus (= 1) occurs only when «*all* the individuals agree as to *all* the issues». This may often be countrinuitive, and not consistent with a real human perception of the very essence of consensus (see, e.g., the citation from a biological context given in the beginning of the paper). Our new degree of consensus can be therefore equal to 1, which stands for full consensus, when, say «*most* of the individuals agree as to *almost all* (of the *relevant*) issues (options)».

Our point of departure is again a set of *individual fuzzy preference relations* which are meant analogously as in Section 3 (see, e.g., (8)).

The degree of consensus is now derived in three steps. First, for each pair of individuals we derive a degree of agreement as to their preferences between all the pair of options, next we pool (aggregate) these degrees to obtain a degree of agreement of each pair of individuals as to their preferences between Q1 (a linguistic quantifier as, e.g., «most», «almost all», much more than 50%», ...) pairs of relevant options, and, finally, we pool these degrees to obtain a degreee of agreement of Q2 (a linguistic quantifier similar to Q1) pairs of important individuals as to their preferences between Q1 pairs of relevant options. This is meant to be the degree of consensus sought. The above derivation process may be formalized by using Zadeh's calculus of linguistically quantified propositions outlined in Section 2.

We start with the degree of *strict agreement* between individuals k1 and k2 as to their preferences between options s_i and s_j

$$v_{ij}(k1, k2) = \begin{cases} 1 & \text{if } r_{ij}^{k1} = r_{ij}^{k2} \\ 0 & \text{otherwise} \end{cases} \tag{30}$$

where here and later on in this section, if not otherwise specified, $k1 = 1, \ldots, m-1$; $k2 = k1 + 1, \ldots, m$; $i = 1, \ldots, n-1$; $j = i + 1, \ldots, n$.

Relevance of options is assumed to be a fuzzy set defined in the set of options such that $\mu_B(s_i) \in [0, 1]$ is a *degree of relevance* of option s_i: from 0 standing for «definitely irrelevant» to 1 for «definitely relevant», through all intermediate values.

Relevance of a pair of options, $(s_i, s_j) \in \mathbf{S} \times \mathbf{S}$, may be defined in various ways among which

$$b_{ij}^{B} = (\mu_B(s_i) + \mu_B(s_j)) / 2 \tag{31}$$

is certainly the most straightforward; obviously, $b_{ij}^{B} = b_{ji}^{B}$, and b_{ii}^{B}'s are irrelevant since they concern the same option.

And analogously for the *importance of individuals*, I, which is defined as a fuzzy set in the set of individuals, with $\mu_I(k) \in [0, 1]$, $k = 1, \ldots, m$, representing the importance of individual k, from definitely important (= 1) to definitely unimportant (= 0) through all intermediate values. Then, the *importance of a pair of individuals*, $b_{k1,k2}^{I} \in [0, 1]$ may also be defined in various ways among which the mean value of type (31) is the most straightforward, and will be used here too.

The degree of agreement between individuals k1 and k2 as to their preferences between all the relevant pairs of options is

$$v_B(k1, k2) = \sum_{i=1}^{n-1} \sum_{j=i+1}^{n} (v_{ij}(k1, k2) \wedge b_{ij}^{B}) / \sum_{i=1}^{n-1} \sum_{j=i+1}^{n} b_{ij}^{B} \tag{32}$$

The degree of agreement between individuals k1 and k2 as to their preferences between Q1 relevant pairs of options is

$$v^{B}_{Q1}(k1, k2) = \mu_{Q1}(v_B(k1, k2)) \tag{33}$$

In turn, the degree of agreement of all the pairs of important individuals as to their preferences between Q1 relevant pairs of options is

$$v^{I,B}_{Q1} = \frac{2}{m(m-1)} \sum_{k1=1}^{m-1} \sum_{k2=k1+1}^{m} (v^{B}_{Q1}(k1, k2) \wedge b^{I}_{k1,k2}) / \sum_{k1=1}^{m-1} \sum_{k2=k1+1}^{m} b^{I}_{k1,k2} \tag{34}$$

and, finally, the degree of agreement of Q2 pairs of important individuals as to their preferences between Q1 relevant pairs of options, called the *degree of Q1/Q2/I/B-consensus*, is

$$\text{con}(Q1, Q2, I, B) = \mu_B(v^{I,B}_{Q1}) \tag{35}$$

Since the strict agreement (30) may be viewed too rigid, we can use the *degree of sufficient agreement* (at least to degree $\alpha \in [0, 1]$) of individuals k1 and k2 as to their preferences between options s_i and s_j, defined by

$$v^{\alpha}_{ij}(k1, k2) = \begin{cases} 1 & \text{if } |r^{k1}_{ij} - r^{k2}_{ij}| \leq 1 - \alpha \leq 1 \\ 0 & \text{otherwise} \end{cases} \tag{36}$$

Then, following the reasoning (31) – (35), we obtain the degree of sufficient agreement (at least to degree α) of Q2 pairs of individuals as to their preferences between Q1 pairs of relevant options (with replacements similar to those in Section 3), called the *degree of α/Q1/Q2/I/B – consensus*, given by

$$\text{con}^{\alpha}(Q1, Q2, I, B) = \mu_{Q2}(v^{\alpha,I,B}_{Q1}) \tag{37}$$

We can also explicitly introduce the strength of agreement into (30), and analogously define the *degree of strong agreement* of individuals k1 and k2 as to their preferences between options s_i and s_j, e.g., as

$$v^{s}_{ij}(k1, k2) = s(|r^{k1}_{ij} - r^{k2}_{ij}|) \tag{38}$$

where $s: [0, 1] \rightarrow [0, 1]$ is some function representing the degree of strong agreements as, e.g.,

$$s(x) = \begin{cases} 1 & \text{for } x \leq 0.05 \\ -10x + 1.5 & \text{for } 0.05 < x < 0.15 \\ 0 & \text{for } x \geq 0.15 \end{cases} \tag{39}$$

such that $x' < x'' \rightarrow s(x') \geq s(x'')$, for all $x', x'' \in [0, 1]$, and $s(x) = 1$ for some $x \in [0, 1]$.

Then, following the reasoning (31) – (35) (with replacements similar to those in Section 3), we obtain the degree of strong agreement of Q2 pairs of important individuals as to their preferences between Q1 pairs of relevant options, called the *degree of s/Q1/Q2/I/B/ – consensus*, as

$$\mathrm{con}_B^s(Q1, Q2) = \mu_{Q2}(v_{Q1,B}^s) \tag{40}$$

Example 4. Suppose that $n = m = 3$, $Q1 = Q2 =$ «most» are given by (3), $\alpha = 0.9$, $s(x)$ is defined by (39), and the individual preference relations are

$$R^1 = [r_{ij}^1] = i \begin{matrix} & j \\ & \begin{matrix} 1 & 2 & 3 \end{matrix} \\ \begin{matrix} 1 \\ 2 \\ 3 \end{matrix} & \begin{bmatrix} 0.0 & 0.1 & 0.6 \\ 0.9 & 0.0 & 0.7 \\ 0.4 & 0.3 & 0.0 \end{bmatrix} \end{matrix}$$

$$R^2 = [r_{ij}^2] = i \begin{matrix} & j \\ & \begin{matrix} 1 & 2 & 3 \end{matrix} \\ \begin{matrix} 1 \\ 2 \\ 3 \end{matrix} & \begin{bmatrix} 0.0 & 0.1 & 0.7 \\ 0.9 & 0.0 & 0.7 \\ 0.3 & 0.3 & 0.0 \end{bmatrix} \end{matrix}$$

$$R^3 = [r_{ij}^3] = i \begin{matrix} & j \\ & \begin{matrix} 1 & 2 & 3 \end{matrix} \\ \begin{matrix} 1 \\ 2 \\ 3 \end{matrix} & \begin{bmatrix} 0.0 & 0.2 & 0.6 \\ 0.8 & 0.0 & 0.7 \\ 0.4 & 0.3 & 0.0 \end{bmatrix} \end{matrix}$$

Now, we assume that $b_i^B = 1/s_1 + 0.6/s_2 + 0.2/s_3$, i.e. $b_{12}^B = 0.8$, $b_{13}^B = 0.6$ and $b_{23}^B = 0.2$, and $b_k^I = 0.8/1 + 1/2 + 0.4/3$, i.e. $b_{12}^I = 0.9$, $b_{13}^I = 0.6$ and $b_{23}^I = 0.7$

Therefore:
con_B(«most», «most», I, B) $\cong 0.35$
$\mathrm{con}^{0.90}$ («most», «most», I, B) $\cong 1.0$
con^s («most», «most», I, B) $\cong 0.75$

For more information on these degrees of consensus, see Fedrizzi and Kacprzyk (1988), Kacprzyk (1987a) and Kacprzyk and Fedrizzi (1986, 1988, 1989). Moreover, the use of Yager's fuzzy – logic – based calculus of linguistically quantified propositions is given in Kacprzyk and Fedrizzi (1989).

5. CONCLUDING REMARKS

In this paper we have tried to show how fuzzy logic with linguistic quantifiers can be used to model a fuzzy majority, and then to define new solution concepts and degrees of consensus based on the fuzzy majority. Fuzzy quantifiers are certainly a natural way of representing a fuzzy majority which cannot practically be adequately represented by conventional formal means. On the other hand, fuzzy logic based calculi of linguistically quantified propositions, in particular the one employed in this paper, offer much simplicity and intuitive appeal, and can help attain more human consistent, hence more adequate and easier implementable group decision making and consensus formation models.

BIBLIOGRAPHY

ARROW K.J. (1963), *Social Choice and Individual Values*, 2nd ed. Yale University Press, New Haven.

BLIN J.M. and A.P. WHINSTON (1973), *Fuzzy sets and social choice*. Journal of Cybernetics, 4, 17 - 22.

BRAYBROOK D. and C. LINDBLOM (1963), *A Strategy of Decision*. Free Press, New York.

CALVERT R. (1986), *Models of Imperfect Information in Politics*. Harwood Academic Publishers, Chur.

FEDRIZZI M. (1986), *Group decisions and consensus: a model using fuzzy sets theory* (in Italian). Rivista per le scienze econ. e soc. A. 9, F. 1, 12 - 20.

FEDRIZZI M. and J. KACPRZYK (1988), *On measuring consensus in the setting of fuzzy preference relations*. In J. Kacprzyk and M. Roubens (Eds.), Non – Conventional Preference Relations in Decision Making. Springer – Verlag, Berlin – New York – Tokyo, 129 – 141.

FEDRIZZI M., J. KACPRZYK and S. ZADROZNY (1988), *An interactive multi – user decision support system for consensus reaching processes using fuzzy logic with linguistic quantifiers*. Decision Support Systems 4, 313 –327.

KACPRZYK J. (1984), *Collective decision making with a fuzzy majority rule*. Proc. WOGSC Congress, AFCET, Paris, 153-159.

KACPRZYK J. (1985a), *Zadeh's commonsense knowledge and its use in multicriteria, multistage and multiperson decision making*. In M.M. Gupta et al. (Eds.), Approximate Reasoning in Expert Systems, North – Holland, Amsterdam, 105-121.

KACPRZYK J. (1985b), *Some «commonsense» solution concepts in group decision making via fuzzy linguistic quantifiers*. In J. Kacprzyk and R.R. Yager (Eds.), Management Decision Support Systems Using Fuzzy Sets and Possibility Theory. Verlag TÜV Rheinland, Cologne, 125-135.

KACPRZYK J. (1985c), *Group decision - making with a fuzzy majority via linguistic quantifiers. Part I: A consensory – like pooling; Part II: A competitive - like pooling.* Cybernetics and Systems: an Int. Journal 16, 119 - 129 (Part I), 131 - 144 (Part II).

KACPRZYK J. (1986 a), *Group decision making with a fuzzy linguistic majority.* Fuzzy Sets and Systems 18, 195 - 118.

KACPRZYK J. (1986b), *Towards an algorithmic/procedural «human consistency» of decision support systems: a fuzzy logic approach.* In W. Karwowski and A. Mital (Eds.), Applications of Fuzzy Sets in Human Factors. Elsevier, Amsterdam, pp. 101 - 116.

KACPRZYK J. (1987a), *On some fuzzy cores and «soft» consensus measures in group decision making.* In J.C. Bezdek (Ed.), The Analysis of Fuzzy Information, Vol. 2. CRC Press, Boca Raton, pp. 119-130.

KACPRZYK J. (1987b), *Towards «human consistent» decision support systems through commonsense – knowledge – based decision making and control models: a fuzzy logic approach.* Computers and Artificial Intelligence 6, 97-122.

KACPRZYK J. and FEDRIZZI M. (1986), *«Soft» consensus measures for monitoring real consensus reaching processes under fuzzy preferences.* Control and Cybernetics 15, 309-323.

KACPRZYK J. and FEDRIZZI M. (1988), *A «soft» measure of consensus in the setting of partial (fuzzy) preferences.* European Journal of Operational Research 34, 315-325.

KACPRZYK J. and FEDRIZZI M. (1989), *A «human–consistent» degree of consensus based on fuzzy logic with linguistic quantifiers.* Mathematical Social Sciences 18, 275-290.

KACPRZYK J. and FEDRIZZI M., Eds. (1990), *Multiperson Decision Making Models Using Fuzzy Sets and Possibility Theory.* Kluwer, Dordrecht – Boston – Lancaster – Tokyo.

KACPRZYK J., FEDRIZZI M. and NURMI H. (1990), *Group decision making with fuzzy majorities represented by linguistic quantifiers.* In J.L. Verdegay and M. Delgado (Eds.): Approximate Reasoning Tools for Artificial Intelligence. Verlag TÜV Rheinland, Cologne, 126-145.

KACPRZYK J. and NURMI H (1989), *Linguistic quantifiers and fuzzy majorities for more realistic and human-consistent group decision making,* in G. Evans, W. Karwowski and M. Wilhelm (Eds.): Fuzzy Methodologies for Industrial and Systems Engineering, Elsevier, Amsterdam, 267-281.

KACPRZYK J. and NURMI H. (1990), *On fuzzy tournaments and their solution concepts in group decision making.* European Journal of Operational Research (forthcoming).

KACPRZYK J. and ROUBENS M., Eds. (1988), *Non – Conventional Preference Relations in Decision Making.* Springer – Verlag, Berlin – New York – Tokyo.

KACPRZYK J. and YAGER R.R. (1984a), *Linguistic quantifiers and belief qualification in fuzzy multicriteria and multistage decision making.* Control and Cybernetics 13, 155-173.

KACPRZYK J. and YAGER R.R. (1984b), *«Softer» optimization and control models via fuzzy linguistic quantifiers.* Information Sciences 34, 157-178.

KACPRZYK J., S. ZADROZNY and M. FEDRIZZI (1988), *An interactive user – friendly decision support system for consensus reaching based on fuzzy logic with linguistic quantifiers*. In M.M. Gupta and T. Yamakawa (Eds.): Fuzzy Computing. Elsevier, Amsterdam, 307-322.

LOEWER B., Guest Ed. (1985), *Special Issue on Consensus*. Synthese 62, No. 1.

LOEWER B. and LADDAGA R. (1985), *Destroying the consensus*. In Loewer (1985), 79-96.

MCKELVEY R.D. (1979), *General Conditions for Global Intransitivities in Formal Voting Models*. Econometrica 47, 1085-1111.

NURMI H. (1981), *Approaches to collective decision making with fuzzy preference relations*. Fuzzy Sets and Systems 6, 249-259.

NURMI H. (1983), *Voting procedures: a summary analysis*. British Journal of Political Science 13, 181-208.

NURMI H. (1987), *Comparing Voting Systems*. Reidel, Dordrecht – Boston – Lancaster – Tokyo.

NURMI H. (1988), *Assumptions on individual preferences in the theory of voting procedures*. In Kacprzyk and Roubens (1988), pp. 142-155.

NURMI H., M. FEDRIZZI and J. KACPRZYK (1990), *Vague notions in the theory of voting*. In J. Kacprzyk and M. Fedrizzi (Eds.): Multiperson Decision Making Models Using Fuzzy Sets and Possibility Theory. Kluwer, Dordrecht – Boston – Lancaster – Tokyo, 43-52.

SCHOFIELD N. (1984), *Existence of Equilibrium on a Manifold*. Mathematics of Operations Research 9, 545-557.

SIMON H.A. (1972), *Theories of Bounded Rationality*. In C.B. McGuire and R. Radner (Eds.): Decision and Organization. North-Holland, Amsterdam.

SAFFERTHWAITE M. (1975), *Strategy-proofness and Arrow's Conditions: Existence and Correspondence Theorem for Voting Procedures and Social Welfare Functions*. Journal of Economic Theory 10, 187-217.

TANINO T. (1988), *Fuzzy preference relations in group decision making*. In Kacprzyk and Roubens (1988), 54-71.

YAGER R.R. (1983a), *Quantifiers in the formulation of multiple objective decision functions*. Information Sciences 31, 107-139.

YAGER R.R. (1983b), *Quantified propositions in a linguistic logic*. International Journal of Man – Machine Studies 19, 195-227.

YAGER R.R. (1984), *General multiple – objective decision functions and linguistically quantified statements*. International Journal of Man – Machine Studies 21, 389 – 400.

YAGER R.R. (1985a), *Reasoning with fuzzy quantified statements: Part I*. Kybernetes 14, 233-240.

YAGER R.R. (1985b), *Aggregating evidence using quantified statements*. Information Sciences 3, 179-206.

YAGER R.R. (1986), *Reasoning with fuzzy quantified statements: Part II*. Kybernetes 15, 111-120.

ZADEH L.A. (1983), *A computational approach to fuzzy quantifiers in natural languages*. Computers and Mathematics with Applications 9, 149-184.

ZADEH L.A. (1985), *Syllogistic reasoning in fuzzy logic and its application to usuality and reasoning with dispositions*. IEEE Transactions on Systems, Man and Cybernetics SMC – 15, 754-763.

14

LEARNING IN UNCERTAIN ENVIRONMENTS

Marco Botta, Attilio Giordana and Lorenza Saitta

Università di Torino
Dipartimento di Informatica
Corso Svizzera 185
10149 TORINO (Italy)
E-mail: saitta@di.unito.it

ABSTRACT

In this paper we briefly survey the problems arising in learning concept descriptions from examples in domains affected by uncertainty and vagueness. A programming environment, called SMART-SHELL, is also presented: it addresses these problems, exploiting fuzzy logic. This is achieved by supplying the learning system with the capability of handling a fuzzy relational database, containing the extensional representation of the acquired logic formulas.

INTRODUCTION

Knowledge acquisition has been recognized as a major problem for the quick and low cost development of expert systems. In fact, knowledge elicitation is a hard and time consuming task, especially in domains where there is a lack or shortage of human experts and/or the knowledge is difficult to be formalized. As a consequence, automated learning methods became appealing and machine learning is now receiving an increasing attention.

Even though the complete automatization of the knowledge acquisition process is beyond the possibilities of the current AI technology, developing tools allowing a substantial part of the necessary knowledge to be first acquired, and, next, maintained and updated, is both a medium-term reachable goal and a very useful one. These tools are likely to become, in the future, a fundamental part of expert systems builders, provided that adequate interfaces towards knowledge engineers and domain expert will be supplied.

Traditionally, machine learning tasks have ranged from acquiring concepts descriptions from examples [1-4] to improving planning heuristics [5-7] and knowledge representation schemes included logical formulas, decision trees (or networks), production rules and semantic networks [8-10,11,31]. Researches on scientific discovery [12-14] and concept formations [7,15-17] received also attention. In all these problems, the notion of learning as a search process, in a space of descriptions or hypotheses, plays a central role [18], especially in inductive approaches.

Recently, new trends emerged, such as the proposal of *chunking* as a general cognitive architecture [19] and the use of deductive methods to performs "justified" learning [5,20-24]. As new, more complex tasks are faced, methods become more refined and integrated models of learning are proposed with the hope of coping with the complexity of real world tasks [25-28].

A great activity is also going on in the field of connectionist models of learning, as it appears, from instance, from [29]. Another interesting approach is also constituted by the genetic algorithm [30], presented as a general-purpose learning method for parallel rule systems.

Unfortunately, many learning systems only work in ideal domains, in which noise in the data and uncertainty in the task are absent. However, the effective use of learning systems in real-world applications substantially depends upon the ability these systems show in handling noise. Some kind of problems arising in real applications are summarized in [31].

Several systems are provided with mechanisms for facing statistical noise, such as the pruning techniques proposed to limit the sizes of decision trees [32-34]. Similar methods have been also proposed for knowledge represented in the form of production rules, as in the AQ15 system, where the initially acquired rules are *truncated* to limit complexity and avoid overfitting [35]. This kind of noise mainly concerns random errors in assigning a value to an attribute or a label to a training event. Several experiments have been performed to investigate the effects of this noise on the effectiveness of the acquired knowledge [36].

However, statistical noise is not the only source of problems; in fact, relevant concepts and relations can be ill-defined and vague. To this purpose, the fuzzy set theory seems the most appropriate tool for handling this type of uncertainty. We have to notice that a continuous-valued semantics, associated to the description language, is a major source of complexity in learning methodologies. Hence, very few systems are able to handle it explicitly and most of these limit themselves to attaching weights to the pieces of acquired knowledge [37,38].

Fuzzy sets occur in symbolic learning methodologies with different roles. In [39], they are used to describe concepts and the varying degrees of typicality of their instances. In [40] the intensional description of a set of classes, to be discriminated from each other, are expressed as fuzzy languages, learned from a set of examples. This approach has been applied to problems in medical diagnosis [41].

Finally, in ML-SMART, a system which learns concept descriptions from examples [42,43] and a domain theory [26,27], the use of fuzzy set theory has proved to be very suited to transform continuous-valued features into a set of categorical attributes and, in general, to define the vague semantics associated with real-world terms and predicates, both in the description of the examples and in the domain theory. ML-SMART is a learning system which uses a full memory approach, supported by the special-purpose shell SMART-SHELL [44,45], especially designed to ease the development of different learning systems. SMART-SHELL mainly consists of a logic programming environment, interfaced toward a relational data-base through a set of operators implementing the basic primitives necessary for a learner. The logic environment has been realized in Common Lisp, whereas the data-base manager has been tailored for the specific class of applications. This data-base differs from the commercial relational data-bases in the sense that many standard features have not been implemented, being not relevant to the particular use it is oriented to, whereas other important aspects have been enhanced: a query language based on full first order logics, including a set of non-standard quantifiers, and the capability of handling continuous-valued semantics.

This paper is organized as follows. Section 2 briefly describes the learning framework used in systems like ML-SMART. Section 3 describes the logic language used for representing both the background knowledge and the acquired knowledge, whereas Section 4 illustrates the behaviour of the basic operators, interfacing the data-base and the learning system. Finally, Section 5 presents some conclusions.

THE LEARNING FRAMEWORK

The learning tasks addressed by the ML-SMART system is that of "learning concept descriptions from examples" [45]. More precisely, the task can be formally defined as follows:

Given: A set H_0 of known concepts (classes),
A set F_0 of classified learning events, represented in an event description language L_E,
A concept description language L.

Find: For every concept $h \in H_0$, a formula $\varphi \in L$ such that $\varphi \rightarrow h$, i.e., φ consists of a set of conditions, sufficient for an example to be an instance of h.

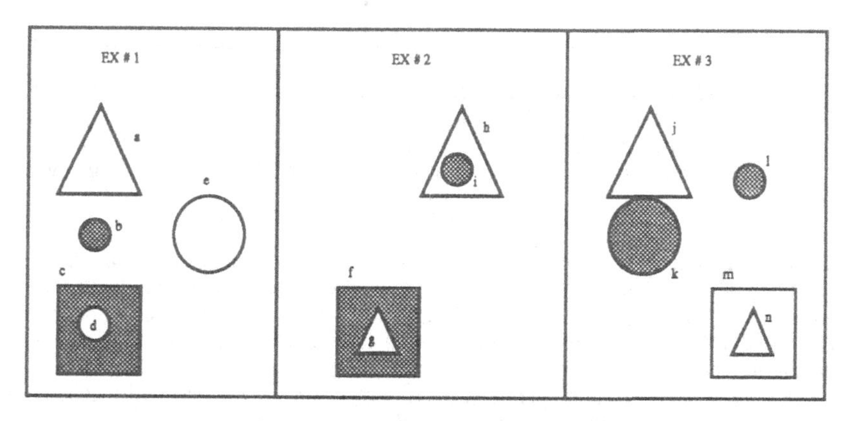

Fig. 1 - Example of instances from a block world domain.

One of the peculiarities distinguishing ML-SMART from other systems, devoted to the same task, consists in the use of a relational data-base as a working memory. In particular, both the learning set F_0 and every inductive hypothesis $\varphi \in \mathcal{L}$, generated during the search, are described extensionally using relations in the data-base. In the following φ^* will denote the extensional representation of the logical formula φ.

For the sake of exemplification, we will use a simple example from a block world domain. Consider the set F_0 of instances reported in Fig. 1; they can be described by means of a set of relations, two of which are reported in Fig. 2(a). All the relations have the homogeneous format <F,H,X1,X2,...,Xn>, where F contains the identifier of the event f, H the correct classification h, and X1,X2,...,Xn are the identifiers of the parts of f (objects) which satisfy the relation; later on, we will slightly extend this basic scheme. By using the standard operators of relational algebra [46], such as natural join, selection and projection, the extension φ^* of a generic formula φ can be computed from the extensions of the predicates occurring in φ; Fig. 2(b) shows the extension of the formula $\varphi = \text{Triangle}(x) \wedge \text{Large}(x)$.

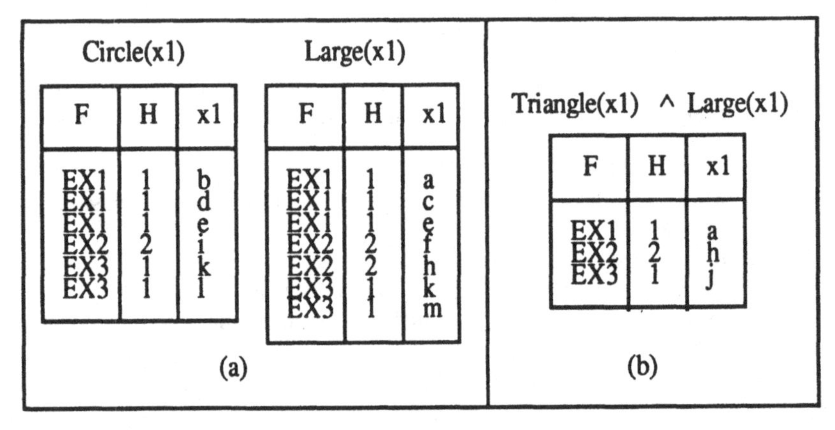

Circle(x1)

F	H	x1
EX1	1	b
EX1	1	d
EX1	1	e
EX2	2	i
EX3	1	k
EX3	1	l

Large(x1)

F	H	x1
EX1	1	a
EX1	1	c
EX1	1	e
EX2	2	f
EX2	2	h
EX3	1	k
EX3	1	m

(a)

Triangle(x1) ∧ Large(x1)

F	H	x1
EX1	1	a
EX2	2	h
EX3	1	j

(b)

Fig. 2 - Example of relations associated to simple formulas, evaluated on the instances of Fig. 1. Relations in (a) are given by the teacher, the one in (b) is computed from the preceding ones. By definition, example 1 and 3 are instances of a concept h_1, whereas example 2 is an instance of another concept h_2.

The learning process can be modcler as a search through the space of formulas which can be generated in this way [18,43]. However, the set of formulas having a non empty extension may be too large and cannot be searched exhaustively. For this reason, ML-SMART uses several strategies for limiting the the number of formulas which are actually created and tested. In particular, it develops a tree of formulas using a set of specialization operators; the root of the tree is the maximally general formula "true", which obviously holds for all the events in F_0, and the leaves are either formulas corresponding to acceptable concept descriptions or formulas which are no more interesting. Three kinds of criteria are used to bias the inductive process in order to limit the size of the tree:

- Simplicity and readability of the formulas.
- Statistical criteria: formulas verified by many examples are preferred.
- If background knowledge is available, formulas which can be deduced from it are preferred. Moreover, formulas contradicting the background knowledge cannot be generated. An extensive description of the methodology can be found in [42,43].

The SMART-SHELL environment provides the basic operators of specialization (and generalization) necessary to implement a problem solver of the type of ML-SMART, as well as a forward/backward inference engine, and the primitives necessary for implementing the search strategies. The relational data-base is a special-purpose one, implemented in such a way to achieve high speed on the most critical operations. On the top of this data-base, a logic environment has been implemented, as well as a user interface, designed to eases the process of supplying the system with the background knowledge and application description.

The tool SMART-SHELL basically consists of three main modules: SMART-CONF, SMART-RUN and SMART-DATABASE which provide the user with a knowledge editor, a set of high level primitives and a data-base manager, respectively. The scheme of the system is reported in Fig. 3.

The module SMART-CONF consists of a user-friendly interface, usable to describe both the background knowledge in input to the learner and other kind of knowledge (control knowledge) which has to be used by the learning strategies. Moreover, it contains a set of compilation procedures which translate this kind of knowledge in a more efficient form, internally used by the other two modules.

SMART-RUN basically implements the logic environment and the learning operators which will be described in the next sections, whereas SMART-DATABASE implements the low level procedures necessary for performing set intersection, natural join, selection and so on. SMART-RUN is embedded in a standard Common Lisp environment, whereas SMART-DATABASE is implemented in C language, for the sake of efficiency.

The learning system is implemented on top of SMART-RUN and basically consists of a set of high level learning strategies which guide the application of the basic deductive and inductive operators. When such strategies are to be very sophisticated, it is a good practice to implement them as knowledge intensive procedures; to this aim the module SMART-CONF turns out to be useful again as a true expert system shell.

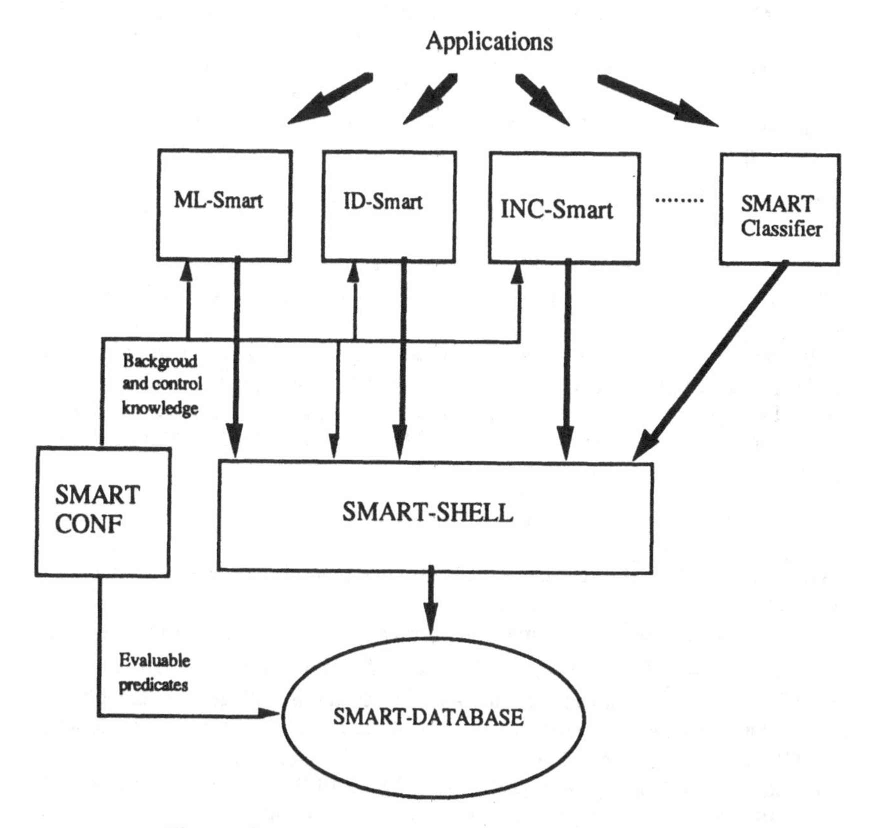

Fig. 3 - Scheme of the SMART-SHELL environment.

KNOWLEDGE REPRESENTATION

In the SMART-SHELL environment, a first order logic language $\mathcal{L}$ is used to describe in a unified form all the knowledge involved in the learning process: the background knowledge (given in input to the learner), and the concept descriptions (the learned knowledge). In addition, in order to facilitate the use of the system by the part of human experts, which may be not familiar with the abstract logic notation, SMART-SHELL offers a frame system, in the style of many standard expert system shells, which allows the knowledge to be also represented in an equivalent object-like paradigm. A compiler automatically performs the conversion from the frame-format representation to the logical one. The frame system can also be used as a tool for implementing the learner itself, as it has been done for ML-SMART [43].

The logic language $\mathcal{L}$ is a Horn clause language, extended with functors, negation and quantifiers. In particular, a well formed formula (wff) of $\mathcal{L}$ takes the form:

$$\varphi(s_1,s_2,....,s_n) \rightarrow p(t_1,t_2,....,t_k) \quad (1)$$

where p is a predicate belonging to a predicate set **P**, the terms $t_1,t_2,....,t_k$ and $s_1,s_2,....,s_n$ can be variables, constants or functions and φ is a logical expression built up using predicates in the set **P**, the connectives $\wedge$ and $\neg$ and the quantifiers ATM, ATL and EX. These quantifiers stand for ATMost, ATLeast and EXactly, respectively, and can be considered as an extension of the standard existential quantifier (similar to the numeric quantifiers used in the system INDUCE [44]). Fuzzy quantifiers are a very important extension to logical languages and have been proposed and deeply analyzed by Zadeh [49]. More precisely, let $\psi(x_1,x_2,...,x_m)$ be a logical expression built up using only the connectives $\wedge$ and $\neg$; then, the expression:

$$\text{ATL } n <y_1,y_2,...,y_k> [\psi(x_1,x_2,...,x_m)] \quad (\{y_1,y_2,...,y_k\} \subseteq \{x_1,x_2,...,x_m\}) \quad (2)$$

is true of a given example f iff there exist **at least** n different bindings, between the variables variables $y_1,y_2,...,y_k$ and the objects occurring in f, satisfying ψ. In an analogous way, ATM n $<y_1,y_2,...,y_k>$ $[\psi(x_1,x_2,...,x_m)]$ and EX n $<y_1,y_2,...,y_k>$ $[\psi(x_1,x_2,...,x_m)]$ require **at most** n and **exactly** n different bindings in order to be satisfied. Notice that, for n = 1, the quantifier ATL corresponds to the existential quantifier $\exists$, whereas, for n = 0, the quantifier ATM n corresponds to $\neg\exists$. Quantifiers can be nested according to the usual rules of the predicate calculus. For instance, the expression :

$$\text{ATL } 1 <x> [\text{EX } 2 <y> [\text{Triangle}(x) \wedge \text{Circle}(y)]] \quad (3)$$

is an example of a wff of the language $\mathcal{L}$ (provided that the predicates Triangle and Circle belong to **P**).

However, some structural restrictions are imposed on the formulas of $\mathcal{L}$. In particular, the set of basic predicates **P** is divided into two disjoint subsets, $\mathbf{P}^{(o)}$ and $\mathbf{P}^{(n)}$. The set $\mathbf{P}^{(o)}$ contains predicates whose extension is evaluable, on the learning set, by means of queries to the data-base manager; examples of predicates belonging to $\mathbf{P}^{(o)}$ are the ones reported in Fig. 2(a). By contrast, predicates in $\mathbf{P}^{(n)}$ are defined by means of implication rules such as (1). We recall that a predicate is evaluable when the data-base manager has a procedure for computing, through a selection operation, its extension from a given relation [47]; examples of standard evaluable predicates are the arithmetic predicates >,< and =. According to the definition given in [26], predicates in $\mathbf{P}^{(o)}$ will be said *operational* and predicates in $\mathbf{P}^{(n)}$ *non-operational*.

Given the set of concepts $\mathbf{H}_0 = \{h_1,h_2,...,h_n\}$, each concept h_i corresponds to a non-operational predicate, which is true of f iff f is an instance of h_i. Concepts descriptions are wffs of the type:

$$\varphi^{(o)} \rightarrow h_i \quad (h_i \in \mathbf{H}_0) \quad (4)$$

where $\varphi^{(o)}$ is a conjunctive wff containing only operational predicates. As, in general, more than one formula (4) is needed to completely define h_i, all these formulas are considered implicitly OR-ed. In the current implementation, the following restrictions on wffs are set:

- Predicates occurring within the scope of a quantifier must belong to the set $P^{(o)}$.
- Negation ($\neg$) can be used only in formulas of the following format:

$$\varphi(x_1,x_2,...,y_1,y_2,...,y_m) \wedge \neg \psi(y_1,y_2,...,y_m) \rightarrow P(...) \qquad (5)$$

Expression (5) states that variables occurring in a negated predicate must also occur in a non-negated one in the same formula. In this way, a simple extension of the SLD-resolution can be used as inference engine.

In order to cope with the vagueness invariably associated with real-world applications, a continuous-valued semantics has been associated to the language $\mathcal{L}$. Each formula $\varphi(x_1,..,x_n) \in \mathcal{L}$ has a corresponding truth degree $\mu \in [0,1]$, computed by combining the truth degrees of the predicates occurring in φ. For this reason, the relation φ^*, associated to the formula φ, has been extended (with respect to the format described in Fig. 2), by adding a new field M, containing the truth value μ of $\varphi(x_1,..,x_n)$ when $x_1,....,x_n$ are bound to the objects specified in the corresponding tuple.

The semantics of an operational predicate can be defined in two ways: extensionally, by giving the corresponding relation on the data-base, or intensionally, by defining a function on attribute values. This two specification forms can both be used in the system. In particular, the implicit form is more compact and efficient but needs an analytic definition, whereas the explicit form can always be given by simply filling up a table when an analytic expression is not available. An example of extensional semantic definition is given in Fig. 4.

F	M	x1	x2
EX1	1.0	d	c
EX2	1.0	g	f
EX2	1.0	i	h
EX3	1.0	n	m

IN (x1, x2)
(Object x1 is inside object x2)

Fig. 4 - Extensional definition of the predicate $IN(x_1,y_2)$.

Furthermore, to ease the writing of semantic functions, the learning events F_0 are usually described by means of a set of numerical and categorical attributes $a_1,...,a_n$; to this aim, a new type of relation has been introduced in SMART-SHELL: the attribute values can be all collected into a unique (n+3)-ary relation, called OBJ. The fields F, H and X contain the identifier f of an event, the classification h of f and the identifier x of a part of f, respectively, whereas the other n columns store the values of the defined attributes for the object x. An an examples, the relation OBJ for the set of instances in Fig. 1 is reported in Fig. 5.

F	H	X	Shape	Area	Texture
EX1	1	a	Triangle	48	Clear
EX1	1	b	Circle	10	Shaded
EX1	1	c	Square	70	Shaded
EX1	1	d	Circle	10	Clear
EX1	1	e	Circle	60	Clear
EX2	2	f	Square	70	Shaded
EX2	2	g	Triangle	15	Clear
EX2	2	h	Triangle	48	Clear
EX2	2	i	Circle	10	Shaded
EX3	1	j	Triangle	48	Clear
EX3	1	k	Circle	60	Shaded
EX3	1	l	Circle	10	Shaded
EX3	1	m	Square	70	Clear
EX3	1	n	Triangle	15	Clear

Fig. 5 - The relation OBJ describing the set of instances of Fig. 1.

The semantic evaluation of a predicate p, depending on the values of the attributes $a_1,a_2,...,a_k$, can be obtained by computing the value of a function:

$$s(a_1,a_2,...,a_k): \quad A_1 \times A_2 \times ... \times A_k \rightarrow [0,1] \tag{6}$$

where A_i is the domain of the attribute a_i. A library of primitive functions has been defined to this aim. For instance, the semantics of a Boolean predicate, such as Triangle(x), can be specified as follows:

if shape(x) = triangle **then** μ=1 **else** μ=0.

Analogously, the continuous-valued semantics of the predicate small(x) can be assigned as a membership function of the object x in the fuzzy set "small". This can be done according to the following syntax:

$$\text{fuzzy}(0,10,30,40,\text{area}(x)) \tag{7}$$

Expression (7) states that the fuzzy set "small" has been defined over the base variable area(x) and has a trapezoidal shape, specified by the four values, 0, 10, 30 and 40 , expressed in same suitable measure units. The corresponding fuzzy set is reported in Fig. 6. What is interesting, in SMART-SHELL, is that the user can give a default semantic for a fuzzy set definition; then the system itself, by analyzing the available examples, can adjust this definition or even learn it from scratch. This facility eases the burden of the domain expert in precisely defining the meaning of the terms he/she uses.

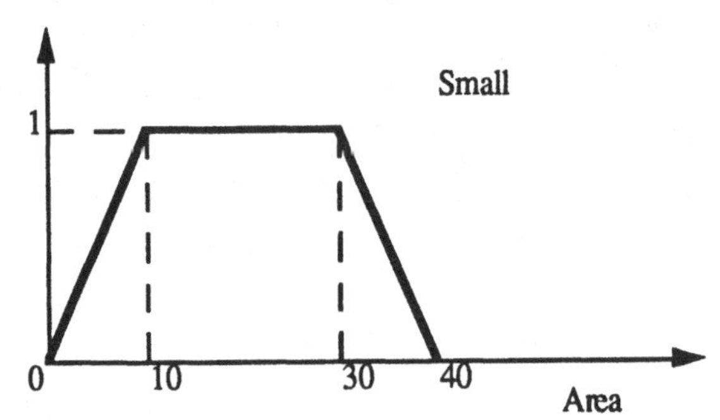

Fig. 6 - Fuzzy set defining the semantics of predicate "small(x)".

Given the truth values u and v of two wffs φ and ψ of $\mathcal{L}$, the semantics of $\varphi \wedge \psi$ and of $\varphi \vee \psi$ is computed according to a pair of corresponding t-norm $\alpha(u,v)$ and t-conorm $\beta(u,v)$; this evaluation reduces to the classical two-valued one in the case of Boolean predicates.

The evaluation of a formula containing a negated predicate (as in formula (5)) is performed by evaluating the function $\alpha(u,1-v)$, where u is the evidence of $\varphi(x_1,x_2,...,y_1,y_2,...,y_m)$ and v the evidence of $\psi(y_1,y_2,...,y_m)$. For what concerns the fuzzy quantifiers, a semantics of the type proposed by Yager in [48] has been adopted. Let us consider the formula :

$$\varphi(\bar{x} - \bar{y}) = \text{ATL } m < \bar{y} > [\psi'(\bar{y}) \wedge \psi''(\bar{x} - \bar{y})] \tag{8}$$

For the sake of simplicity, the vectors $\bar{x}$ and $\bar{y}$ denote, in (8), sets of variables. Given an example f, let $b_1,...,b_r$ be the different bindings between the variables in $\bar{y}$ and the objects in f such that $\psi'(\bar{y})$ is true on f. Let, moreover, $\mu_j(\psi')$ be the evaluation of ψ' for the binding b_j $(1 \leq j \leq r)$. Let us now sort the μ_j's in a non-increasing order :

$$\mu_1 \geq \mu_2 \geq ... \geq \mu_r$$

Then, the evidence of the quantified formula (8) is computed as follows :

$$\mu(\varphi(x)) = \begin{cases} \beta(\alpha(\mu_1,\mu_2,...,\mu_m),\mu_{m+1},...,\mu_r) & \text{if } m \leq r \wedge \bar{y} \equiv \bar{x} \\ \alpha(\beta(\alpha(\mu_1,\mu_2,...,\mu_m),\mu_{m+1},...,\mu_r),\mu(\psi'')) & \text{if } m \leq r \wedge \bar{y} \subset \bar{x} \\ 0 & \text{otherwise} \end{cases} \tag{9}$$

The evaluation of the other two fuzzy quantifiers can be derived from the following relationships :

$$\text{ATM } m < \bar{y} > [\psi(\bar{x})] = \neg \text{ ATL } (m+1) < \bar{y} > [\psi(\bar{x})] \tag{10}$$

$$\text{EX } m < \bar{y} > [\psi(\bar{x})] = \text{ATL } m < \bar{y} > [\varphi(\bar{x})] \wedge \text{ATM } m < \bar{y} > [\psi(\bar{x})] \tag{11}$$

For efficiency reasons, the procedures for evaluating the predicate semantics and for updating the evidence of the formulas are handled by the data-base manager program.The truth evaluation of a non-operational predicate activates a deductive procedure which builds a corresponding operational formula, evaluable as described above.

THE LEARNING OPERATORS

The description of the learning methodology is out of the scope of this paper; we will focus, instead, on the mechanisms used to handle the fuzzy relations associated to the formulas generated during the search. To understand how these mechanisms work, it is sufficient to know what basic specialization and generalization operators can be applied to candidate formulas. The system SMART-SHELL provides the basic primitives necessary to search for concept descriptions in the space of formulas belonging to the language $\mathcal{L}$. In particular it provides inductive operators, namely, specialization and generalization operators, and a deductive operator.

The inductive operators

The basic operations available in the inductive part of the system are *specialization* and *generalization*. Each one of them can be performed by applying different operators, as described in the following.

Specialization operators

Specialization by detailing. Given a formula $\varphi(x_1,x_2,...,y_1,y_2,...,y_n)$, one way of obtaining from it a more specific formula ψ is by adding to it a predicate containing a subsets of the variables occurring in φ:

$$\psi(x_1,x_2,...,y_1,y_2,...,y_n) = \varphi(x_1,x_2,...,y_1,y_2,...,y_n) \wedge p(y_1,y_2,...,y_n)$$

In this way, the original description is enriched with some new details on the same objects considered before. Given the relation φ^*, the extension ψ^* of ψ is built up by selecting from φ^* those tuples satisfying the predicate $p(y_1,...,y_n)$. The relation ψ^* will have the same number of columns as φ^*. An example is given in Fig. 7.

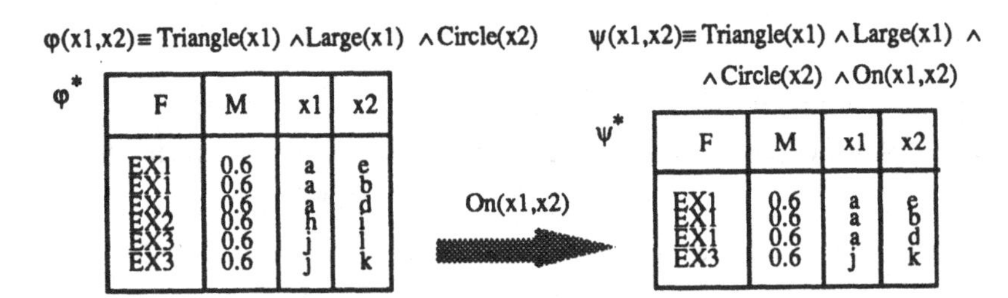

F	M	x1	x2
EX1	0.6	a	e
EX1	0.6	a	b
EX1	0.6	a	d
EX2	0.6	h	i
EX3	0.6	j	l
EX3	0.6	j	k

F	M	x1	x2
EX1	0.6	a	e
EX1	0.6	a	b
EX1	0.6	a	d
EX3	0.6	j	k

Fig. 7 - Example of specialization by detailing.

Specialization by negation. Let $\varphi(x_1,...,x_k,y_1,...,y_n)$ and $\rho(x_1,...,x_k)$ be two formulas. Then, the new formula

$$\psi(x_1,...,x_k,y_1,...,y_n) = \varphi(x_1,...,x_k,y_1,...,y_n) \wedge \neg\rho(x_1,...,x_k)$$

is obtained by negating the assertion ρ for the objects bound to $<x_1,...,x_k>$ in φ. This operator is based on the negation as failure paradigm : given the extensions φ^* and ρ^*, the resulting relation is obtained from φ^* by removing those tuples which do not verify ρ. An example is given in Fig. 8.

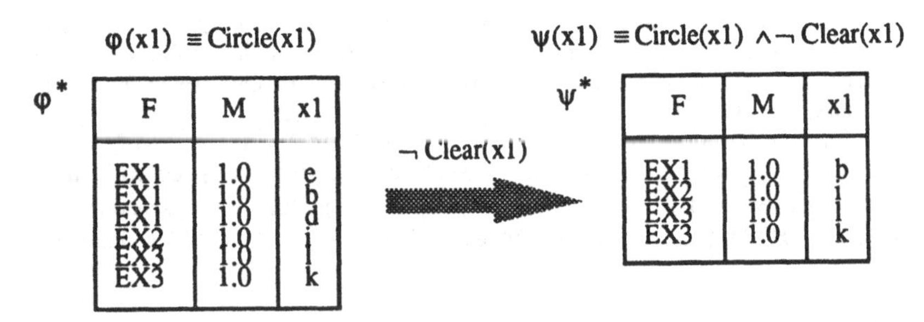

F	M	x1
EX1	1.0	e
EX1	1.0	b
EX1	1.0	d
EX2	1.0	i
EX3	1.0	l
EX3	1.0	k

F	M	x1
EX1	1.0	b
EX2	1.0	i
EX3	1.0	l
EX3	1.0	k

Fig. 8 - Example of specialization by negation.

<u>Specialization by conjunction</u>. Let $\varphi(x_1,...,x_k)$ and $\psi(y_1,...,y_n)$ be two formulas; the formula $\rho(x_1,...,x_k,y_1,...,y_n) = \varphi(x_1,...,x_k) \wedge \psi(y_1,...,y_n)$ is more specific than both φ and ψ. A natural join is performed between the two relations φ^* and ψ^*. The resulting relation, an example of which is reported in Fig. 9, will have k+n+2 columns.

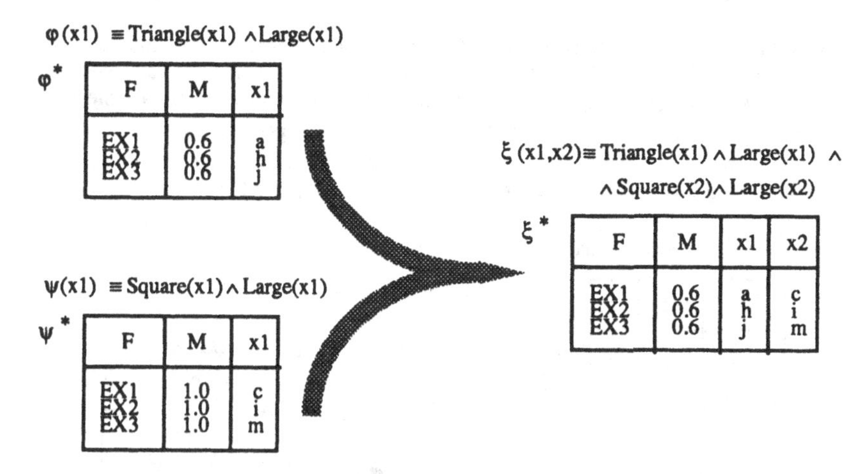

Fig. 9 - Example of specialization by conjunction.

<u>Specialization by quantification</u>. Let $\varphi(x_1,...,x_i,...,x_k)$ be a formula; we can build up the quantified expression $\psi = q\ n <x_1,x_2,..., x_i> [\varphi(x_1,...,x_i,...,x_k)]$, where $q \in \{EX,ATM,ATL\}$. The formula ψ is closed with respect to the variables $<x_1,x_2,..., x_i>$ and, then, the tuples in the relation ψ^* contain only the variables $<x_{i+1},...,x_k>$. While the quantifiers ATL and EX are implemented as data-base operators, ATM is handled by a higher level procedure, that uses both the quantification operators implemented and a set-difference operator described in the following.

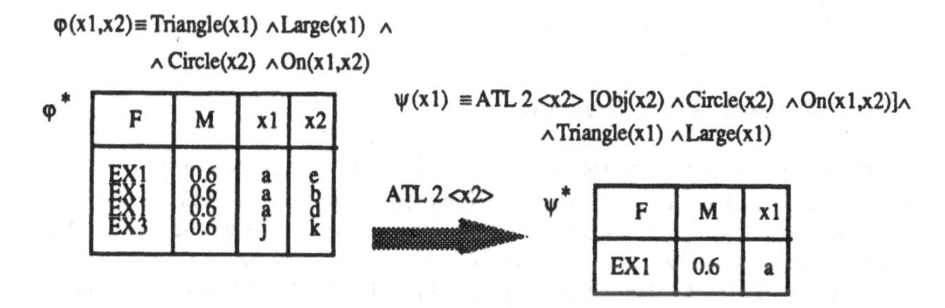

Fig. 10 - Example of specialization by quantifying.

Quantification is similar to a counting operator: for each example in a relation, it counts the number of tuples having different bindings to the variables $<x_1,...,x_i>$ which exist in that relation; then, it projects over $<x_{i+1},...,x_k>$ the relation, discarding those tuples whose count is not q n. An example of this operation is reported in Fig. 10. According to this procedure, the result of the ATM quantifier is neither a more specific nor a more general formula; in fact, the resulting extension is not comparable with the original one (it is not always possible to say which is a subset of the other one).

Generalization operators

Only one basic generalization operator has been considered, i.e., the one that performs a disjunction of two formulas having the same number of variables. Let $\varphi(x_1,...,x_k)$ and $\psi(x_1,...,x_k)$ be two formulas with the same number of variables; the formula $\rho(x_1,...,x_k) = \psi(x_1,...,x_k) \vee \varphi(x_1,...,x_k)$ is a generalization of both. A merging operator, similar to the union operator of relational algebra, is used for implementing this operation. In Fig. 11 an example is reported.

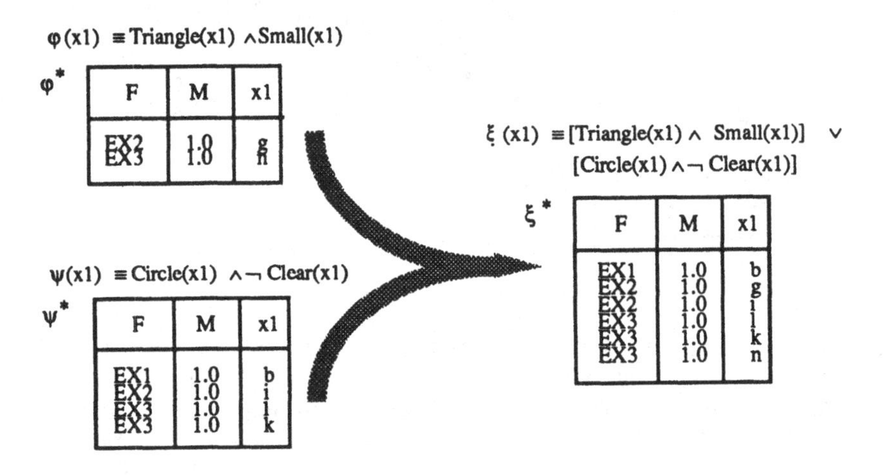

Fig. 11 - Example of generalization by disjunction.

Using the described basic operators, any kind of formula in the relational calculus, extended with the above defined non standard quantifiers, can be built up.

Basic mechanisms for deduction

As mentioned in Section 3, SMART-SHELL allows the background knowledge to be described as a Horn clause theory, in which the non-operational predicates $P^{(n)}$ can be defined by means of implication rules (1). We will now briefly describe how the standard SLD-resolution mechanism can be extended in order to perform deduction using all the learning examples belonging to F_0 at the same time. Let us consider a goal expressed in conjunctive form:

$$g = p_1^{(o)} \wedge p_2^{(o)} \wedge ... \wedge p_1^{(n)} \wedge p_2^{(n)} \wedge ...$$

and the implication rule:

$$p_1^{(n)} \leftarrow q_1^{(o)} \wedge q_2^{(o)} \wedge q_1^{(n)} \qquad (12)$$

Variables are omitted for the sake of simplicity. By applying the basic step of Horn clause resolution, the goal g can be reduced to the goal g':

$$g' = p_1^{(o)} \wedge p_2^{(o)} \wedge ... \wedge q_1^{(o)} \wedge q_2^{(o)} \wedge q_1^{(n)} \wedge p_2^{(n)} \wedge ...$$

where the non-operational predicate $p_1^{(n)}$ has been replaced by the body of the rule (13), after applying the unification with the terms occurring in g. However, suppose we know the extension g* on F_0 of the operational subformula $p_1^{(o)} \wedge p_2^{(o)} \wedge ...$ in the goal g; then, the extension g'* can be easily computed by specializing g* with the formula $q_1^{(o)} \wedge q_2^{(o)}$, i.e., by using the specialization operators defined in the previous section.

This basic deductive step corresponds to the one used in deductive data-bases, which utilize the method of "queries and subqueries" [50]; in particular, SMART-SHELL incorporates a deductive data-base of this form, which has been obtained by extending Robinson's LOGLISP [51].

Inductive specialization and deductive steps can be also easily interleaved, realizing an effective integration of analytical and empirical learning [27]: in this framework, specialization steps allow one to modify the partial operational descriptions obtained from the theory, thus improving their classification performance. On the other hand, the deductive use of background knowledge supplies a skeleton for the inductive process, limits the search space and gives structural meaning to the obtained concept descriptions.

Finally, as the semantics of the predicates can be freely defined by the user, he is also allowed to change it dynamically, in the sense that the shell provides a mechanism to firstly define non-operational predicates through a set of Horn clauses and then move them to an operational state, by deducing their operational form in a context free environment. In this case the predicates' semantics is given extensionally, by means of the relations built up during the former process.

CONCLUSIONS

In this paper we have described the tool SMART-SHELL, designed to ease the development of learning systems oriented to classification and diagnostic expert systems. The learning framework is based on an integrated paradigm allowing empirical learning (i.e. induction) and analytic learning (i.e explanation-based learning) to be the interleaved. This paradigm, that proved very effective in practice, can be easily implemented using a deductive data-base. Then, the environment SMART-SHELL can be considered as a special-purpose deductive data-base, extended in order to support the development of knowledge-intensive learners. An important feature of the system is the capability of handling fuzzy relations.

So far, SMART-SHELL has been used to develop four families of learners, the best known being ML-SMART; they have been applied in several real-world domains, such as pattern recognition [43] and fault diagnosis of electromechanical equipments [45] among others. In these applications SMART-SHELL proved to be reliable enough and usable even by peoples who did not participate to the implementation of the tool itself (one version of ML-SMART has been develop by the SOGESTA s.p.a). The help obtained for speeding up the prototyping time has been evaluated excellent, when used by a trained programmer; several prototypes have been developed in few weeks.

The facility of handling fuzzy logic was also a key for the success, especially in diagnostic problems, where coping with the vagueness of the terms used by a human expert and with the approximation of the measures cues was a must.

Moreover, the possibility of automatically acquiring the required fuzzy set definitions greatly enhance the system's usefulness.

REFERENCES

1. F. Hayes-Roth and J. McDermott, "An Interference Matching Technique for Inducing Abstractions," *Communications of the ACM*, vol. 21, no. 5, pp. 401-410, 1978.
2. R. S. Michalski, "Pattern Recognition as Rule-guided Inductive Inference," *IEEE Transactions on Pattern Analysis and Machine Intelligence*, vol. 2, pp. 349-361, 1980.
3. S. A. Vere, "Induction of Concepts in the Predicate Calculus," in *Proc. of the Fourth IJCAI*, pp. 281-287, Tbilisi, USSR, 1975.
4. P. H. Winston, "Learning Structural Descriptions from Examples," in *The Psychology of Computer Vision*, ed. P.H. Winston, McGraw Hill, New York, 1975.
5. S. Minton, J.G. Carbonell, "Strategies for Learning Search Control Rules: An Explanation-Based Approach," in *Proc IJCAI-87*, pp. 228-235, Milano, Italy, 1987.
6. L. Rendell, "A General Framework for Induction and a Study of Selective Induction," *Machine Learning*, vol. 1, pp. 177-226, 1986.
7. D.B. Lenat, "The Role of Heuristics in Learning by Discovery: Three Case Studies," in *Machine Learning, An Artificial Intelligence Approach*, ed. R. S. Michalski, J. G. Carbonell, T. M. Mitchell, pp. 243-306, Tioga Publishing Company, 1983.
8. R. S. Michalski, J. Carbonell, and T. Mitchell, *Machine Learning, An Artificial Intelligence Approach*, Vol. 1, Tioga Publishing Company, Palo Alto, CA, 1983.
9. R. S. Michalski, J. Carbonell, and T. Mitchell, *Machine Learning, An Artificial Intelligence Approach*, Vol. 2, Morgan Kaufmann, Los Altos, CA, 1985.
10. R. S. Michalski and Y. Kodratoff, eds. : "Machine Learning: An Artificial Intelligence Approach", vol. 3, Morgan Kaufmann, Palo Alto, CA, 1988.
11. **Artificial Intelligence,** Special Issue on Machine Learning, J. Carbonell (Ed.), **40**, N. 1-3 (1989)
12. P. Langley, G.L. Bradshaw, and H.A. Simon, "Rediscovering Chemistry with the Bacon System," in *Machine Learning, An Artificial Intelligence Approach*, ed. R. S. Michalski, J. G. Carbonell, T. M. Mitchell, pp. 307- 330, Tioga Publishing Company, 1983.
13. P. Langley, J.M. Zytkow, H.R. Simon, and G.L. Bradshaw, "The Search of Regularities: Four Aspect of Scientific Discovery," in *Machine Learning, An Artificial Intelligence Approach*, Vol. 2, ed. R. S. Michalski, J. G. Carbonell, T. M. Mitchell, pp. 425-470, Morgan Kaufmann, Los Altos, CA, 1985.
14. B.C. Falkenhainer, R.S. Michalski, "Integrating Quantitative and Qualitative Discovery: The ABACUS System," *Machine Learning*, no. 1-4, pp. 367-402, 1986.
15. R.S. Michalski and R.E. Stepp, "Learning from Observation: Conceptual Clustering," in *Machine Learning, An Artificial Intelligence Approach*, ed. R. S. Michalski, J. G. Carbonell, T. M. Mitchell, pp. 331-364, Tioga Publishing Company, 1983.
16. M. Lebowitz, "Experiments with Incremental Concept Formation: UNIMEM," *Machine Learning*, no. 2-2, pp. 103-138, 1987.
17. D.H. Fisher, "Knowledge Acquisition Via Incremental Conceptual Clustering," *Machine Learning*, no. 2-2, pp. 139-162, 1987.

18. T. M. Mitchell, "Generalization as Search," *Artificial Intelligence*, vol. 18, pp. 203-226, 1982.
19. J.E. Laird, P.S. Rosenbloom, and A. Newell, "Chunking in Soar: The Anatomy of a General Learning Mechanism," *Machine Learning*, no. 1-1, pp. 11-46, 1986.
20. T. M. Mitchell, R. M. Keller, and S. J. Kedar-Cabelli, "Explanation-based Generalization: a Unifying View," *Machine Learning*, vol. 1, pp. 47-80,1986.
21. G. Dejong and R. Mooney, "Explanation-Based Learning: An Alternative View," *Machine Learning*, vol. 1, pp. 145-176.
22. M.J. Pazzani, "Explanation-based learning for knowledge-based systems," *Int. J. of Man-Machine Studies*, pp. 413-424, 1987.
23. R. Keller, "Defining Operationality for Explanation Based Learning," in *Proc. AAAI-87*, pp. 482-487, Seattle (WS), 1987.
24. S. Rajamoney, G. DeJong, "The Classification, Detection and Handling of Imperfect Theory Problems," in *Proc IJCAI-87*, pp. 205-207, Milano, Italy, 1987.
25. M. Lebowitz, "Not The Path to Perdition: The Utility of Similarity Based Learning," in *Proc IMAL-86*, Les Arc, 1986.
26. F. Bergadano and A. Giordana : "A knowledge Intensive Approach to Concept Induction", *Proc. of the Fifth International Conference on Machine Learning*, Ann Harbor (1988).
27. F. Bergadano, A. Giordana and L. Saitta : "Concept Acquisition in an Integrated Ebl and Sbl Environment", *Proc. European Conf. on Artificial Intelligence*, (Munich, Germany, 1988) pp. 363-368.
28. A. Danyluk: "The Use of Explanations for Similarity Based Learning", *Proc. IJCAI-87* (Milano, Italy, 1987), pp. 274-279.
29. IAAA, "Neural Network," in *Proc of the International Conference on Neural Networks*, S. Diego, CA, 1987.
30. J.H. Holland, "Escaping Brittleness: The Possibility of General-Purpose Learning Algorithms Applied to Parallel Rule Based Systems," in *Machine Learning, An Artificial Intelligence Approach*, Vol. 2, ed. R. S. Michalski, J. G. Carbonell, T. M. Mitchell, pp. 593-624, Tioga Publishing Company, 1983.
31. Y. Kodratoff, M. Manago, "Generalization and Noise", *Int. J. of Man-Machine Studies*, 27, 181-204 (1987).
32. J.R. Quinlan, "Induction of Decision Trees," *Machine Learning*, no. 1-1, pp. 81-106, 1986.
33. P. Clark, T. Niblett: "Induction in Noisy Domains", in *Progress in Machine Learning*, I. Bratko and N. Lavrac (Eds.), Sigma Press (Wilmslow, UK, 1987), pp.11-30.
34. J. Mingers: "An Empirical Comparison of Pruning Methods for decision Trees Induction", *Machine Learning* , 4, 227-243 (1989).
35. R. S. Michalski, Igor Mozetic, Jiarong Hong, and Nada Lavrac, "The AQ15 Inductive Learning System: An Overview and Experiments," in *Proc. of the International Meeting on Advances in Learning - IMAL*, 1986.
36. J. C. Schlimmer and R. H. Granger, Jr., "Incremental Learning from Noisy Data," *Machine Learning*, vol. 1, pp. 317-354, 1986.
37. P. Politakis, S. Weiss, "Using Empirical Analysis to Refine Expert System Knowledge Bases," *Artificial Intelligence*, no. 22, pp. 23-48, 1984.
38. R. Rada, "Gradualness Facilitates Knowledge Refinement," *IEEE Trans. on Pattern Analysis and Machine Intelligence*, vol. PAMI-7, pp. 523-530, 1985.
39. A. Ralescu, J. Baldwin: "Concept Learning from Examples and Counter-Examples", *Int. J. of Man-Machine Studies*, 30, 329-354 (1989)
40. R. De Mori, L. Saitta: "Automatic Learning of Fuzzy Naming Relations over Finite Languages", *Information Sciences*, 21, 93-139 (1980).

41. L. Lesmo, L. Saitta, P. Torasso: "Learning of Fuzzy Production Rules in Medical Diagnosis", *Invited paper* in M. Gupta and E. Sanchez (Eds.), *'Approximate Reasoning in Decision Analysis'*, North-Holland Publ. Co. (1982), pp. 249-260.
42. F. Bergadano, A. Giordana, and L. Saitta, "Learning from Examples in Presence of Uncertainty," in *Approximate Reasoning in Intelligent Systems Decision and Control*, ed. E. Sanchez and L. Zadeh, pp. 105-124, Pergamon Press , 1986.
43. F. Bergadano, A. Giordana, and L. Saitta, "Concept Acquisition in Noisy Environment," *IEEE Trans. on Pattern Analysis and Machine Intelligence*, vol. PAMI-10, 1988.
44. F. Bergadano, R. Gemello, A. Giordana and L. Saitta : "Smart : A Problem Solver for Learning from Examples", *Fundamenta Informaticae*, 12, pp. 29-50 (1989).
45. R. S. Michalski, "A Theory and Methodology of Inductive Learning," *Artificial Intelligence*, 20, pp. 111-161, 1983.
46. J. D. Ullman : *Principles of Database Systems*, Computer Science Press (1983).
47. J. D. Ullman : "Implementation of Logical Query Languages for Database", *ACM Trans. on Database Systems*, 10, pp. 289-321 (1985).
48. R. Yager : "Quantified Propositions in a Linguistic Logic", *International J. of Man-Machine Studies*, 19, 195-227 (1983).
49. L. Zadeh : "A Computational Approach to Fuzzy Quantifiers in Natural Languages", *Computer and Mathematics with Applications*, 9, 149-184 (1983).
50. L. Vieille : "Recursive Axioms in Deductive Database : the Query Sub-query Approach", *Proc. of the 1st Int. Conf. on Expert Database Systems*, (Charlston, SC, 1986).
51. J. A. Robinson and E. E. Siebert : "LOGLISP : An Alternative to Prolog", *Machine Intelligence*, 10, J. E. Hayes and D. Michie (eds), 399-419 (1982).
52. F. Bergadano, F. Brancadori, A. Giordana, L. Saitta: " A System that Learns Diagnostic Knowledge in a Data-Base Framework", *Proc. IJCAI Workshop on Knowledge Discovery in Databases* (Detroit, MI, 1989), pp. 4-15.

15

EVIDENTIAL REASONING UNDER PROBABILISTIC AND FUZZY UNCERTAINTIES

J. F. BALDWIN
Engineering Mathematics Dept
University of Bristol
Bristol BS8 1TR
England

1. INTRODUCTION

1.1 GENERAL KNOWLEDGE AND EVIDENCE

An expert's knowledge of an application is concerned with general tendencies, what is likely to be the case, frequent conjunctions, rules of thumb and other forms of statistical statements. An investigator may know that a certain type of crime is common among criminals of a certain type, an insurance company may know that a person with certain characteristics is a good risk, a doctor knows that certain symptoms almost always means the person is suffering from a given disease. The conclusion in each of these cases comes from studying tendencies in a population of relevant cases and using these to infer something about an individual case.

Rules of thumb such as

"most tall persons wear large shoes"

can be expressed as the rule

person X wears large shoes IF person X is tall : very likely

X is a variable which can be instantiated to any member assumed to be drawn at random from the population of persons. This says that the head of the rule is very likely given the body of the rule is true. It makes a vague statement about the conditional probability Pr(large shoes | tall). This probability represents the proportion of persons who wear large shoes in some population of tall persons. It is a statement about the population as a whole rather than a statement about any particular individual person. If an individual person is known to be tall then one can infer that the probability of this person wearing large shoes is very likely. In this sense the variable in the rule is universally quantified,
i.e $\forall x$ Pr(x wears large shoes | x is tall) = very likely.

The use of "very likely" rather than a point probability value further complicates matters. As a first approximation we might equate "very likely" with the interval [0.9, 1]. This means that the Pr(x wears large shoes | x is tall) lies in the interval [0.9, 1]. We could further express this as the necessary support in favour of (x wears large shoes | x is tall) is 0.9 and the necessary support in favour of (x does not wear large shoes | x is tall) is 0. The term necessary support can be replaced with the term "belief". We can also express this in the form of a mass assignment over the power set of

{(x wears large shoes | x is tall), (x does not wear large shoes | x is tall)}

namely,
(x wears large shoes | x is tall) : 0.9
(x does not wear large shoes | x is tall) : 0
{(x wears large shoes | x is tall), (x does not wear large shoes | x is tall)} : 0.1
where an assignment of mass m to set Y means m is the probability associated with exactly Y but not to any subset of Y. The meaning of these various terms will be expanded upon later in the paper. In order to more adequately capture the true semantics of the vague statement "very likely" we require to model this linguistic term using a fuzzy set, [ZADEH 1965]

1.2 An Example

Consider the following simple example. A bag contains 70% red balls and 30% blue balls. Each ball is either large or small. 60% of the red balls are large and 40% of the blue balls are large.

Problem 1a
What is the probability that a ball drawn randomly from the bag is large?
Of course this is a very elementary problem and can be solved by fusing the pieces of information concerning the balls in the bag to calculate this probability. If {y1, y2, y3, y4} stand for the probabilities {Pr(rl), Pr(rs), Pr(bl), Pr(bs)} respectively and r, l signifies "red", "large" respectively then

$y1 + y2 = 0.7$; $y3 + y4 = 0.3$
$y1 / (y1 + y2) = 0.6$; $y3 / (y3 + y4) = 0.4$
so that $y1 = 0.42$, $y2 = 0.28$, $y3 = 0.12$ and $y4 = 0.18$
from which $Pr(l) = y1 + y3 = 0.54$.
This is simply a probability logic problem. In the sequel this fusion of probabilistic information will be done by means of a general assignment method.

Problem 1b
A ball drawn at random from the population is known to be large. What is the probability that it is red?
The solution is given by $y1 / (y1 + y3) = 0.7778$ and comes from fusing the given information using probability logic and calculating the required conditional probability.

Of course $y1 / (y1 + y3) = Pr(l \mid r)Pr(r) / Pr(l)$ which is Bayes theorem applied to this problem. This can therefore be viewed as an updating problem in which the apriori distribution {y1, y2, y3, y4} is updated using the certain information that the ball in question is large.

Problem 2

The balls in the bag are shown, as a black and white image on a screen, one by one to an observer. The observer is then asked if the third ball shown was red. The observer believes that the third ball was large but is not certain of this fact. He expresses this belief as Pr(third ball shown is large) = 0.8. He does not have information about the colours of the balls shown. What should be his belief that it is red?

One possible answer to this problem is obtained by using Jeffrey's rule, [JEFFREY 1967] namely

Pr(third ball is r) = Pr(r | l)Pr(third ball is l) + Pr(r | s)Pr(third ball is s)

$= 0.7778 * 0.8 + \{y2 / (y2 + y4)\} * 0.2$

$= 0.7778 * 0.8 + 0.6087 * 0.2 = 0.744$

It looks as if we have used the theorem of total probabilities, namely,

Pr(third ball is r) = Pr(third ball is r |third ball is l)Pr(third ball is l)

+ Pr(third ball is red | third ball is small)Pr(third ball is s)

with the assumption that

Pr(third ball is r |third ball is l) = Pr(r | l) and

Pr(third ball is r | third ball is s) = Pr(r | s).

We do not have to make this assumption if the following philosophy is accepted. The apriori distribution over the labels {rl, rs, bl, bs} is {y1, y2, y3, y4}. This is to be updated using the specific information P'r(l) = 0.8, where the ' is used to signify that this is not the proportion of large balls in the population but a belief in one particular ball being large. We could update to {y'1, y'2, y'3, y'4} by choosing the {y'i} such that the relative information

$$I = \sum y'i \, Ln\,(y'i / yi)$$

is minimised. This will be discussed further later. This forms the basis of the iterative assignment method to be discussed in detail in a later section.

1.3 Rule Form of Knowledge Representation

Various forms of knowledge representation can be used to express the general tendencies and specific information discussed above. The assignment methods mentioned above can be used with various forms of knowledge representation but in this paper we will concentrate on a rule form like that used by Prolog but extended to allow for uncertainties of both a probabilistic and a fuzzy kind to be expressed. The methods developed in this paper are extensions of those used in the AI language FRIL, [BALDWIN 1986, 1987] and [BALDWIN et al 1987].

Consider the following Prolog program.

```
married(X) :-  middle_aged(X), has_children(X).
middle_aged(mary).
has_children(mary).
```

This says that any middle aged person who has children is married and that Mary has children. We can conclude from this that Mary is married. In Prolog we ask the query

?- married(mary)

to which we get the reply

yes.

Suppose it is known that at least 70% and at most 90% of middle aged persons who have children are married. We will define the fuzzy term "middle aged" using the fuzzy set middle_aged with membership function

$$\chi_{middle\_aged}(x) = \begin{cases} 1/5.x - 7 \text{ for } 35 \le x \le 40 \\ 1 \text{ for } 40 \le x \le 50 \\ -1/5.x + 11 \text{ for } 50 \le x \le 55 \\ 0 \text{ elsewhere} \end{cases}$$

We also define the fuzzy term "about_35" using the fuzzy set about_35 with membership function

$$\chi_{about\_35}(x) = \begin{cases} 1/5.x - 6 \text{ for } 30 \le x \le 35 \\ -1/5.x + 8 \text{ for } 35 \le x \le 40 \\ 0 \text{ elsewhere} \end{cases}$$

Suppose it is known that Mary is about 35 and it is believed with a probability of at least 0.8 that Mary has children.

A more realistic program is

```
married(X) :- age(X, middle_aged), has_children(X) : [0.7, 0.9].
age(mary, about_35)
has_children(mary) : [0.8, 1].
```

which is interpreted as saying that the conditional probability that someone is married if the person is middle aged and has children lies between 0.7 and 0.9 and that the person Mary is about 35 years old and the probability that Mary has children lies between 0.8 and 1.

We can now ask how do we answer the query

?- married(mary)

We would expect the answer to take the form of an interval containing the probability that Mary is married.

The methods of inference developed below will allow us to answer this query for this program. In deriving this interval both the probabilistic and fuzzy types of uncertainty must be taken into account. For example, the rule talks about middle aged persons while the age of Mary is given as "about 35". From a syntactic point of view it would appear that the rule has no relevance to Mary but from a semantic point of view it does since someone who is "about 35" is to some degree middle aged. This degree depends on the definitions of the fuzzy sets "middle aged" and "about 35". In order to answer the query given it is necessary to determine an interval containing the conditional probability Pr{age(mary, middle_aged) | age(mary | about_35). We term this process "semantic unification", [BALDWIN 1990a].

When the second argument of the age predicate is always a crisp set then this interval is
[0, 0], [1, 1] or [0, 1].
For example,
Pr{age(mary, [40, 50]) | age(mary |[35, 39])} = 0 and
Pr{age(mary, [35, 45]) | age(mary |[37, 42])} = 1 and
Pr{age(mary, [40, 50]) | age(mary | [45, 55])} is contained in the interval [0, 1].
Semantic unification extends this to the case of fuzzy sets.

A program can also contain universally quantified facts. For example
middle_aged(X) : [0.4, 0.5]
would be interpreted as saying that between 40% and 50% of the relevant population of objects being considered are middle aged. While X can be instantiated to the object mary, the statement
middle_aged(X) : [0.4, 0.5] with X = mary
is not interpreted in the same way as
middle_aged(mary) : [0.4, 0.5].
The former is a statement about the population as a whole and if Mary is an object drawn at random from this population then Pr{middle_aged(mary)} lies in [0.4, 0.5]. The latter statement makes no reference to the population but is a statement made about the object Mary by inspecting this object independently of any population statistics. The first statement is concerned with a general tendency while the second statement is specific to the object concerned.

1.4 AIMS OF PAPER

In this paper we will discuss methods for answering queries of the types given above from a knowledge base expressed in rule form containing statements representing general tendencies and also specific facts. The specific facts expressed in probabilistic terms will be used as evidences to update the family of possible apriori distributions obtained from the relevant general statements and this update used to provide the answer to the given query.The rules and facts can contain both probabilistic and fuzzy uncertainties.

The inference method for processing the knowledge base to answer queries will use the following three methods
(1) general assignment method
(2) iterative assignment method
(3) semantic unification.

In special cases the inference process simplifies to using the theorem of total probabilities if only general statements from the same population are used or Jeffrey's rule when both general statements and specific evidences are used. This is the inference mechanism of the AI language FRIL.

Each of these methods requires the information in the form of a mass assignment over a frame of discernment whose elements are labels formed from the information. We discuss this more fully in the appropriate sections that follow. A general treatment will not be given here and the reader is expected to generalise for him/herself from those cases discussed. Other aspects can be found in [BALDWIN 1990b, 1990c]

2. MASS ASSIGNMENTS, SUPPORT MEASURES AND SUPPORT PAIRS

2.1 LABELS AND FRAME OF DISCERNMENT

A knowledge base statement, either in the form of a rule or a fact, is converted into the form of a mass assignment over a set of labels. Each label is a concatenation of instantiations of the proposition variables and the proposition variables come from all the information of the knowledge base relevant to answering the given query. The following example will illustrate this. A general theory in terms of inference diagrams and logic proof paths can be given but space does not allow this to be included here.
Consider the knowledge base

```
fly(X) :- bird(X) : [0.9, 0.95].
bird(X) :- penguin(X).
fly(X) :- penguin(X) : [0, 0].
penguin(obj) : [0.4, 0.4].
bird(obj) : [0.9, 1].
etc
```

and the query

```
?- fly(obj)
```

To answer this query we form the following inference diagram

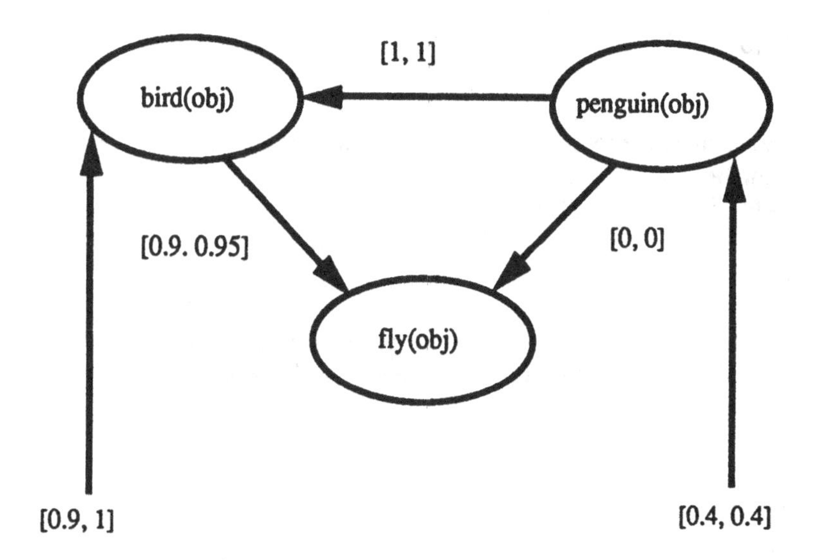

from which we extract the propositional variables
B = bird(obj) ; F = fly(obj) ; P = penguin(obj)
Each of these variables can be instantiated to "true" or "false". We represent the set of instantiations for a propositional variable X by $\{x, \neg x\}$. A label is a possible instantiation of the concatenation of the three variables B, P and F written as BPF. Thus the possible set of labels corresponding to the frame of discernment is
$L = \{\neg b \neg p \neg f, \neg b \neg p f, \neg b p \neg f, \neg b p f, b \neg p \neg f, b \neg p f, b p \neg f, b p f\}$.

For any knowledge base consisting of facts and rules and for any query, a frame of discernment can be established using this method of constructing an inference diagram and extracting the propositional variables. The inference diagram is obtained by using the unification and backtracking mechanisms of Prolog with extensions to include semantic unification as discussed later.

2.2 Mass Assignments

A mass assignment over a finite frame of discernment X is a function
$m : P(X) \rightarrow [0, 1]$ where $P(X)$ is the power set of X
such that
$m(\emptyset) = 0$ and

$$\sum_{A \,\varepsilon\, P(X)} m(A) = 1$$

and corresponds to the basic probability assignment function of the Shafer / Dempster theory of evidence, [SHAFER 1976]. m(A) represents a probability mass assigned exactly to A. It does not include any masses assigned to subsets of A.

As an example consider the specific evidences

penguin(obj) : [0.4, 0.4].
bird(obj) : [0.9, 1].
given above.
The first is equivalent to the following mass assignment over the set of labels L
m({¬bp¬f, ¬bpf, bp¬f, bpf}) = 0.4
m({¬b¬p¬f, ¬b¬pf, b¬p¬f, b¬pf}) = 0.6
which can be written as
{_ p _} : 0.4
{_ ¬p _} : 0.6
where _ can be instantiated to the appropriate proposition or its negation.
Similarly the second evidence is equivalent to the mass assignment
{b _ _} : 0.9
{_ _ _} : 0.1

Consider the general statements
fly(X) :- bird(X) : [0.9, 0.95].
bird(X) :- penguin(X).
fly(X) :- penguin(X) : [0, 0].
The second of these clauses say that the labels {¬bpf, ¬bp¬f} are not possible. The third says that {bpf} are not possible. The combined two statements says that the labels {¬bpf, ¬bp¬f, bpf} are not possible. We can therefore express the first clause as a mass assignment over the reduced set of labels
L'= {¬b¬p¬f, ¬b¬pf, b¬p¬f, b¬pf, bp¬f}, by combining the following two evidences, each expressed as a mass assignment over L'
(1) {b¬p _ , bp¬f} : k , {¬b¬p _ } : 1-k
(2) {b¬pf} : 0.9k , {¬b¬p _ , b _ ¬f} : 1-0.9k
corresponding to
Pr{bird(obj)} = k , Pr{¬ bird(obj)} = 1-k
and
Pr{bird(obj) ∧ fly(obj)} = 0.9k Pr{¬ (bird(obj) ∧ fly(obj))} = 1-0.9k
The combination of these two evidences, using the general assignment method defined below gives the mass assignment over L' as
{b¬pf} : 0.9k
{¬b¬p _ } : 1-k
{b _ ¬f} : 0.1k
for the combined relevant general statements in the knowledge base. The conditional statements of rules can always be treated in this way. Pure logic rules simple reduce the set of possible labels.

2.3 SUPPORT PAIRS

We use the concept of belief and plausibility measures of [SHAFER 1976] to define necessary support and possible support measures. Names are changed to be consistent with the notation used in support logic programming, [BALDWIN 1986] and the FRIL language [Baldwin et al 1987], and to avoid confusion with conclusions and derived results based on the use of the Dempster rule of combining evidences. The methods given here do not use the Dempster rule and the necessary and possible supports are more in keeping with upper and lower probabilities, [DUBOIS, PRADE 1986].

A necessary support measure is a function

$$Sn : P(X) \rightarrow [0, 1]$$

where X is a set of labels and $P(X)$ is the power set of X that satisfies the following axioms

Axiom 1 (boundary condition). $Sn(\emptyset) = 0$ and $Sn(X) = 1$ where $\emptyset$ is the empty set

Axiom 2 : $$Sn(A_1 \cup A_2 \cup ... \cup A_n) \geq \Sigma_i Sn(A_i) - \Sigma_{i<j} Sn(A_i \cap A_j) + ... + (-1)^{n+1} Sn(A_1 \cap A_2 \cap ... \cap A_n)$$

for every collection of subsets of X.

For each $A \in P(X)$, Sn(A) is interpreted as the necessary support, based on available evidence, that a given label of X belongs to the set A of labels.

When the sets $A_1, A_2, ..., A_n$ in axiom 2 are pairwise disjoint ie.

$(A_i \cap A_j) = \emptyset$ for all $i, j \in \{1, 2, ..., n\}$ such that $i \neq j$

the axiom requires that the necessary support associated with the union of the sets is not smaller than the sum of the necessary supports pertaining to the individual sets. The basic axiom of necessary support measures is thus a weaker version of the additivity axiom of probability theory.

It is easy to show that axiom 2 above implies that for every $A, B \in P(X)$, if $A \subseteq B$, then $Sn(A) \leq Sn(B)$

and also that

$Sn(A) + Sn(\overline{A}) \leq 1$

Possible Support Measure

Associated with each necessary support measure is a possible support measure Sp, defined by the equation

$Sp(A) = 1 - Snl(\overline{A})$

for all $A \in P(X)$.

Similarly

$Sn(A) = 1 - Sp(\overline{A})$

Necessary support measures and possible support measures are therefore mutually dual and it is easy to show that

$Sp(A) + Sp(\overline{A}) \geq 1$

Given a basic probability assignment m, a necessary support measure and possible support measure are uniquely determined by the formulae

$$Sn(A) = \sum_{B \subseteq A} m(B)$$

and

$$Sp(A) = \sum_{A \cap B \neq \emptyset} m(B)$$

which are applicable for all A ε *P* (X).

Focal Elements

Every set A ε *P* (X) for which m(A) > 0 is called a focal element of m. We can represent the mass assignment as (m, F) where F is the set of focal elements.

Total ignorance is expressed in terms of the mass assignment by
m(X) = 1 and m(A) = 0 for all A ≠ X.
Using the formula above for Snl in terms of m, we can therefore also express total ignorance as
Sn(X) = 1 and Sp(A) = 0 for all A ≠ X
The total ignorance in terms of the possible support measure is
Sp(Ø) = 0 and Sp(A) = 1 for all A ≠ Ø.

A support pair for A ε *P* (X) is given by [MIN Sn(A), MAX Sp(A)] and this defines an interval containing the Pr(A) where the MIN and MAX are over the set of values of any possible parameters that Sn(A) and Sp(A) may depend on. This will be illustrated later.

3. GENERAL ASSIGNMENT METHOD

3.1 COMBINING MASS ASSIGNMENTS

Let m1 and m2 be two mass assignments over the power set *P* (X) where X is a set of labels. Evidence 1 and evidence 2 are denoted by (m1, F1) and (m2, F2) respectively, where F1 and F2 are the sets of focal elements of *P* (X) for m1 and m2 respectively.

Suppose F1 = {L1k} for k = 1, ..., n1 and F2 = {L2k} for k = 1, ..., n2
then Lij is a subset of *P* (X) for which mi(Lij) ≠ 0.

Let (m F) be the evidence resulting from combining evidence 1 with evidence 2 using the general assignment method. This is denoted as
(m, F) = (m1, F1) ⊕ (m2, F2)
where
F = {L1i ∩ L2j | m(L1i ∩ L2j) ≠ 0}

$$m(Y) = \sum_{ij : L1i \cap L2j = Y} m'(L1i \cap L2j) \qquad \text{for any } Y \in F$$

$m'(L1i \cap L2j)$ for $i = 1, ..., n1$; $j = 1, ..., n2$ satisfies

$$\sum_{j} m'(L1i \cap L2j) = m1(L1i) \qquad \text{for } i = 1, ..., n1$$

$$\sum_{i} m'(L1i \cap L2j) = m(L2j) \qquad \text{for } j = 1, ..., n2$$

$m'(L1i \cap L2j) = 0$ if $L1i \cap L2j = \emptyset$ the empty set ; for $i = 1,..., n1$; $j = 1, ..., n2$

The problem of determining the mass assignment m is an assignment problem as depicted in the following diagram

If there are more than two evidences to combine then they are combined two at a time. For example to combine (m1, F1), (m2, F2), (m3, F3) and (m4, F4) use
$(m, F) = (((m1, F1) \oplus (m2, F2)) \oplus (m3, F3)) \oplus (m4, F4)$

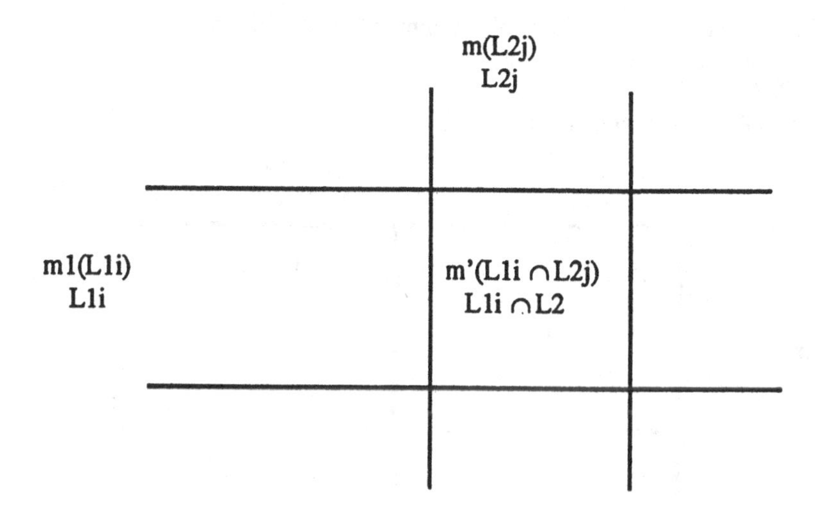

The labels in a cell is the intersection of the subset of labels of evidence 1 associated with the row of the cell and the subset of labels of evidence 2 associated with the column of the cell.

The mass assignment entry in a cell is 0 if the intersection of the subset of labels of evidence 2 associated with the column of the cell and the subset of labels of evidence 1 associated with the row of the cell is empty.

The mass assignment in a cell is associated with the subset of labels in the cell.

The sum of the cell mass assignment entries in a row must equal the mass assignment associated with m1 in that row.

The sum of the cell mass assignment entries in a column must equal the mass assignment associated with m2 in that column.

If there are no loops, where a loop is formed by a movement from a non zero assignment cell to other non zero assignment cells by alternative vertical and horizontal moves returning to the starting point, then the general assignment problem gives a unique solution for the mass assignment cell entries. If a loop exists then it is possible to add and subtract a quantity from the assignment values around the loop without violating the row and column constraints and the solution will not then be unique. If a non-unique solution exists then the family of solutions can be parametrised with known constraints on the parameter values. These possible parameter values must be taken into account when determining support pairs from the necessary and possible support measures.

3.2 The Bird Penguin Example revisited

Consider the example discussed above of combining the two evidences
(1) {b¬p _ , bp¬f} : k , {¬b¬p _ } : 1-k
(2) {b¬pf} : 0.9k , {¬b¬p _ , b _ ¬f} : 1-0.9k
Using the general assignment method we obtain

	0.9k {b¬pf}	1-0.9k {¬b¬p _ , b _ ¬f}
k {b¬p _ , bp¬f}	{b¬pf} 0.9k	{b _ ¬f} 0.1k
1 - k {¬b¬p _ }	Ø 0	{¬b¬p _ } 1-k

giving the result quoted previously.

Consider combining the specific evidences expressed as mass assignments over the set
X = {bp, b¬p, ¬bp, ¬b¬p}
in the example above, namely
penguin(obj) : [0.4, 0.4].
bird(obj) : [0.9, 1].
We have
(1) {_ p } : 0.4 , {_ ¬p } : 0.6
(2) {b _ } : 0.9 , {_ _ } : 0.1
Using the general assignment method we obtain

	0.9 b	0.1 {b, ¬b}
0.4 p	bp 0.4 - x	{_ p} x
0.6 ¬p	b¬p 0.5 + x	{_ ¬p} 0.1 - x

where $0 \leq x \leq 0.1$
An abbreviated form of labelling is used for convenience.

The necessary and possible supports for the various elements of X are given by
Sn(bp) = 0.4 - x ; Sp(bp) = 0.4
Sn(b¬p) = 0.5 + x ; Sp(b¬p) = 0.6
Sn(¬bp) = 0 ; Sp(¬bp) = x
Sn(¬b¬p) = 0 ; Sp(¬b¬p) = 0.1 - x
from which we can calculate the support pairs
bp : [0.3, 0.4] ; b¬p : [0.5, 0.6] ; ¬bp : [0, 0.1] ; ¬b¬p : [0, 0.1]

3.3 Another Example

Consider the example:

90% of birds can fly ------------------------------ (1)
No penguins can fly ------------------------------ (2)
All penguins are birds --------------------------- (3)

70% of objects exhibited are birds -------------- (4)
5% of objects exhibited are penguins ----------- (5)
which in program form would be

fly(X) :- bird(X) : [0.9, 0.9].
fly(X) :- penguin(X) : [0, 0].
bird(X) :- penguin(X) : [1, 1].
bird(X) : [0.7, 0.7].
penguin(X) : [0.5, 0.5].
We can ask the following question. What is the probability that an object drawn at random can fly.

To answer the query we combine the following mass assignments over the label set L' using the general assignment method.
(1) $\{b\neg p\neg f, b\neg pf, bp\neg f\} : 0.7$; $\{\neg b\neg p\neg f, \neg b\neg pf\} : 0.3$ using (4)
(2) $\{bp\neg f\} : 0.05$; $\{\neg b\neg p\neg f, \neg b\neg pf, b\neg p\neg f, b\neg pf\} : 0.95$ using (5)
(3) $\{b\neg pf\} : 0.9k$; $\{\neg b\neg p\neg f, \neg b\neg pf, b\neg p\neg f, bp\neg f\} : 1 - 0.9k$ using (1)
where k is the probability assigned to $\{b\neg p\neg f, b\neg pf, bp\neg f\}$

For this particular example the solution is easily found by elementary analysis to be
$\{b\neg pf\} : 0.63$
$\{bp\neg f\} : 0.05$
$\{b\neg p\neg f\} : 0.02$
$\{\neg b\neg pf, \neg b\neg p\neg f\} : 0.3$
so that the answer to the query is that the Pr{(x can f)} lies in the interval [0.63, 0.93] since the 0.3 associated with $\{\neg b\neg pf, \neg b\neg p\neg f\}$ could be all be associated with $\neg b\neg pf$ although this is not necessarily the case. This conclusion is expressed in the form of a support pair. We can obtain this result using the general assignment as follows:

Combining (1) and (2) gives

Evidence 1 \ Evidence 2	0.05 $\{bp\neg f\}$	0.95 $\{\neg b\neg p\_, b\neg p\_\}$
0.7 $\{b\neg p\_, bp\neg f\}$	$\{bp\neg f\}$ 0.05	$\{b\neg p\_\}$ 0.65
0.3 $\{\neg b\neg p\_\}$	Ø 0	$\{\neg b\neg p\_\}$ 0.3

The mass assignment resulting from combining(1) and (2) is thus
$\{bp\neg f\} : 0.05$

$\{b\neg pf, b\neg p\neg f\} : 0.65$
$\{\neg b\neg pf, \neg b\neg p\neg f\} : 0.3$
This is now combined with (3) as follows

Evidence 1, 2 \ Evidence 3	0.9k $b\neg pf$	1 - 0.9k $\{\neg b\neg p\_, b\_\neg f\}$
$0.05 : \{bp\neg f\}$	$\emptyset$ 0	$\{bp\neg f\}$ 0.05
$0.65 : \{b\neg p\_\}$	$\{b\neg pf\}$ 0.9k	$\{b\neg p\neg f\}$ 0.65 - 0.9k
$0.3 : \{\neg b\neg p\_\}$	$\emptyset$ 0	$\{\neg b\neg p\_\}$ 0.3

where $0.9k + 0.05 + (0.65 - 0.9k) = k$
so that $k = 0.7$ giving the combined mass assignment

$\{bp\neg f\} : 0.05$
$\{b\neg pf\} : 0.63$
$\{b\neg p\neg f\} : 0.02$
$\{\neg b\neg pf, \neg b\neg p\neg f\} : 0.3$

4. ITERATIVE ASSIGNMENT METHOD

4.1 UPDATING PROBLEM

Suppose an apriori mass assignment m_a is given over the focal set A whose elements are subsets of the power set P (X) where X is a set of labels. This assignment represents general tendencies and is derived from statistical considerations of some sample space or general rules applicable to such a space.

Suppose we also have a set of specific evidences {E1, E2, ..., En} where for each i, Ei is (mi, Fi) where Fi is the set of focal elements of P (X) for Ei and mi is the mass assignment for these focal elements. These evidences are assumed to be relevant to some object and derived by consideration of this object alone and not influenced by the sample space of objects from which the object came from.

We wish to update the apriori assignment m_a with {E1, ..., En} to give the updated mass assignment m such that the minimum information principle concerned with the relative information of m given m_a is satisfied.

4.2 ALGORITHM FOR SIMPLE CASE

The minimum information principle
Let p be an apriori distribution defined over the set of labels X.

Let specific evidences E1, ..., En be given where Ei expresses a probability distribution over a partition of X.

Let p' be a distribution such that

$$\sum_{x \varepsilon X} p'(x) \operatorname{Ln} (p'(x) / p(x))$$

is minimised subject to the constraints
E1, ..., En

p' is said to satisfy the minimum information principle for updating the distribution p over X with specific evidences E1, ..., En where each Ei is expressed as a distribution over a partition of X.

The sequential iterative assignment algorithm updates p using E1 to obtain the update p1 satisfying the minimum information principle which is similarly updated using E2 to obtain p2 which ... in turn is updated using En to obtain pn. At each stage of this process only the evidence used for updating is necessarily satisfied and previous evidences used will no longer be necessarily satisfied. We therefore replace the apriori with pn and repeat the process. The iteration is continued until a pn is found which satisfies all the evidences E1, ..., En.

This iterative process in fact converges to the solution which satisfies the minimum information principle of minimising the relative information with respect to the apriori p subject to the constraints E1, ..., En. The multi constraint optimisation problem is therefore solved by a succession of single constraint optimisation problems and iterating.
The single constraint optimisation problem has a particularly simple algorithm for its solution which we will now consider.

Let p(r-1) = p, say, be updated to p(r) = p', say, using Er with the following algorithm which gives a p' such that

$$\sum_{x \,\varepsilon\, X} p'(x).\mathrm{Ln}\{p'(x) / p(x)\}$$

is minimised subject to the constraint Er being satisfied.

Let the partition for evidence Er be {X1, ..., Xk} with probability distribution {Pr(Xi)} given.

$$\text{Let } Ki = \sum_{x \,\varepsilon\, Xi} p(x)$$

for i = 1, ..., k
then
p'(x) = p(x)Pr(Xi) / Ki for x ε X IF x ε Xi for x a label of X, for all labels of X

The algorithm is particularly simple when the apriori is expressed as a probability distribution over the set of labels and the evidences as a distribution over a partition of the set of labels.

We can give a pictorial representation for this algorithm

	Pr(Xj) Xj ={...., li, ...}	
pi : li	li : Kj.pi.Pr(Xj)	,Ø : 0 for all other cells in this row
	Ø : 0 if label in row is not in Xj	
	$Kj = \dfrac{1}{\sum_{k : lk \,\varepsilon\, Xk} pk}$	

4.3 <u>GENERALISATION FOR EVIDENCES EXPRESSED AS MASS ASSIGNMENTS</u>

If the evidences are expressed as mass assignments over X with the apriori assignment still being a probability distribution over the set of labels X then a more complicated case must be considered.

Let the p(r-1) = p be denoted as in 4.2 and let Er be the mass assignment
{Xrk : mrk, for k = 1, ..., nr}
where Xrk is a subset of X for all k and mrk is the mass assignment given to Xrk.

$$p'(x) = \sum_{k\,:\,x\,\varepsilon Xrk} p(x).mrk.\,Kk \qquad ;\ \text{for x any label, for all labels}$$

where

$$Kk = \frac{1}{\sum_{s\,:\,ls\,\varepsilon Xrk} ps}$$

We can express this in pictorial form as:

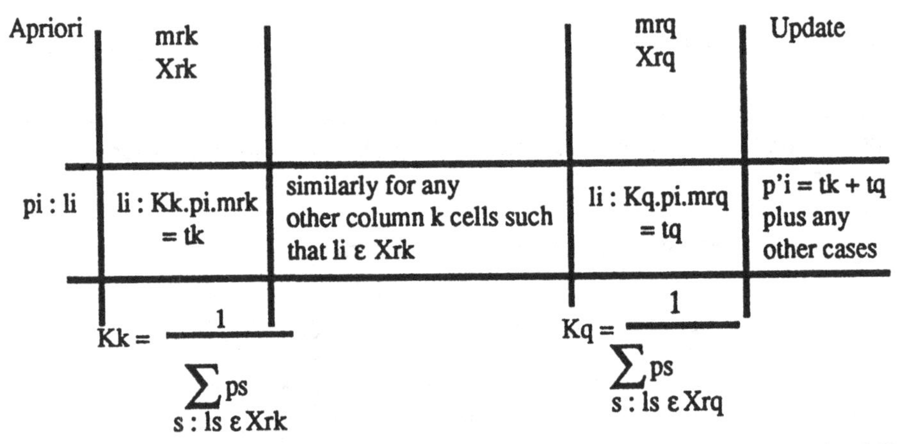

Apriori
mrk
Xrk
mrq
Xrq
Update
pi : li
li : Kk.pi.mrk
= tk
similarly for any
other column k cells such
that li ε Xrk
li : Kq.pi.mrq
= tq
p'i = tk + tq
plus any
other cases
Kk = 1 / Σ ps
s : ls ε Xrk
Kq = 1 / Σ ps
s : ls ε Xrq

Each column constraint is satisfied. The labels in the tableau cells are all labels of X since the focal elements of Er when intersected with a label in the apriori gives the apriori label. The update of this label is the sum of all the cell assignments associated with this label. A cell in a row where the apriori label is not a member of the cell column focal element of Er has a zero mass assignment associated with the empty set.
In this case the update solution p' satisfies the following relative information optimisation problem.

is minimised subject to the constraints
$Sn(Y) \leq p'(Y) \leq Sp(Y)$ for all subsets Y of *P* (X)
where Sn(Y) and Sp(Y) are determined from the mass assignment (mr , Fr)

4.4 Generalisation for both apriori and evidences expressed as mass assignments

In this case the intersection of the row subset of labels of the apriori assignment with the column subset of labels of the evidence assignment for a given cell of the tableau is a subset of P (X). In the case when this intersection is the empty set the mass assignment for that cell is zero, When the intersection is not empty then the mass assignment is the product of the row apriori assignment and the evidence column assignment scaled with the K multiplier for that column. The K multiplier for a column is the sum of the apriori row assignments corresponding to those cells of the tableau in the column which have non-empty label intersections. The update is a mass assignment over the set of subsets of labels of the cells of the tableau. The update can therefore be over a different set of subsets of labels to that of the apriori. Iteration proceeds as before and convergence will be obtained with a mass assignment over P (X) which will correspond to a family of possible probability distributions over X.

We can give a pictorial view of the algorithm:

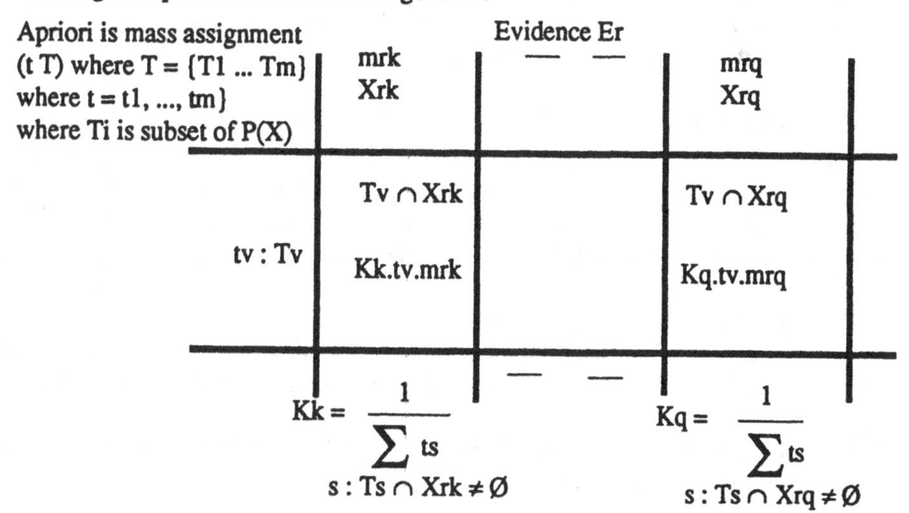

The apriori assignment was also a family of possible probability distributions. For any one of these, the calculation is that of 4.3 and satisfies the minimum information principle where the constraints are in terms of necessary and possible supports from the evidence assignments. A member of the apriori set of possible probability distributions over the label set X will be updated with the evidences E1, ..., Er to a final probability distribution over X satisfying the minimum information principle. Each member of the set of possible apriori probability distributions will be updated, in general to a different final probability distribution. The final set of distributions can be expressed as an assignment over P (X). This is what the algorithm described above does in this case. The calculation is no more involved as for the other more simple cases apart from having to determine the intersections for finding the subset of labels for each cell and taking note of these in the final update.

The examples which follow will illustrate the method.
In this case the updating tableau can contain loops in a similar manner to the general

assignment case. The loop can be treated in exactly the same way as for the general assignment method. This will be illustrated in the examples that follow. Each solution of the loop satisfies the minimum information principle.

5. EXAMPLES OF USE OF ITERATIVE ASSIGNMENT METHOD

5.1 A SIMPLE RULE SYSTEM

```
a(X) :- b(X), c(X) : [0.9, 0.9], [0, 0]. ------------------------------------  (1)
c(X) :- d(X) : [0,.85 1], [0, 0]. -------------------------------------------  (2)
b(mary) : [0.8, 0.8]. -------------------------------------------------------  (3)
d(mary) : [0.95, 0.95]. -----------------------------------------------------  (4)
```

This is a simple FRIL type program. X is a variable and a, b, c, d are predicates. The first two sentences are rules which express general statements about persons and the third and fourth are facts about a specific person mary.

If a rule contains one list after the colon then this gives the interval containing the probability of the head of the rule given the body of the rule is true. If the list after the colon contains two lists then the first gives the probability of the head of the rule given the body of the rule is true and the second gives the interval for the probability of the head of the rule given the body of the rule is false.

The first rule says that for any person X the probability that (X is a) given that (X is b) and (X is c) is 0.9. This expresses the fact that 90% of persons who are both b and c are also a. It also says that (X is a) cannot be true unless both b and c are satisfied.

The second rule says that at least 85% of persons who are d are also c while no person who is not a d can be a c.

We can ask the query in FRIL
?- a(mary).
to determine the support pair for (Mary is a)

In this example the rules are used to determine a family of apriori assignments over the two sets of labels

{ABC} and {CD}

where A, B, C, D denotes a or $\neg$ a, b or $\neg$ b, c or $\neg$ c, d or $\neg$ d respectively. Rule (2) can be used to construct a family of apriori assignments for {CD} which can be updated using (4) for the specific person Mary and from this a support pair for Pr{c(mary)} determined. This can be used with (3) to update a family of apriori assignments determined from (1) for the set of labels {ABC}. From this update the Pr{a(mary)} can be determined.

Alternatively we could update the family of apriori assignments over the set of labels

{ABCD} constructed using rules (1) and (2) with specific evidences (3) and (4) and determine Pr{a(mary)} from the final update.

These two approaches are equivalent and the first approach decomposes the problem of finding Pr{a(mary)} into two sub problems. Decomposition will not be discussed in detail in this paper but it is important to reduce the computational burden associated with updating over a large set of labels.

The computation of the first approach is shown below

Labels with apriori assignment

y1	cd
y2	c¬d
y3	¬cd
y4	¬c¬d

rule (2) constrains {yi} such that

$y1 / (y1 + y3) = [0.85, 1]$ and $y2 / (y2 + y4) = 0$

so that

$y1 = [0.85, 1]k$ where $k = y1 + y3$ for $0 < k \leq 1$

$y2 = 0$

Thus

{_d} : k , {¬c¬d} : 1-k -- (5)

{cd} : 0.85k , {¬c_} : 1 - k , {cd, ¬c_} : 0.15k ---------- (6)

(5) and (6) can be combined using the general assignment method to form a family of apriori assignments over this set of labels

	0.85k {cd}	1-k {¬c_}	0.15k {cd, ¬c_}
k : {_d}	**{cd} : 0.85k**	**{¬cd} : 0+x**	**{_d} : 0.15k-x**
1-k : {¬c¬d}	**Ø : 0**	**{¬c¬d} : 1-k-x**	**{¬c¬d} : 0+x**

where $0 \leq x \leq MIN\{0.15k, 1-k\}$

giving apriori assignment

{cd} : 0.85k , {¬cd} : x , {¬c¬d} : 1-k-x , {_d} : 0.15k-x

which can be updated using

{_d} : 0.95 , {_¬d} : 0.05

by the iterative assignment method as follows

Apriori	0.95 {_d}	0.05 {_¬d}	Update
0.85k : {cd}	**0.8075**	**0**	**0.8075**
x : {¬cd}	**0.95x / k**	**0**	**0.95x / k**
1-k : {¬c¬d}	**0**	**0.05**	**0.05**
0.15k-x : {_d}	**0.1425 - (0.95x / k)**	**0**	**0.1425 - (0.95x / k)**
K's	1/k	1/(1-k)	

From this we calculate
Pr{c(mary)} ε [0.8075, MAX{0.95 - (0.95x / k)}]
The upper limit is maximum for x = 0, so that
Pr{c(mary)} ε [0.8075, 0.95]
which is to be used for the next stage.

Labels and Apriori

y1	abc	
y2	¬ abc	
y3	¬ ab¬ c	{ab¬ c, a¬ bc, a¬ b¬ c} : 0 from rule 1
y4	¬ a¬ bc	
y5	¬ a¬ b¬ c	

Also y1 / (y1 + y2) = 0.9
so that y1 =0.9k where k = y1 + y2
ie we must combine
1. {abc} : 0.9k , {¬ a _ _ } : 1-0.9k
2. {_ bc} : k , {¬ ab¬ c, ¬ a¬ b _ } : 1-k

The general assignment method gives

	0.9k {abc}	1-0.9k {¬ a _ _ }
k : {_ bc}	**{abc} : 0.9k**	**{¬ abc} : 0.1k**
1-k : {¬ ab¬ c, ¬ a¬ b_}	**Ø : 0**	**{¬ ab¬ c, ¬ a¬ b _} : 1-k**

giving the apriori family of assignments over the labels {ABC} as
0.9k {abc}
0.1k {¬ abc}
1-k {¬ ab¬ c, ¬ a¬ b _ }
which is updated using
Specific Evidence 1 :- { _ b _ } : 0.8 , { _ _ _ } : 0.2
Specific Evidence 2 :- { _ _ c} : 0.8075 , { _ _ ¬ c} : 0.05 , { _ _ _ } : 0.1425

0.9k	{abc}	UPDATE USING	{abc}
0.1k	{¬ abc}	Specific Evidence 1	{¬ abc}
1-k	{¬ ab¬ c, ¬ a¬ b _ }	TO GIVE UPDATE	{¬ ab¬ c}
			{¬ a¬ b _ }

{abc}	UPDATE USING	{abc}
{¬ abc}	Specific Evidence 2	{¬ abc}
{¬ ab¬ c}	TO GIVE UPDATE	{¬ ab¬ c}
{¬ a¬ b _ }		{¬ a¬ bc}
		{¬ a¬ b¬ c}
		{¬ a¬ b _ }

$\{abc\}$	UPDATE USING	$\{abc\}$	
$\{\neg abc\}$	Specific Evidence 1	$\{\neg abc\}$	
$\{\neg ab\neg c\}$	TO GIVE UPDATE	$\{\neg ab\neg c\}$	
$\{\neg a\neg bc\}$		$\{\neg a\neg bc\}$	
$\{\neg a\neg b\neg c\}$		$\{\neg a\neg b\neg c\}$	I
$\{\neg a\neg b\_\}$		$\{\neg a\neg b\_\}$	T
			E
			R
			A
$\{abc\}$	UPDATE USING	$\{abc\}$	T
$\{\neg abc\}$	Specific Evidence 2	$\{\neg abc\}$	E
$\{\neg ab\neg c\}$	TO GIVE UPDATE	$\{\neg ab\neg c\}$	
$\{\neg a\neg bc\}$		$\{\neg a\neg bc\}$	
$\{\neg a\neg b\neg c\}$		$\{\neg a\neg b\neg c\}$	
$\{\neg a\neg b\_\}$		$\{\neg a\neg b\_\}$	

For all values of k this will give an interval for {abc} and thus a. We leave the actual calculation to the reader. The result of this calculation is

$$a : (0.6675 \quad 0.72)$$

so that Pr{a(mary)} ε [0.6675, 0.72].

The interval for bc can also be calculated and this is [0.6075, 0.8]. It should be noted that if only the answer for a(mary) is required the last 5 rows of the last two tables can be collapsed into one row with the assignment for this row equal to the sum of the assignments of the five rows. This simplifies the calculation process.

In this example each stage of the process retains the information given by the appropriate rule. For example in the final table Pr{a(mary) | b(mary), c(mary)} = 0.9. This simply means that there is a member of the family of apriori assignments which can satisfy the specific evidences. For this example several steps are required for the final iteration to converge. This is because of the imprecision found for Pr{c(mary)}. If a point value was used for Pr{c(mary)} then the iteration would have converged in one step. In a later section we will deal with the non-monotonic logic case in which the specific evidences are inconsistent with the family of apriori assignments.

5.2 THREE CLOWNS EXAMPLE

Three clowns stood in line. Each clown was either a man or a woman. The audience was asked to vote on each of the first and last clown being male. 90% voted that the first clown, the one on the left, was a man and 20% thought the third clown, the one on the far right, was a man. Nothing was recorded about the middle clown. What is the probability that a male clown stands next to a female clown with the male on the left?

If it were known for sure that the first was male and the third was female then a male would certainly be standing next to a female with the male on the left. This problem can be expressed in first order logic and the theorem proved by case analysis. The refutation resolution method popular in computer theorem proving programs could also be used but is much more cumbersome. The problem posed above is a probabilistic version of this.

The set of labels for this problem is
{mmm, mmf, mfm, mff, fmm, fmf, ffm, fff}
Two evidences have been supplied:-
Evidence 1:- {mmm, mmf, mfm, mff} : 0.9
Evidence 2:- {mmm, mfm, fmm, ffm} : 0.2

We can combine these two evidences using the general assignment method

	Evidence 2	
Evidence 1	0.2 {__m}	0.8 {__f}
0.9 : {m__}	{m_m} x	{m_f} 0.9 - x
0.1 : {f__}	{f_m} 0.2 - x	{f_f} x - 0.1

$0 \le x \le 0.2$

Therefore the support pair for the statement S = "clowns of opposite sex stand next to each other with a male on the left of the pair = [MIN(0.9 - x), MAX(0.8 + x)] = [0.7, 1].

We now consider this example using the iterative assignment method. Above we used specific information about the three clowns in line. We did not use any apriori information concerning clowns in general. In fact the apriori information we assumed was of the form
{mmm, mmf, mfm, mff, fmm, fmf, ffm, fff} : 1
This mass assignment could be used with the iterative assignment method using the specific information given for updating as follows

		0.9 {m__}	0.1 {f__}
1 : {m__, f__}		{m__} : 0.9	{f__} : 0.1
	K's	1	1

		0.2	0.8
		{_ _ m}	{_ _ f}
0.9 : {m _ _ }		{m _ m} : 0.18	{m _ f} : 0.72
0.1 : {f _ _ }		{f _ m} : 0.02	{f _ f} : 0.08
	K's	1	1

Final Update is
0.18 : {m _ m}
0.72 : {m _ f}
0.02 : {f _ m}
0.08 : {f _ f}
since {m _ _} : 0.9 is satisfied so that both updating evidences are satisfied.

The loop in this final mass assignment means that we can add and subtract around the loop without destroying the constraints and all these solutions satisfy the minimum relative entropy criteria with respect to some apriori assignment in the set of all possible apriori assignments. The solution which is produced by the iterative assignment method before any adding and subtracting around the loop is performed is that corresponding to the maximum entropy apriori assignment, ie. that member of the set of possible apriori assignments corresponding to maximum entropy.

To obtain the necessary support for the statement S we must minimise the assignment given to {m _ f} so that we use the assignment
0.2 : {m _ m}
0.7 : {m _ f}
0 : {f _ m}
0.1 : {f _ f}
since 0.02 is the maximum value we can subtract from 0.72 since otherwise the entry in the cell with assignment 0.02 would go negative.

The possible support for S is obtained by maximising the assignment given to {m_ m, m _ f, f _ f} ie minimising the assignment given to {f _ m}. This is also satisfied by this last assignment giving the support pair [0.7 1] for S.

The above analysis is equivalent to using the iterative assignment with all apriori distributions over the label set {mmm, mmf, mfm, mff, fmm, fmf, ffm, fff} which will allow both specific evidences to be retained when using the iterative assignment method. For example
The apriori {0 0 0 .25 0 .25 0 0 0.25 0.25} will give 0.9
The apriori {0.25 0.25 0 0 0.25 0.25 0 0} will give 0.8
The apriori {0 0 0.1 0.4 0 0 0.25 0.25} will give 0.9
The apriori {0 0.7 0.2 0 0 0.1 0 0} gives 1
The apriori {0.2 0.4 0 0.3 0 0 0 0.1} gives 0.7

6. NON_MONOTONIC REASONING

6.1 WHY SHOULD THERE BE A PROBLEM?

Consider the following example. Population statistics tell us that a thirty year old Englishman has a very high probability of living another 5 years. The statistics also tell us that a thirty year old Englishman who has lung cancer only has a small probability of living another 5 years. We are told that John is a thirty year old Englishman. We can conclude that it is very probable that he will live another 5 years. If we are later told that he has lung cancer then we conclude he has little chance of living another 5 years. What we could conclude before this additional piece of information was given we can no longer conclude. From a logic point of view it appears that we have a situation in which we can approximate the modelling of this situation by replacing propositions with high probabilities with those propositions and propositions with low probabilities with their negation. Thus we have

$\forall x$ {Englishman (x) $\wedge$ InThirties (x)} $\supset$ Live5yrsmore (x)
$\forall x$ {Englishman (x) $\wedge$ InThirties (x) $\wedge$ Cancer (x)} $\supset \neg$ Live5yrsmore (x)
If Englishman (John) $\wedge$ InThirties (John)
then we conclude Live5yrsmore (John)
If Englishman (John) $\wedge$ InThirties (John) $\wedge$ Cancer (John)
then we conclude $\neg$ Live5yrsmore (John)
showing a nonmonotonic behaviour.
Thus situations like the above seem to make difficulties if we try to model them using first order predicate logic.

From a probabilistic point of view there is no problem. We are told that
Pr{Live5yrsmore (x) | Englishman (x) $\wedge$ InThirties (x)} is high ;
for any x ------------ (1)
Pr{Live5yrsmore (x) | Englishman (x) $\wedge$ InThirties (x) $\wedge$ Cancer (x)} is low ;
for any x --------- (2)
This will not lead to any form of inconsistency. In fact if it is known that
Pr{Live5yrsmore (John) | Englishman (John) $\wedge$ InThirties (John)} is high
this will tell us nothing about
Pr{Live5yrsmore (John) | Englishman (John) $\wedge$ InThirties (John) $\wedge$ Cancer (John)}
which can take any value in the range [0, 1].

We make inferences by selecting the correct sample space using the given specific information and determine the desired probability using this. In the case of John who is known to be an Englishman in his thirties the answer for the probability of him living another 5 years will be "high" if this is all we know about him. If we also know that he has lung cancer then a different sample space is used and the answer is "low".

In terms of the iterative assignment method, the general statements (1) and (2) above are used to determine a family of apriori assignments which are updated with the specific evidences concerning John. These specific evidences could be uncertain in some sense, ie probabilistic statements, in this case.

The next example illustrates this.

6.2 PENGUIN EXAMPLE

We reconsider this example which was discussed above.

fly(X) :- bird(X)) : [0.9, 0.9]. -- (1)
bird(X) :- penguin(X). -- (2)
fly(X) :- penguin(X). -- (3)
penguin(obj) : [0.4, 0.4] -- (4)
bird(obj) : [0.9, 0.9]. -- (5)

The rules (1), (2) and (3) define the family of apriori assignments. (2) and (3) eliminate certain possible labels as discussed above. The labels are:

y1 $\neg b \neg p \neg f$
y2 $\neg b \neg p f$
y3 $b \neg p \neg f$
y4 $b \neg p f$
y5 $b p \neg f$

so that $y4 / (y3 + y4 + y5) = 0.9$ and $y1 + y2 + y3 + y4 + y5 = 1$
If we let
$k = y3 + y4 + y5$ -- (6)
then
$y4 = 0.9k$ -- (7)
and the family of apriori for a given k, $0 < k \leq 1$ is
$\{b \neg pf\} : 0.9k$
$\{\neg b \neg p\_\} : 1\text{-}k$
$\{b\_\neg f\} : 0.1k$
determined by combining (6) and (7) using the general assignment method.
This family of assignments are updated using the specific evidences (4) and (5) with the iterative assignment method using the scheme

0.9k	$\{b \neg pf\}$	UPDATE USING	$\{b \neg pf\}$
0.1k	$\{b\_\neg f\}$	$Pr\{(b)\} = 0.9$	$\{b\_\neg f\}$
1-k	$\{\neg b \neg p\_\}$	TO GIVE UPDATE	$\{\neg b \neg p\_\}$

{b¬pf} {b_¬f} {¬b¬p_}	UPDATE USING Pr{(p)} = 0.4 TO GIVE UPDATE	{bp¬f} {b¬pf} {b¬p¬f} {¬b¬p_}	
{bp¬f} {b¬pf} {b¬p¬f} {¬b¬p_}	UPDATE USING Pr{(b)} = 0.9 TO GIVE UPDATE	{bp¬f} {b¬pf} {b¬p¬f} {¬b¬p_}	I T E R
{bp¬f} {b¬pf} {b¬p¬f} {¬b¬p_}	UPDATE USING Pr{(p)} = 0.4 TO GIVE UPDATE	{bp¬f} {b¬pf} {b¬p¬f} {¬b¬p_}	A T E

From the final family of assignments we can determine the support pair for f(obj)
f(k) : [assignment for {b¬ pf}, assignment for {b¬ pf} + assignment for {¬ b¬ p _ }]
= [0.45, 0.55]
This final support pair is in actual fact independent of k, so that this is the actual support pair for "f"

ie.Pr{fly(obj)} ε [0.45, 0.55]

6.3 Complete Model for Penguin Example

Consider the program
bird(X) : [0.7, 0.7].
fly(X) :- bird(X) : [0.9, 0.9].
fly(X) :- bird(X), penguin(X) : [0, 0][0.95, 0.95][_ _][0.1, 0.1].
bird(X) :- penguin(X) : [1, 1].
fly(X) :- penguin(X) : [0, 0)].
penguin(obj) : [0.4, 0.4].
bird(obj) : [0.9, 0.9)].

This program says that
The proportion of birds in the relevant population of objects is 70%. 90% of the birds can fly. No object which is a bird and penguin can fly, 95% of birds which are not penguins can fly, 10% of objects which are not birds can fly. All penguins are birds. No penguin can fly. The 4 support pairs associated with the third rule correspond to Pr{fly(X) | bird(X), penguin(X)}, Pr{fly(X) | bird(X), ¬ penguin(X)}
Pr{fly(X) | ¬ bird(X), penguin(X)}, Pr{fly(X) | ¬ bird(X), ¬ penguin(X)} respectively.

It also gives specific information about the object obj, namely that there is a probability of 0.9 that obj is a bird and a probability of 0.4 that obj is a penguin.

This information allows the following unique distribution over the relevant labels to be constructed:

Apriori

$y1 = 0.27$	$\neg b \neg p \neg f$
$y2 = 0.03$	$\neg b \neg p f$
$y3 = 0.0332$	$b \neg p \neg f$
$y4 = 0.63$	$b \neg p f$
$y5 = 0.0368$	$b p \neg f$

using

$y4 / (y3 + y4) = 0.95$; $y2 / (y1 + y2) = 0.1$; $y4 / (y3 + y4 + y5) = 0.9$

$y3 + y4 + y5 = 0.7$; $y1 + y2 + y3 + y4 + y5 = 1$

The iterative assignment update then gives (fly Mary) : (0.485 0.485).

Intuitive solution

In this problem we are presented with two pieces if information:

1. Object Mary came from a population with statistics

$\neg b \neg p \neg f$	0.27
$\neg b \neg p f$	0.03
$b \neg p \neg f$	0.0332
$b \neg p f$	0.63
$b p \neg f$	0.0368

so that

IF object obj has properties bp then Pr(obj can fly) = 0

IF object obj has properties $b \neg p$ then Pr(obj can fly) = 0.63 / 0.6632 = 0.95

IF object obj has properties $\neg b \neg p$ then Pr(obj can fly) = 0.03 / 0.3 = 0.1

2. Object properties

2(a) $b : 0.9$; $\neg b : 0.1$

2(b) $p : 0.4$; $\neg p : 0.6$

2(c) obj cannot be a penguin and not a bird

Combining 2(a) and 2(b) taking account of 2(c) by allowing only the set of labels $\{\neg b \neg p, b \neg p, bp\}$

using the general assignment method gives

	p :0.4	$\neg p$: 0.6
0.9 b	bp 0.4	$b \neg p$ 0.5
0.1 $\neg b$	$\neg b p$ (not allowed) 0	$\neg b \neg p$ 0.1

giving
bp : 0.4 ; b¬p : 0.5 ; ¬b¬p : 0.1

Expected value of Pr(obj can fly) = 0.5*0.95 + 0.1*0.1 = 0.485
the value given by the iterative assignment method.

We can write
P'r(f) = Pr(f | bp)P'r(bp) + Pr(f | b¬p)P'r(b¬p) + Pr(f | ¬b¬p)P'r(f | ¬b¬p)
where Pr(.) signifies a probability determined from the population statistics, information 1, and P'r(.) signifies a probability determined from the specific information, information 2 and set of possible labels.

This is a form of Jeffrey's rule.

7. SEMANTIC UNIFICATION

7.1 NESTED SETS, POSSIBILITY DISTRIBUTIONS AND MASS ASSIGNMENTS

Let X = {x1, x2, ... , xn}
Let A1, A2, ... , An be nested subsets of X such that

A1 ⊂ A2 ... ⊂ An where Ai = {x1, ... , xi}

Let m be a mass assignment over these nested sets

$$m(Ai) \geq 0 \text{, all } i$$

$$\sum_{i} m(Ai) = 1$$

Let the necessary support and possible support measures for this special case of nested sets be called necessary and possibility measures denoted by N(.) and P(.) respectively. It is easy to show that
P(A ∪ B) = MAX{P(A), P(B)}
N(A ∩ B) + MIN{N(A), N(B)}
for all A, B ε *P*(X)
[ZADEH 1978], [KLIR, FOLGER 1988].

Let p_f be a function

$p_f : X \rightarrow [0, 1]$

called the possibility distribution of f over X.

Let ρi = p_f(xi) for all xi ε X and ordered such that

$$\rho_1 \geq \rho_2 \geq ... \geq \rho_n$$

We will only consider normalised possibility distributions corresponding to $\rho_1 = 1$.

Define a mass assignment over the nested sets
A1 = {x1}, A2 = {x1, x2}, ... , An = X
as
mi = m(Ai) where mi = $\rho_i - \rho_{i+1}$ with $\rho_{n+1} = 0$
so that

$$P(Ai) = \sum_{Ai \cap Ak \neq \emptyset} m(Ak) \qquad = \underset{xi \, \varepsilon \, Ai}{MAX} \; \rho_i$$

and more specifically

$$P(\{xi\}) = \sum_{k=i}^{n} mk \; = \rho_i$$

Corresponding to a possibility distribution p_f for f over X there is a unique mass assignment m over the subsets {Ai} given by the set of formulae above.

Let f be a normalised fuzzy set
f = x1 / χ_1 + x2 / χ_2 + ... + xn / χ_n
where χ_1=1 and $\chi_1 \geq \chi_2 \geq ... \geq \chi_n$
This induces a possibility distribution p_f over X given by
$p_f(xi) = \rho_i = \chi_i$
with an associated mass assignment over the nested sets
A1 = {x1}, A2 = {x1, x2}, ... , An = {x1, x2, ..., xn}
given by
m(A1) = 1 - χ_1 ; m(A2) = $\chi_2 - \chi_3$; ... ; m(An) = χ_n
This mass assignment represents the family of possible probability distributions over X induced by the fuzzy set f.

We can generalise this to the case of continuous fuzzy sets like those discussed in the introduction but we will not do this in this paper. The continuous case can always be treated by approximating the continuous fuzzy set f with membership function χ_f defined over R by a discrete set of pairs {xi / χ_i} where $\chi_f(xi) = \chi_i$ and the interval R is approximated by the set of points {x1, x2, ..., xn}.

7.2 Examples

We can associate with the fuzzy set defined on {a, b, c, d, e}
f1 = a / 0.2 + b / 0.4 + c / 0.8 + d / 1
the mass assignment
{d} : 0.2 ; {c, d} : 0.4 ; {b, c, d} : 0.2 ; {a, b, c, d} : 0.2

We give an example with repeated membership levels:
Associated with the fuzzy set
f2 = a / 0.1 + b / 0.3 + c / 0.3 + d / 0.7 + e / 1 + f / 1
is the mass assignment
{e, f} : 0.3 ; {d, e, f} : 0.4 ; {b, c, d, e, f} : 0.2 ; {a, b, c, d, e, f} : 0.1

7.3 VOTING MODEL INTERPRETATION

We will use a voting model with constant thresholds to interpret the meaning of a fuzzy set. Consider the fuzzy set "tall" defined on the height space [4ft, 8ft] by means of the membership function χ_{tall}. How can we interpret χ_{tall}(5ft 10")? Consider a representative population sample of persons, S say. We ask each member of S to accept or reject the height 5ft 10" as satisfying the concept "tall". Each member must accept or reject ; there is no allowed abstention. χ_{tall}(5ft 10") is put equal to the proportion of S who accept.

We can therefore interpret the fuzzy set
f1 = a / 0.2 + b / 0.4 + c / 0.8 + d / 1
as
20% of S accept a as f1
40% of S accept b as f1
80% of S accept c as f1
100% of S accept d as f1
100% of S reject e as f1

One possible voting pattern of acceptances is

1	2	3	4	5	6	7	8	9	10
a	a								
b	b	b	b						
c	c	c	c	c	c	c	c		
d	d	d	d	d	d	d	d	d	d

An alternative pattern is

1	2	3	4	5	6	7	8	9	10
a		a							
b	b		b		b				
c		c		c	c	c	c	c	c
d	d	d	d	d	d	d	d	d	d

The first pattern is more reasonable than the second. In the second pattern voter 3 accepts a which has a low membership level but doesn't accept b which has a higher membership level. It seems that anyone who accepts a member with a certain membership level will accept all members with a higher membership level. This we call the **constant threshold assumption.** The first pattern satisfies the constant

threshold assumption. From the first pattern we can deduce
20% of S give acceptance to exactly {d}
40% of S give acceptance to exactly {c, d}
20% of S give acceptance to exactly {b, c, d}
20% of S give acceptance to exactly {a, b, c, d}
and this defines a mass assignment over the nested sets
{d}, {c, d}, {b, c, d}, {a, b, c, d}
namely
{d} : 0.2 ; {c, d} : 0.4 ; {b, c, d} : 0.2 ; {a, b, c, d} : 0.2

We can interpret this mass assignment in the following way. If the population S is told Z has property f1 and a member of the population drawn at random is asked what the value of this property taken from {a, b, c, d, e} for Z is, the answer would be a family of distributions over {a, b, c, d, e} deduced from the mass assignment above.

This interpretation is consistent with the general method given above.

This interpretation is not valid if the fuzzy set is non- normalised since the constant threshold model cannot be satisfied.

7.4 SEMANTIC UNIFICATION

We discussed the need to determine an interval containing the conditional probability Pr{age(mary, middle_aged) | age(mary | about_35) for the example given in the introduction. This we term semantic unification.

Consider the statements
X is f1
a is f2
where f1 and f2 are fuzzy sets defined on the universe of discourse F. Then we are interested in determining Pr{a is f1 | a is f2}.

We can associate the mass assignments (m1, F1) and (m2, F2) with f1 and f2 respectively where F1 and F2 are the focal elements and are nested sets.

For any member s1i of F1 and any member s2j of F2 we can determine the support pair for s1i | s2j from the set {[0, 0], [1, 1], [0, 1]}. Let this be [Sn(s1i | s2j), Sp(s1i | s2j)]

Let m1 = {m1i} and m2 = {m2j}

Therefore the expected value of Pr(a is f1 | a is f2) is contained in the support pair [Sn(a is f1 | a is f2), Sp(a is f1 | a is f2)]
where

$$Sn(a \text{ is } f1 \mid a \text{ is } f2) = \sum_{i,j} m1i.m2j.Sn(s1i \mid s2j)$$

$$Sn(a \text{ is } f1 \mid a \text{ is } f2) = \sum_{i,j} m1i.m2j.Sp(s1i \mid s2j)$$

7.5 EXAMPLE

Consider the fuzzy sets defined on {a, b, c, d, e}
f1 = a / 0.2 + b / 0.4 + c / 0.8 + d / 1
f2 = a /1 + b / 0.8 + c / 0.1

The corresponding associated mass assignments are
{d} : 0.2 ; {c, d} : 0.4 ; {b, c, d} : 0.2 ; {a, b, c, d} : 0.2
and
{a} : 0.2 ; {a, b} : 0.7 ; {a, b, c} : 0.1
respectively.

Therefore
S({d} | {a}) = [0, 0]
S({c, d} | {a}) = [0, 0]
S({b, c, d} | {a}) = [0, 0]
S({a, b, c, d} | {a}) = [1, 1]
and
S({d} | {a, b}) = [0, 0]
S({c, d} | {a, b}) = [0, 0]
S({b, c, d} | {a, b}) = [0, 1]
S({a, b, c, d} | {a, b}) = [1, 1]
and
S({d} | {a, b, c}) = [0, 0]
S({c, d} | {a, b, c}) = [0, 1]
S({b, c, d} | {a, b, c}) = [0, 1]
S({a, b, c, d} | {a, b, c}) = [1, 1]
so that
Sn(a is f1 | a is f2) = 0.2*0.2 + 0.2*0.7 + 0.2*0.1 = 0.2
Sp(a is f1 | a is f2) = 0.2*0.2 + 0.2*0.7 + 0.2*0.7 + 0.4*0.1 + 0.2*0.1 + 0.2*0.1 = 0.4

so that
The support pair for the unification of f1 given f2 is
a is f1 | a is f2 : [0.2, 0.4]

7.6 A SPECIAL CASE

Let F = {e1, e2, ... , e10}

and
f = e1/0.1 + e2/0.2 + e3/0.3 + e4/0.4 + e5/0.5
+ e6/0.6 + e7/0.7 + e8/0.8 + e9/0.9 + e10/1.0
then
f1 | f1 : [0.55, 1]

Let F = {e1, e2, ... , e10}
and

f = e1/0 + e2/0.1 + e3/0.2 + e4/0.3 + e5/0.4
+ e6/0.5 + e7/0.6 + e8/0.7 + e9/0.8 + e10/0.9
then
f | f : [0.45, 1]

These are two approximations for determining f|f where f is a ramp fuzzy set on an interval R. In the limit as more and more points in R are used we obtain
f | f : [0.5, 1].

This illustrates how we can deal with continuous fuzzy sets.

7.7 Iterative Assignment Method with Semantic Unification

Consider the program discussed in the introduction

married(X) :- age(X, middle_aged), has_children(X) : [0.7, 0.9].
age(mary, about_35)
has_children(mary) : [0.8, 1].

We can ask the query
?- married(mary)

To answer this query we determine the support pair [x, y] below by the method in the last section applied to
middle_aged | about_35

age(mary, middle_aged) :- age(mary, about_35) : [x, y].

We then solve

married(X) :- age(X, middle_aged), has_children(X) : [0.7, 0.9].
age(mary, middle_aged) : [x, y].
has_children(mary) : [0.8, 1].

using the iterative assignment method as described previously.

8. CONCLUSIONS

This paper provides a general approach to evidential reasoning when the knowledge representation is in the form of rules and fact with both probabilistic and fuzzy uncertainties included. The methods provided can be used for other forms of knowledge representation, for example, Bayesian networks [PEARL 1988], Moral Graphs [LAURITZEN, SPIEGELHALTER 1988] with extensions to the case of uncertain specific information and valuation-based languages for expert systems [SHENOY 1989]. The non-monotonic case is not seen to be a problem. Without decomposition methods the approach given here could easily become computationally excessive. Decomposition methods have only been lightly touched on in this paper although expressing knowledge in the form of rules provides a natural decomposition. Inference diagrams mcan be used to construct a decomposition from a group of rules. For special cases the decomposition allows the calculus of support logic programming used in FRIL to be used for answering queries. The methods given here extends FRIL to cases which cannot be treated by the present version. The next version will take account of these extensions.

9. REFERENCES

Baldwin J.F, (1986), "Support Logic Programming", in: A.I.Jones et al,. Eds., Fuzzy Sets Theory and Applications, (Reidel, Dordrecht-Boston).

Baldwin J.F, (1987), "Evidential Support Logic Programming", Fuzzy Sets and Systems, 24, pp 1-26.

Baldwin J.F. et al, (1987), "FRIL Manual", Fril Systems Ltd, St Anne's House, St Anne's Rd, Bristol BS4 4A, UK

Baldwin J.F., (1990a), "Computational Models of Uncertainty Reasoning in Expert Systems", Computers Math. Applic., Vol. 19, No 11, pp 105-119.

Baldwin J.F., (1990b), "Combining Evidences for Evidential Reasoning", Int. J. of Intelligent Systems, To Appear.

Baldwin J.F., (1990c), "Towards a general theory of intelligent reasoning", 3rd Int. Conf IPMU, Paris, July 1990

Dubois D., Prade H., (1986), "On the unicity of Dempster Rule of Combination", Int. J. of Intelligent Systems, 1, no. 2, pp 133-142

Jeffrey R., (1965) "The Logic of Decision", McGraw-Hill, New York

Klir G.J., Folger T.A., (1988), Fuzzy Sets, Uncertainty, and Information, Prentice-

Hall

Lauritzen S.L, Spiegelhalter D.J., (1988), "Local computations with probabilities on graphical structures and their application to expert systems", J. Roy. Stat. Soc. Ser. B 50(2), 157-224

Pearl J., (1988), "Probabilistic reasoning in Intelligent Systems", Morgan Kaufmann Pub. Co.

Shafer G., (1976), "A mathematical theory of evidence" , Princeton Univ. Press

Shenoy P.P., (1989), "A Valuation - Based Language for Expert Systems, Int. J. of Approx. Reasoning, V01 3, No. 5.

Zadeh L, (1965), "Fuzzy sets", Information and Control, **8**, pp 338-353.

Zadeh L, (1978), "Fuzzy Sets as a basis for a theory of Possibility", Fuzy Sets and Systems 1, 3-28

16

PROBABILISTIC SETS
PROBABILISTIC EXTENSION OF FUZZY SETS

Kaoru Hirota
Dept. of Instrument & Control Engineering College of Engineering, Hosei University 3-7-2 Kajino-cho, Koganei-city, Tokyo 184, Japan

Introduction

In the field of pattern recognition or decision making theory, the following complicated problems have been left unsolved: (1)ambiguity of objects, (2)variety of character, (3)subjectivity of observers, (4)evolution of knowledge or learning. With regard to each problem, however, there are several general theories: many-valued logic, fuzzy set theory (in connection with (1)and(2)), modal logic (in conjunction with (2)and(4)), and subjective probability (in relation to (3)). It seems, however, that there are few carefully thought-out investigations by paying attention to all problems mentioned above. In this paper we would like to give our opinion about these problems and to introduce a new concept called 'probabilistic sets'.

By giving examples in comparison with fuzzy set theory, the background idea of probabilistic sets is explained in Section 2. In probabilistic sets, it is essential to regard the value of membership functions of fuzzy sets as a random variable. A probabilistic set A on a total space X is defined by a defining function $\mu_A(x,\omega)$, which is a point (i.e. $x \in X$)-wise (B, B_c)-measurable function from a parameter space (Ω, B, P) to a characteristic space (Ω_c, B_c). The parameter space (Ω, B, P) is a probability space and is closely related with subjectivity, personality, and evolution of knowledge. The characteristic space (Ω_c, B_c) is a measurable space usually adopt ([0,1],Borel sets) as (Ω_c, B_c) . Section 3 describes definitions of probabilistic sets from a measure-theoretical viewpoint. The concept of probabilistic sets includes the concept of classical fuzzy sets. Some other properties are important results

is that the family of all probabilistic sets constitutes a complete pseudo-Boolean algebra. In Section 5, some new concepts are shown such as moment analysis and expected cardinal numbers. The possibility of moment analysis is an essential feature of probabilistic sets and it is a great advantage in applications.

The background idea of probabilistic sets

Digital computers have been widely used in the field of pattern recognition, decision making theory, artificial intelligence and so on. It should be noted, however that they involved the following complicated problems to be solved:

(1) ambiguity of objects,
(2) variety of property or character,
(3) subjectivity of observers,
(4) evolution of knowledge or learning.

In order to take up these problems, several general studies have been made such as fuzzy set theory, many-valued logic, modal logic, quantum logic, subjective probability. In particular, fuzzy set theory has been widely studied. (More than one thousand papers have been published since L.A. Zadeh presented fuzzy set theory [14].) We have also been studying these problems especially by paying attention to inherent and special characteristics of pattern recognition and decision making theory. Giving an example, we shall deal with the background of our idea 'probabilistic sets' and shall compare it with fuzzy sets.

Let all real numbers be a total space X . Consider all numbers nearly equal to one and all numbers nearly equal to minus one. In fuzzy set theory, their membership functions are shown as in Fig.1.

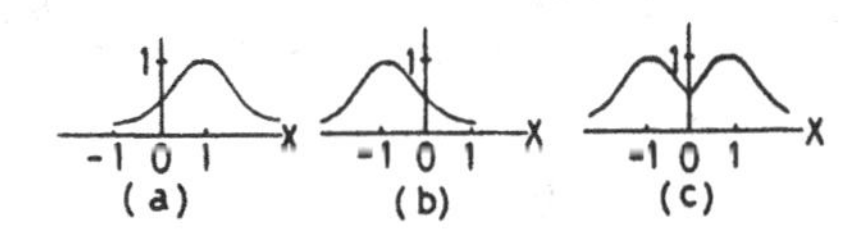

Fig. 1. Fuzzy sets; (a) numbers near one, (b) numbers near minus one, (c) the union (numbers near one or minus one).

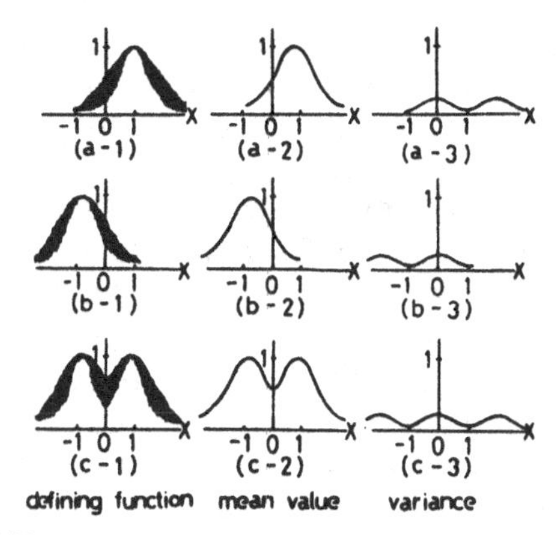

Fig. 2. Probabilistic sets; (a) numbers near one, (b) numbers near minus one, (c) the union (numbers near one or minus one).

In this situation, however, the following discussion may be possible: If the degree of ambiguity were accurately given, it would no longer be ambiguous. Although mean value or variance may be determined and a rough tendency nay be given, it is impossible in general to assign definite [0,1]-values. To make the matters worse, the tendency varies according to observers' subjectivity, situations and so on. Hence we shall introduce a probability space (Ω, B, P) , called a parameter space, whose element represents a standard of judgments. It is assumed that if a standard $\omega(\in\Omega)$ is fixed, the degree of ambiguity of considered objects (i.e. elements of the total space X) can be definitely determined. A set of all degrees of ambiguity will be called a characteristic space (Ω_c, B_c) . We usually adopt ([0,1],Borel sets) as the characteristic space, because it is an infinite totally ordered set with a maximum element 1 and a minimum element 0 and because it is in harmony with characteristic functions of ordinary sets and membership functions of fuzzy sets. A probabilistic set on a total space X is defined by giving a (Ω_c, B_c)-valued random variable on (Ω, B, P) for each object $x(\in X)$, and this correspondence will be called a defining function of the probabilistic set. The corresponding probabilistic sets of Fig.1 are shown in Fig.2(a-1),(b-2),(c-1). The parameter space (Ω, B, P) is expected to be adopted suitably according to each situation, hence in general no restrictions are added to the parameter space except it is a probability space. For example, in the case of Fig.2, the parameter space might exist in observers' subconscious and might be changed according to estimate it by a statistical method. One of the most important facts in probabilistic set theory is a possibility of moment analysis by using a probabilistic measure P of the parameter space. For instance, in Fig.2, mean values and variances are shown the parameter space. For instance, in Fig.2, mean values and variances are shown in (a-2),(b-2),(c-2),and (a-3),(b-3),(c-3), respectively. The mean value indicates the first approximation of probabilistic sets and might be considered to be the same one as a membership function of 'union' generally has a continuous but non-smooth (i.e. non-differentiable) point as shown in Fig.1(c) (at a point of $x=0$). It will be natural to expect a smooth curve like Fig.2(c-2) as the first approximation of 'numbers nearly equal to one or minus one'. The variance provides the second information and it indicates a disordered degree of judgments. Higher moments can be considered in the same way. Moreover, it can be shown theoretically that the nth moment around mean value tends to zero as n tends to infinity (cf. Proposition 9). Hence, from a practical viewpoint, it is sufficient to consider only the lower moments, i.e. mean value and variance. If we consider a probabilistic set with variance zero, it could be identified with a fuzzy set. In this sense, it can be concluded that the concepts of probabilistic sets include classical fuzzy concepts.

To make sure our option, we shall give several comments. A distinction between the total space X and the parameter space (Ω, B, P) is very important in probabilistic set theory. The concept of probabilistic sets differs intrinsically from Zadeh's way of thinking [15] on this point.

A notion of fuzzy set of type 2(Mizunoto and Tanaka [12]) is introduced in order to resolve the difficulty of settling a definite ambiguous degree. Fuzzy set of type n is also characterized by n step recursively defined ambiguity. However, the number of steps (i.e. n) has no upper bound and, to make matters worse, realistic meanings decline as n increases. In probabilistic sets, the ambiguity is arranged on the parameter space and realistic meanings are made clear in connection with the subjectivity of observers.

A family of probabilistic sets constitutes a complete pseudo-Boolean algebra(cf. Theorem 1). A pseudo-Boolean algebra is a subclass of distributive lattices(Fig.3).

Hence, from a lattice theoretical viewpoint (cf.[1]), probabilistic set theory takes its position between L-fuzzy set theory [3] and Boolean algebra valued set theory [13].

In probabilistic set theory, the parameter space (Ω, B, P) plays an important role, but it has no restriction except it is a probabilistic measure space. The most important task in applications is a choice of a suitable parameter space, especially an establishment of probabilistic measure P. Finally, we would like to add that there is no need to recollect a probabilistic randomness like casting a dice in spite of a diction 'probabilistic' sets.

Definitions of probabilistic sets

The above discussion is informal from a mathematical point of view. The strict definitions are shown in this section. The mathematical foundation of this theory is measure theory and some well-known facts in measure theory will be used(cf.[5]).

First, we would like to define the following three terms. (The meanings were discussed in a previous section.)

Definition 1.

(Ω, B, P) is a parameter space, (Ω_c, B_c)=([0,1],Borel sets) is a characteristic space, $M=\{\mu \mid \mu: \Omega \rightarrow \Omega_c (B, B_c)$ - measurable function} is a family of characteristic variables.

It is easily shown that M satisfies the following properties.

Proposition 1.

For arbitrary μ_i's ($\mu_i \in M, i=1,2,\ldots$ at most countably infinite), the following properties are satisfied.

$$\min(\mu_1, \mu_2) \in M, \tag{1}$$

$$\max(\mu_1, \mu_2) \in M, \tag{2}$$

$$\mu = c \in M \quad \textit{where } c \in \Omega_c = [0, 1] \; (\mu: \textit{constant fn.}), \tag{3}$$

$$|\mu_1 - \mu_2| \in M, \tag{4}$$

$$\lambda\mu_1 + (1-\lambda)\mu_2 \in M \qquad \text{where } 0 \leqslant \lambda \leqslant 1, \tag{5}$$

$$\mu_1^{\alpha} \in M \qquad \text{where } \alpha \geqslant 0, \tag{6}$$

$$\mu_1\mu_2 \in M, \tag{7}$$

$$\inf_{i \geqslant 1} \mu_i \in M, \tag{8}$$

$$\sup_{i \geqslant 1} \mu_i \in M, \tag{9}$$

$$\underline{\lim}_{i \to \infty} \mu_i = \sup_{i \geqslant 1} \inf_{j \geqslant i} \mu_j \in M, \tag{10}$$

$$\overline{\lim}_{i \to \infty} \mu_i = \inf_{i \geqslant 1} \sup_{j \geqslant i} \mu_j \in M. \tag{11}$$

The fundamental definition of probabilistic sets will be given as follows. Here a total space $X=\{x\}$, which represents a set of all objects discussed in each situation, is arbitrarily fixed.

Definition 2.
A probabilistic set A on X is defined by a defining function μ_A

$$\begin{array}{c} \mu_A : X \times \Omega \rightarrow \Omega_{c'} \\ \quad\ \ \Psi \qquad\quad \Psi \\ (x, \omega) \mapsto \mu_A(x, \omega) \end{array} \tag{12}$$

where $\mu_A(x, \cdot)$ is the (B, B_c)-measurable function for each fixed $x (\in X)$.

For any two probabilistic sets A and B, whose defining function are $\mu_A(x, \omega)$ and $\mu_B(x, \omega)$, respectively, A is said to be included in $B (A \subset B)$ if for each $x (\in X)$ there exists $E (\in B)$ which satisfies

$$P(E) = 1, \tag{13}$$

$$\mu_A(x, \omega) \leqslant \mu_B(x, \omega) \qquad \text{for all } \omega \in E.. \tag{14}$$

In this situation we will sometimes use a brief notation as follows,

$$\mu_A(x,\omega) \leq \mu_B(x,\omega) \qquad \text{for all } x \in X \text{ and a.e. } \omega \in \Omega. \tag{15}$$

If both $A \subset B$ and $B \subset A$ are satisfied, A and B are said to be equivalent ($A \equiv B$). (Indeed this relation $\equiv$ satisfies an equivalence relation; i.e. reflexivity, symmetricity, and transitivity.) All equivalent probabilistic sets are considered to be the same one and are not distinguished. All probabilistic sets on X is said to be a family of probabilistic sets and is denoted by $\mathscr{P}(X)$.

Note.
An element of $\mathscr{P}(X)$ represents an equivalence class of M by the equivalence relation $\equiv$ for each $x (\in X)$.
The inclusion relation in $\mathscr{P}(X)$ satisfies reflexivity, anti-symmetricity, and transitivity, hence $(\mathscr{P}(X), \subset)$. constitutes a poset (partially ordered set).

In the following, several operations in $\mathscr{P}(X)$ will be defined. A fundamental operation in $\mathscr{P}(X)$ is 'union', however, it is a little complicated. Let A_γ ($\gamma \in \Gamma, \Gamma$:possibly infinite) be probabilistic sets on X whose defining functions are $\mu_{A_\gamma}(x,\omega)$ respectively. The union of $\{A_\gamma\}_{\gamma\in\Gamma}$, which is denoted by $\bigcup A_\gamma$, is defined by a defining function $\mu_{\bigcup A_\gamma}(x,\omega)$ which will be given by the following procedure. For the time being, consider a case where each $x (\in X)$ is arbitrarily fixed. Then $\mu_{A_\gamma}(x,\cdot)$ may be regarded as a function of $\omega \in \Omega$ (i.e. an element of M). Since $\mu_{A_\gamma}(x,\cdot)$ is a $\Omega_c = [0,1]$-valued measurable function, and since the total measure is finite (i.e. $P(\Omega) = 1$), $\mu_{A_\gamma}(x,\cdot)$ is always P -integrable,

$$0 \leq \int_\Omega \mu_{A_\gamma}(x,\omega) \cdot \mathrm{d}P(\omega) \leq 1. \tag{16}$$

For arbitrarily fixed n indices $\gamma_1, \gamma_2, \ldots, \gamma_n (\in \Gamma)$, a function max $\{\mu_{A_{\gamma_i}}(x,\cdot) \mid 1 \leq i \leq n\}$ is also an element of M (see Proposition 1(2)). Hence it is also P -integrable,

$$0 \leq \int_\Omega \max\{\mu_{A_{\gamma_i}}(x,\omega) \mid 1 \leq i \leq n\} \cdot \mathrm{d}P(\omega) \leq 1. \tag{17}$$

The selection of $\gamma_1, \gamma_2, \ldots, \gamma_n$ from Γ is varied. The least upper bound, denoted by $a(x)$, can be calculated,

$$a(x)=\sup\left\{\int_{\Omega} \max\{\mu_{A_{\gamma_i}}(x,\omega)\mid 1\leq i\leq n\}\mathrm{d}P(\omega)\mid n\in N \text{ (natural numbers)}, \gamma_i\in\Gamma\right\}, \tag{18}$$

$$0\leq a(x)\leq 1. \tag{19}$$

Since $a(x)$ is a 'least upper bound', there exists a countably infinite subsequence

$$\{\max\{\mu_{A_{\gamma_i}}(x,\omega)\mid 1\leq i\leq n_j\}\mid n_j\in N,\ \gamma_i\in\Gamma\}_{j=1}^{\infty}$$

such that

$$\lim_{j\to\infty}\int_{\Omega} \max\{\mu_{A_{\gamma_i}}(x,\omega)\mid 1\leq i\leq n_j\}\cdot \mathrm{d}P(\omega)=a(x). \tag{20}$$

Although an element $x(\in X)$ was arbitrarily fixed, this procedure could be done for each $x(\in X)$. We shall define the defining function $\mu_{\bigcup A_\gamma}(x,\omega)$ by

$$\mu_{\bigcup A_\gamma}(x,\omega)=\sup\{\max\{\mu_{A_{\gamma_i}}(x,\omega)\mid 1\leq i\leq n_j\}\mid 1\leq j<\infty\} \tag{21}$$

The justification of this definition will be ensured by the following Proposition 2.

Proposition 2.
(1)The union $\bigcup A_\gamma$ is determined uniquely by (21), i.e. if there exists another countably infinite subsequence which satisfies (20), the result given by the same equation as (21) also belongs to the same equivalence class of M (for each $x\in X$) in the sense of Definition 2.
(2) For all $\gamma\in\Gamma$, we have $A_\gamma\subset\bigcup A_\gamma$.
(3) If there exists an A which satisfies $A_\gamma\subset A$ for all $\gamma\in\Gamma$, then we have $\bigcup A_\gamma\subset A$.

The proof is omitted here, since it requires some results in measure theory and a rather long description (cf.[7]).
Although the above stated procedure of union is rather complicated, it can be simplified in a case that the index set Γ is at most countably infinitc. For oxample, the union of A and B (whose defining functions are $\mu_A(x,\omega)$ and $\mu_B(x,\omega)$, respectively) may be defined by

$$\mu_{A\cup B}(x,\omega)=\max\{\mu_A(x,\omega),\mu_B(x,\omega)\} \tag{22}$$

for each $x \in X$ and each $\omega \in \Omega$, and the union of $\{A_n\}_{n=1}^{\infty}$ may be defined by

$$\mu_{\bigcup A_n}(x, \omega) = \sup\{\mu_{A_n}(x, \omega) \mid 1 \leq n < \infty\} \tag{23}$$

for each $x \in X$ and each $\omega \in \Omega$. The complexity in a general case arises from the fact that M is not always closed by more than countably infinite operations (see Proposition 1).

The 'intersection' of $\{A_\gamma\}_{\gamma \in \Gamma}$, which is denoted by $\bigcap A_\gamma$, is a dual concept of 'union' $\bigcup A_\gamma$, and it is defined as follows. Put

$$b(x) = \inf\left\{\int_\Omega \min\{\mu_{A_{\gamma_i}}(x, \omega) \mid 1 \leq i \leq n\} \, \mathrm{d}P(w) \mid n \in N, \ \gamma_i \in \Gamma\right\}, \tag{24}$$

$$0 \leq b(x) \leq 1, \tag{25}$$

and choose a countably infinite subsequence

$$\{\min\{\mu_{A_{\gamma_i}}(x, \omega) \mid 1 \leq i \leq n_j\} \mid n_j \in N, \ \gamma_i \in \Gamma\}_{j=1}^{\infty} \tag{26}$$

such that

$$\lim_{j \to \infty} \int_\Omega \min\{\mu_{A_{\gamma_i}}(x, \omega) \mid 1 \leq i \leq n_j\} \, \mathrm{d}P(\omega) = b(x), \tag{27}$$

and define

$$\mu_{\bigcap A_\gamma}(x, \omega) = \inf\{\min\{\mu_{A_{\gamma_i}}(x, \omega) \mid 1 \leq i \leq n_j\} \mid 1 \leq j < \infty\}. \tag{28}$$

The justification of this definition will also be ensured by the same proposition as Proposition 2. (Change symbols $\bigcup$, $\subset$ to $\bigcap$, $\supset$ respectively in Proposition 2.)

Some other useful concepts or operations on $\mathcal{P}(X)$ could be defined. They will be summarized as follows. The justification of these definitions is also ensured by Proposition 1.

Definition 3.

Total set X

$$\mu_X(x, \omega) = 1 \qquad \text{for all } x \in X \text{ and a.e. } \omega \in \Omega. \tag{29}$$

(This notation will be omitted until (36).)

Void set (or null set) ϕ

$$\mu_{\phi}(x, \omega) = 0. \tag{30}$$

Complement of A A^c

$$\mu_{A^c}(x, \omega) = 1 - \mu_A(x, \omega). \tag{31}$$

Difference $A - B$

$$\mu_{A-B}(x, \omega) = \max\{0, \mu_A(x, \omega) - \mu_B(x, \omega)\}. \tag{32}$$

Symmetric difference $A \Delta B$

$$\mu_{A \Delta B}(x, \omega) = |\mu_A(x, \omega) - \mu_B(x, \omega)|. \tag{33}$$

Algebraic sum $A \oplus B$

$$\mu_{A \oplus B}(x, \omega) = \mu_A(x, \omega) + \mu_B(x, \omega) - \mu_A(x, \omega)\mu_B(x, \omega). \tag{34}$$

λ sum $A \overset{+}{_\lambda} B$ (where $0 \leqslant \lambda \leqslant 1$)

$$\mu_{A \overset{+}{_\lambda} B}(x, \omega) = \lambda \mu_A(x, \omega) + (1 - \lambda)\mu_B(x, \omega). \tag{35}$$

α power A^α (where $\alpha \geqslant 0$)

$$\mu_{A^\alpha}(x, \omega) = \mu_A(x, \omega)^\alpha. \tag{36}$$

Superior limit of $\{A_n\}_{n=1}^{\infty}$

$$\overline{\lim_{n \to \infty}} A_n = \bigcap_{n=1}^{\infty} \bigcup_{k=n}^{\infty} A_k. \tag{37}$$

Inferior limit of $\{A_n\}_{n=1}^{\infty}$

$$\underline{\lim_{n \to \infty}} A_n = \bigcup_{n=1}^{\infty} \bigcap_{k=n}^{\infty} A_k. \tag{38}$$

An ordered pair $(\mu_A(x, \omega), \mu_B(x, \omega))$ is said to be a direct product of A and B, and is denoted by $A \times B$.

A_y is said to be an one point probabilistic set at $y \in X$, if its defining function $\mu_{A_y}(x, \omega)$ satisfies

$$\int_\Omega \mu_{A_y}(x, \omega) \cdot \mathrm{d}P(\omega) \begin{cases} = 0 & x \neq y, \\ > 0 & x = y. \end{cases} \tag{39}$$

A_y is said to be a full one point probabilistic set at $y \in X$, if its defining function $\mu_{A_y}(x,\omega)$ satisfies

$$\int_\Omega \mu_{A_y}(x,\omega)\cdot \mathrm{d}P(\omega) = \begin{cases} 0 & x \neq y, \\ 1 & x = y. \end{cases} \tag{40}$$

Some properties of probabilistic sets

Some properties of probabilistic sets can be characterized from a lattice theoretical viewpoint (cf.[17]).

A family of probabilistic sets $(\mathscr{P}(X), \subset)$ constitutes a poset(see the note of Definition 2). For arbitrary $A, B (\in \mathscr{P}(X))$, there exist a supremum $A \cup B$ and an infimum $A \cap B$ with respect to this partial order $\subset$ (see Proposition 2 (2)and(3)). Hence the poset $(\mathscr{P}(X), \subset)$ forms a lattice and the following proposition holds. (Note that a set of the following properties is a necessary and sufficient condition of being a lattice.)

Proposition 3.

For arbitrary $A, B, C (\in \mathscr{P}(X))$, we have commutativity

$$A \cup B = B \cup A, \tag{41}$$

$$A \cap B = B \cap A, \tag{42}$$

associativity

$$(A \cup B) \cup C = A \cup (B \cup C), \tag{43}$$

$$(A \cap B) \cap C = A \cap (B \cap C), \tag{44}$$

absorption law

$$A \cup (A \cap B) = A, \tag{45}$$

$$A \cap (A \cup B) = A. \tag{46}$$

It is also possible to show that there exist pseudo complements in $\mathscr{P}(X)$. Let A and B be arbitrarily fixed two probabilistic sets whose defining functions are $\mu_A(x,\omega)$ and $\mu_B(x,\omega)$, respectively. Consider the following equation,

$$\mu_{A'}(x,\omega) = \begin{cases} 1 & \text{if } \mu_A(x,\omega) \leqslant \mu_B(x,\omega), \\ \mu_B(x,\omega) & \text{if } \mu_A(x,\omega) > \mu_B(x,\omega). \end{cases} \tag{47}$$

For each $x(\in X)$, $\mu_{A'}(x,\cdot)$ is a (B, B_c)-measurable function and is an element of M, since a set $\{\omega \mid \mu_B(x,\omega)-\mu_A(x,\omega)\geqslant 0\}$ belongs to B (i.e. this set is measurable). Hence it is possible to define a probabilistic set A' by (47). It is also clear that A' is the largest probabilistic set of those C's which satisfy $A \cap C \subset B (C \in \mathscr{P}(X))$. In this sense, A' is said to be a pseudo complement of A relative to B. Hence $(\mathscr{P}(X), \subset)$ constitutes a pseudo-Boolean algebra (cf. Fig.3). (A pseudo-Boolean algebra is a relative complemented lattice with a minimum element. In this case the minimum element is ϕ.)

Moreover, for arbitrary $\{A_\gamma\}_{\gamma\in\Gamma}(\subset \mathscr{P}(X))$ (Γ: possibly infinite), the existence of $\bigcup A_\gamma$ and $\bigcap A_\gamma$ was shown in a previous section and they played a role of a supremum and an infimum with respect to the order $\subset$. Hence it is proved that the lattice $(\mathscr{P}(X), \subset)$ is complete, and so we can conclude the following Theorem 1 from a lattice theoretical viewpoint.

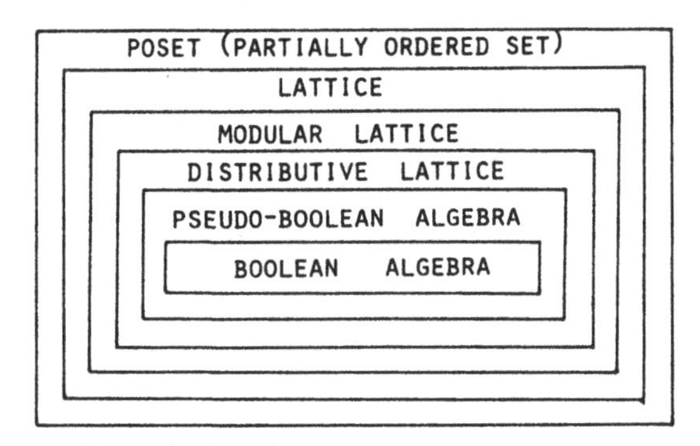

Fig. 3. An inclusion diagram of various lattices.

Theorem 1.

A family of probabilistic sets $(\mathscr{P}(X), \subset)$ constitutes a complete pseudo-Boolean algebra.

Note.

In ordinary set theory, a family of all subsets constitutes a complete Boolean algebra. The difference between the two is the lack of a complemented law (i.e. $A \cup A^c \neq X, A \cap A^c \neq \phi$). In probabilistic set theory, it is essential to consider ambiguous states, so we can not get any definite information if we know that the considered object is not in one state. In ordinary set theory, however, we information that it is not in one state. (Note that ordinary set theory can be considered to be a two-valued logic.) Hence the lack of complemented law is unavoidable in probabilistic set theory.

Since the notion of pseudo-Boolean algebra is included in that of distributive lattice (see Fig.3), a distributive law holds in $(\mathscr{P}(X), \subset)$. Moreover, in connection with its completeness, we can generalize commutative law, associative law, distributive law, and de-Morgan's law as follows. (Proofs are omitted here.)

Proposition 4.

For arbitrary subfamilies of probabilistic sets $\{A_\gamma\}_{\gamma\in\Gamma}$ and $\{B_\lambda\}_{\lambda\in\Lambda}$,
we have
generalized associative law

$$\left(\bigcup_{\gamma\in\Gamma} A_\gamma\right) \cup \left(\bigcup_{\lambda\in\Lambda} B_\lambda\right) = \bigcup_{\gamma,\lambda} (A_\gamma \cup B_\lambda), \tag{48}$$

$$\left(\bigcap_{\gamma\in\Gamma} A_\gamma\right) \cap \left(\bigcap_{\lambda\in\Lambda} B_\lambda\right) = \bigcap_{\gamma,\lambda} (A_\gamma \cap B_\lambda), \tag{49}$$

generalized distributive law

$$\left(\bigcup_{\gamma\in\Gamma} A_\gamma\right) \cap \left(\bigcup_{\lambda\in\Lambda} B_\lambda\right) = \bigcup_{\gamma,\lambda} (A_\gamma \cap B_\lambda), \tag{50}$$

$$\left(\bigcap_{\gamma\in\Gamma} A_\gamma\right) \cup \left(\bigcap_{\lambda\in\Lambda} B_\lambda\right) = \bigcap_{\gamma,\lambda} (A_\gamma \cup B_\lambda), \tag{51}$$

generalized de-Morgan's law

$$\left(\bigcup_{\gamma\in\Gamma} A_\gamma\right)^c = \bigcap_{\gamma\in\Gamma} A_\gamma^c, \tag{52}$$

$$\left(\bigcap_{\gamma\in\Gamma} A_\gamma\right)^c = \bigcup_{\gamma\in\Gamma} A_\gamma^c. \tag{53}$$

Some other important properties in $\mathscr{P}(X)$ will be mentioned in the following without proofs.

proposition 5.

For arbitrary $A, B, C(\in\mathscr{P}(X))$, we have
idempotent law

$$A \cup A = A, \tag{54}$$

$$A \cap A = A, \tag{55}$$

involution law

$$A^{cc} = A, \tag{56}$$

elimination law

$$\left.\begin{array}{l} A \cup B = A \cup C \\ A \cap B = A \cap C \end{array}\right\} \rightarrow B = C, \tag{57}$$

identity law

$$A \cup X = X, \tag{58}$$

$$A \cap X = A, \tag{59}$$

$$A \cup \phi = A, \tag{60}$$

$$A \cap \phi = \phi. \tag{61}$$

Proposition 6.

For arbitrary $\{A_n\}_{n=1}^{\infty}(\subset \mathscr{P}(X))$, we have

$$\underline{\lim}_{n\to\infty} A_n \subset \overline{\lim}_{n\to\infty} A_n, \tag{62}$$

$$\overline{\lim}_{n\to\infty} A_n^c = \left(\underline{\lim}_{n\to\infty} A_n\right)^c. \tag{63}$$

If $A_1 \subset A_2 \subset \cdots \subset A_n \subset \cdots$, then we have

$$\underline{\lim}_{n\to\infty} A_n = \overline{\lim}_{n\to\infty} A_n = \bigcup_{n=1}^{\infty} A_n. \tag{64}$$

If $A_1 \supset A_2 \supset \cdots \supset A_n \supset \cdots$, then we have

$$\underline{\lim}_{n\to\infty} A_n = \overline{\lim}_{n\to\infty} A_n = \bigcap_{n=1}^{\infty} A_n. \tag{65}$$

If $A_{2n+1} = A$ and $A_{2n} = B$, then we have

$$\underline{\lim}_{n\to\infty} A_n = A \cap B \quad \textit{and} \quad \overline{\lim}_{n\to\infty} A_n = A \cup B. \tag{66}$$

Proposition 7.

Each of $(\mathscr{P}(X), \cup), (\mathscr{P}(X), \cap), (\mathscr{P}(X), \cdot)$, and $(\mathscr{P}(X), \oplus)$ constitutes a commutative monoid (i.e. a commutative semigroup with a unit) and, for arbitrary $A, B, C(\in \mathscr{P}(X))$, we have

$$A \Delta B = (A - B) \cup (B - A), \tag{67}$$
$$A \oplus B = (A^c B^c)^c, \tag{68}$$
$$A \cdot B \subset A \cap B \subset A \underset{\wedge}{+} B \subset A \cup B \subset A \oplus B. \tag{69}$$

Note.
In ordinary set theory, it is possible to define sixteen different kinds of binary operations. (Because the total space X can be divided into four regions for arbitrary subsets A and B, hence there exist $2^4 = 16$ combinations.) Among these sixteen binary operations, symmetric difference $A \Delta B$ has a very good property from an algebraic viewpoint, namely, it constitutes an Abelian group. In probabilistic set theory, however, $(\mathscr{P}(X), \Delta)$ doesn't satisfy such a good property. On the contrary, it doesn't satisfy the associative law.

Proposition 8.
Let $X_\gamma (\gamma \in \Gamma)$ be total spaces (possibly infinite), then we have

$$\bigcup_{\gamma \in \Gamma} \mathscr{P}(X_\gamma) \subset \mathscr{P}\left(\bigcup_{\gamma \in \Gamma} X_\gamma\right), \tag{70}$$

$$\bigcap_{\gamma \in \Gamma} \mathscr{P}(X_\gamma) = \mathscr{P}\left(\bigcap_{\gamma \in \Gamma} X_\gamma\right). \tag{71}$$

Some extended concepts of probabilistic sets

1.Probabilistic mappings
A mapping f from X to Y is usually defined as a correspondence from an element $x (\in X)$ to an element $y (\in Y)$. There also exist some variations such as a set function (a correspondence from a subset $A (\subset X)$ to an element $y (\in Y)$) and a multivalued mapping (a correspondence from an element $x (\in X)$ to a subset $B (\subset Y)$). The concepts of set functions and multivalued mappings play an important role in the fields of measure theory and functional analysis, respectively. In the field of pattern recognition or learning theory, it is essential to consider an ambiguous correspondence (i.e. a probabilistic mapping) which will be defined as follows.

Definition 4.
A probabilistic mapping f from X to Y on a parameter space (Ω_m, B_m, P_m) is defined by

$$f: X \times \Omega_m \to Y, \quad (x, \omega_m) \mapsto f(x, \omega_m). \tag{72}$$

Some extended concepts can be defined in connection with probabilistic mappings, such as induced images and induced inverse images of probabilistic sets by a probabilistic mapping, and some properties are also investigated. However, all of them are omitted here.

2.Moment analysis
The parameter space (Ω, B, P) is a (probabilistic) measure space and plays an essential role in applications of probabilistic set theory. By using the measure P of this parameter space, we can carry out moment analysis. The possibility of a moment analysis is one of the most important features in probabilistic set theory and can not be found in other theories.

Definition 5.
Let A be a probabilistic set on X whose defining function is $\mu_A(x, \omega)$. For each fixed $x(\in X)$, mean value $E(\mu_A)(x)$, variance $V(\mu_A)(x)$, standard deviation $\sigma(\mu_A)(x)$, nth moment $M^n(\mu_A)(x)$, nth moment around mean value $M_0^n(\mu_A)(x)$, nth absolute moment around mean value $\bar{M}_0^n(\mu_A)(x)$ are defined as follows.

$$E(\mu_A)(x) = \int_\Omega \mu_A(x, \omega) \cdot \mathrm{d}P(\omega)(\underset{\Delta}{=} M^1(\mu_A)(x)), \tag{73}$$

$$V(\mu_A)(x) = \int_\Omega (\mu_A(x, \omega) - E(\mu_A)(x))^2 \, \mathrm{d}P(\omega)(\underset{\Delta}{=} M_0^2(\mu_A)(x)), \tag{74}$$

$$\sigma(\mu_A)(x) = \sqrt{V(\mu_A)(x)}, \tag{75}$$

$$M^n(\mu_A)(x) = \int_\Omega \mu_A(x, \omega)^n \cdot \mathrm{d}P(\omega) \qquad (n \in N), \tag{76}$$

$$M_0^n(\mu_A)(x) = \int_\Omega (\mu_A(x, \omega) - E(\mu_A)(x))^n \cdot \mathrm{d}P(\omega), \tag{77}$$

$$\bar{M}_0^n(\mu_A)(x) = \int_\Omega |\mu_A(x, \omega) - E(\mu_A)(x)|^n \cdot \mathrm{d}P(\omega). \tag{78}$$

The justification of above stated definitions is ensured by Proposition 1, and the following properties follow from these definitions.

Proposition 9.
In the situation of Definition 5, we have

$$0 \leqslant E(\mu_A)(x) \leqslant 1 \qquad \textit{for all } x \in X, \tag{79}$$

$$\begin{aligned} 0 &\leqslant \cdots \leqslant M^3(\mu_A)(x) \leqslant M^2(\mu_A)(x) \leqslant M^1(\mu_A)(x) \\ &= E(\mu_A)(x) \leqslant \{M^2(\mu_A)(x)\}^{\frac{1}{2}} \\ &\leqslant \{M^3(\mu_A)(x)\}^{\frac{1}{3}} \leqslant \cdots \leqslant 1 \qquad \textit{for all } x \in X, \end{aligned} \tag{80}$$

$$V(\mu_A)(x) = M^2(\mu_A)(x) - (E(\mu_A)(x))^2 \qquad \textit{for all } x \in X, \tag{81}$$

$$n \geqslant m(\geqslant 1) \rightarrow 0 \leqslant \bar{M}_0^n(\mu_A)(x) \leqslant \bar{M}_0^m(\mu_A)(x) \leqslant 1 \qquad \textit{for all } x \in X, \tag{82}$$

$$\lim_{n\to\infty} M_0^n(\mu_A)(x) = \lim_{n\to\infty} \bar{M}_0^n(\mu_A)(x) = 0 \qquad \textit{for all } x \in X. \tag{83}$$

Definition 6.

Let A and B be probabilistic sets on X . For each fixed $x \in X$,covariance $C(\mu_A, \mu_B)(x)$ and correlation coefficient $r(\mu_A, \mu_B)(x)$ are defined by

$$\begin{aligned} C(\mu_A, \mu_B)(x) = \int_\Omega & (\mu_A(x, \omega) - E(\mu_A)(x)) \\ & \cdot (\mu_B(x, \omega) - E(\mu_B)(x)) \cdot \mathrm{d}P(\omega), \end{aligned} \tag{84}$$

$$r(\mu_A, \mu_B)(x) = C(\mu_A, \mu_B)(x) / \sqrt{V(\mu_A)(x) \cdot V(\mu_B)(x)}. \tag{85}$$

(If $V(\mu_A)(x) \cdot V(\mu_B)(x) = 0$, $r(\mu_A, \mu_B)(x)$ is not defind.)

Proposition 10.
In the situation of Definition 6, we have

$$0 \leqslant |C(\mu_A, \mu_B)(x)| \leqslant V(\mu_A)(x) \cdot V(\mu_B)(x) \leqslant 1, \tag{86}$$

$$0 \leqslant |r(\mu_A, \mu_B)(x)| \leqslant 1, \tag{87}$$

$$C(\mu_A, \mu_A)(x) = V(\mu_A)(x), \tag{88}$$

$$C(\mu_A, \mu_B)(x) = E(\mu_A, \mu_B)(x) - E(\mu_A)(x) \cdot E(\mu_B)(x), \tag{89}$$

$$\begin{aligned} r(\mu_A, \mu_B)(x) = \pm 1 \rightarrow & \textit{ there exist real numbers } a \textit{ and } b \textit{ such that} \\ & \mu_A(x, \omega) = a \cdot \mu_B(x, \omega) + b \textit{ for a.e. } \omega \in \Omega. \end{aligned} \tag{90}$$

Definition 7.

Let $A_1, A_2, \ldots, A_n$ be probabilistic sets on X whose defining functions are $\mu_{A_1}(x, \omega), \mu_{A_2}(x, \omega), \ldots, \mu_{A_n}(x, \omega)$, respectively. For arbitrary $x, y (\in X)$, moment matrix $M(x, y)$ and variable-covariance matrix $V(x, y)$ of $A_1, A_2, \ldots, A_n$ are defined by

$$M(x, y) = [m_{i,j}]_{1 \leq i,j \leq n} \qquad \text{where } m_{i,j} = \int_{\Omega} \mu_{A_i}(x, \omega) \cdot \mu_{A_j}(x, \omega) \cdot \mathrm{d}P(\omega), \tag{91}$$

$$V(x, y) = [v_{i,j}]_{1 \leq i,j \leq n} \qquad \text{where } v_{i,j} = \int_{\Omega} (\mu_{A_i}(x, \omega) - E(\mu_{A_i})(x)) \cdot (\mu_{A_j}(x, \omega) - E(\mu_{A_j})(x)) \cdot \mathrm{d}P(\omega). \tag{92}$$

3.Expected cardinal number

In ordinary set theory a notion of the cardinal number of a finite set is defined as the number of elements of the set. This concept can be extended to probabilistic set theory as follows.

Definition 8.

Let A be a probabilistic set on X whose defining function is $\mu_A(x, \omega)$. The expected support of A is defined as the following(ordinary) subset of X,

$$\operatorname{supp} A = \left\{ x \in X \mid E(\mu_A)(x) = \int_{\Omega} \mu_A(x, \omega) \, \mathrm{d}P(\omega) > 0 \right\}, \tag{93}$$

and the expected cardinal number of A, denoted by $\#A$, is defined by

$$\#A = \begin{cases} \sum_{x \in \operatorname{supp} A} \left(\int_{\Omega} \mu_A(x, \omega) \, \mathrm{d}P(\omega) \right) & \text{if } \# \operatorname{supp} A \leq \chi_0, \\ \# \operatorname{supp} A & \text{if } \# \operatorname{supp} A > \chi_0. \end{cases} \tag{94}$$

Conclusions

The background idea of probabilistic sets was discussed in comparison with fuzzy sets, and its mathematical structure was explained without proofs. Main results are (1)a family of probabilistic sets constitutes a pseudo-Boolean algebra; (2)the possibility of moment analysis is a great advantage in applications(cf. [9]).

The concepts of probabilistic sets presented in this

paper seems to provide a new mathematical foundation in the field of pattern recognition or provide a new mathematical foundation in the field of pattern recognition or decision making theory. Several studies are being done in these fields [9]. We will be glad if our idea is any help to the people concerned.

References

[1]G.Birkhoff, Lattice Theory,Am.Math.Soc.Colloq.Publ. (Am.Math.Soc.,Mew York,1969).

[2]J.G.Brown, A note on Fuzzy sets, Information and Control 18(1971) 32-39.

[3]J.A.Geguen, L-fuzzy sets, J.Math.Anal.Appl. 18(1967) 145-174.

[4]P.R.Halmos, Naive Set Theory(Van Nostrand,New York, 1960).

[5]P.R.Halmos, Measure Theory (Van Nostrand, New York, 1960).

[6]K.Hirota, Kakuritsu-Shugoron to sono Oyourei (Probabilistic sets and its applications), Presented at the Behaviormetric Society of Japan 3rd Conference(1975)(in Japanese).

[7]K.Hirota, Kakuritsu-Shugoron (Probabilistic set theory), in: Fundamental research works of fuzzy system theory and artificial intelligence, Research Reports of Scientific Research fund from the Ministry of Education in Japan(1976) 193-213(in Japanese).

[8]K.Hirota, Concepts of probabilistic sets,IEEE Conf. on Decision and Control (New Orleans)(1977) 1361-1366.

[9]K.Hirota, Extended fuzzy expression of probabilistic sets-Analytical expression of ambiguity and subjectivity in pattern recognition, Presented at Seminar on Applied Functional Analysis(July,1978) 13-18.

[10]K.Hirota er al., A decision making model-A new approach based on the concepts of probabilistic sets, Presented at Int.Conf.on Cybernetics and Society 1978, Tokyo(Nov.1978) 1348-1353.

[11]K.Hirota et al., The bounded variation quantity (B.V.Q) and its application to feature extractions, Presented at the 4th Int.Conf. on Pattern Recognition,

Kyoto(Nov.1978) 456-461.
[12]M.Mizumoto and K.Tanaka,Some properties of fuzzy sets of type 2, Information and Control 31(1976) 312-340.
[13]K.Nanba, shugo-ron(Set Theory) (Science-sha Publ., 1975) (in Japanese).
[14]L.A.Zadeh, Fuzzy sets, Information and Control 81965) 228-353.
[15]L.A.Zadeh, Probabilistic measure of fuzzy events, J. Math. Ana.. Appl.23(1968) 421-427.
[16]L.A.Zadeh et al., Fuzzy Sets and their Applications to Cognitive and Decision Theory (Academic Press, New York,1975).

INDEX